Communications and Control Engineering

Ioan D. Landau and Gianluca Zito

Digital Control Systems

Design, Identification and Implementation

With 238 Figures

Ioan D. Landau, PhD
Gianluca Zito, PhD
Lab. d'Automatique de Grenoble (INPG/CNRS)
ENSIEG
BP 46
38402 Saint Martin d'Heres
France

Series Editors
E.D. Sontag · M. Thoma · A. Isidori · J.H. van Schuppen

British Library Cataloguing in Publication Data
Landau, Ioan D., 1938-
    Digital control systems : design, identification and
    implementation. - (Communications and control engineering)
    1. Digital control systems 2. Digital control systems -
    Design and construction
    I. Title II.Zito, Gianluca
    629.8'9

Communications and Control Engineering Series ISSN 0178-5354
ISBN-13: 978-1-84996-551-4   e-ISBN 978-1-84628-056-6   Printed on acid-free paper

To Lina, Vlad, Carla, Maria Luisa, Francesco

"Ce qui est simple est toujours faux
Ce qui ne l'est pas est inutilisable."

(Paul Valery)

Preface

The extraordinary development of digital computers (microprocessors, microcontrollers) and their extensive use in control systems in all fields of applications has brought about important changes in the design of control systems. Their performance and their low cost make them suitable for use in control systems of various kinds which demand far better capabilities and performances than those provided by analog controllers.

However, in order really to take advantage of the capabilities of microprocessors, it is not enough to reproduce the behavior of analog (PID) controllers. One needs to implement specific and high-performance *model based* control techniques developed for computer-controlled systems (techniques that have been extensively tested in practice). In this context identification of a plant *dynamic model* from data is a fundamental step in the design of the control system.

The book takes into account the fact that the association of books with software and on-line material is radically changing the teaching methods of the control discipline. Despite its interactive character, computer-aided control design software requires the understanding of a number of concepts in order to be used efficiently. The use of software for illustrating the various concepts and algorithms helps understanding and rapidly gives a *feeling* of the various phenomena. Complementary information and material for teaching and applications can be found on the book website:

http://landau-bookic.lag.ensieg.inpg.fr

The Aim of the Book
The aim of this book is to give the necessary knowledge for the comprehension and implementation of digital techniques for system identification and control design. These techniques are applicable to various types of process. The book has been written taking into account the needs of the designer and the user of such systems. Theoretical developments that are not directly relevant to the design have been omitted. The book also takes into account the availability of dedicated control software. A number of useful routines have been developed and they can be freely

downloaded from the book website. Details concerning effective implementation and on-site optimization of the control systems designed have been provided.

An important feature of the book, which makes it different from other books on the subject, is the fact that equal weight has been given to *system identification* and *control design*. This is because both techniques are equally important for design and optimization of a high-performance control system. A control engineer has to possess a balance of knowledge in both subjects since identification cannot be dissociated from control design. The book also emphasizes control robustness aspects and controller complexity reduction, both very important issues in practice.

The Object of Study

The closed loop control systems studied in this book are characterized by the fact that the control law is implemented on a digital computer (microprocessor, microcontroller). This type of system is sketched in Figure 0.1.

The continuous-time plant to be controlled is formed by the set of actuator, process and sensor. The continuous-time measured output *y(t)* is converted into a sequence of numbers {*y(k)*} by an analog-to-digital converter (ADC), at sampling instants *k* defined by the synchronization clock. This sequence is compared with the reference sequence{*r(k)*} and the resulting sequence of errors is processed by the digital computer using a *control algorithm* that will generate a control sequence {*u(k)*}. By means of a digital-to-analog converter (DAC), this sequence is converted into an analog signal, which is usually maintained constant between the sampling instants by a zero-order hold (ZOH).

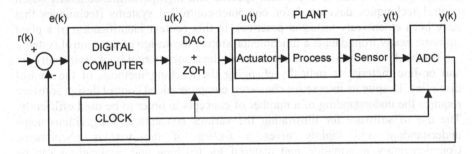

Figure 0.1. Digital control system

The Main Stream

Figure 0.2 summarizes the general principles for controller design, implementation and validation.

For design and tuning of a good controller one needs:

1. To specify the desired control loop performance and robustness
2. To know the dynamic model of the plant to be controlled
3. To possess a suitable controller design method making it possible to achieve the desired performance and robustness specifications for the corresponding plant model

4. To implement the resulting controller taking into account practical constraints
5. To validate the controller performance on site and, if necessary, to re-tune it

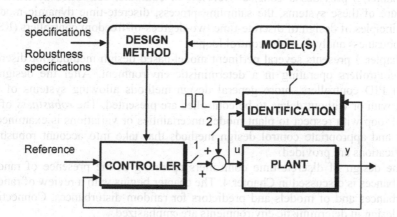

Figure 0.2. Principle of controller design and validation

In order to obtain a relevant dynamic plant model for design, system identification techniques using input/output measurements (switch 1 is off, switch 2 is on) should be considered. The methodology for system identification is presented in the book together with dedicated algorithms implemented as software tools.

Once the system model is available, the book provides a set of methods (and the corresponding software tools) for the design of an appropriate controller.

The implementation of the controller should take into account aspects related to data acquisition, switching from open loop to closed loop, and saturation of the actuator as well as constraints on the complexity of the controller. These aspects are examined in detail in the book.

Expected Audience

The book represents a course reference for Universities and Engineering Schools offering courses on applied computer-controlled systems and system identification.

In addition to its academic audience, *Digital Control Systems* is aimed at practising engineers wishing to acquire the concepts and techniques of system identification, control design and implementation using a digital computer. The industrial references for the techniques presented in the book and the various applications described provide useful information for those directly involved in the real-world uses of control.

Readers who are already familiar with the basics of computer-controlled systems will find in this book a clear, application oriented, methodology for system identification and the design of various types of controllers for single-input, single-output (SISO) systems.

The Content

Chapter 1 briefly reviews the continuous-time control techniques which will be used later on as a reference for the introduction of basic concepts for computer control.

Chapter 2 provides a concise overview of computer-controlled systems: the structure of these systems, the sampling process, discrete-time dynamic models, the principles of design of discrete-time two-degrees-of-freedom controllers (RST), and robustness analysis of the control loops.

Chapter 3 presents several pertinent model-based design methods for discrete-time controllers operating in a deterministic environment. After the design of digital PID controllers, more general design methods allowing systems of any order, with or without delay, to be controlled are presented. The *robustness* of the closed loop with respect to plant model uncertainties or variations is examined in detail and appropriate control design methods that take into account robustness specifications are provided.

The design of discrete-time controllers operating in the presence of random disturbances is discussed in Chapter 4. The chapter begins with a review of random disturbances and of models and predictors for random disturbances. Connections with design in deterministic environments are emphasized.

The basics of system identification using a digital computer are presented in Chapter 5. Methods that are used for the identification of discrete-time models, and model validation techniques as well as techniques for order estimation from input/output data are described in Chapter 6.

Chapter 7 discusses the practical aspects of system identification using data from several applications: air heater, distillation column, DC motor, and flexible transmission.

The main goal of this work, the use of control design methods and system identification techniques in the implementation of a digital controller for a specific application, is discussed in Chapter 8. Implementation aspects are reviewed and several applications presented (air heater, speed and position control of a DC motor, flexible transmission, flexible arm, and hot-dip galvanizing).

For on-site optimization and controller re-tuning a plant model should be obtained by identification in closed loop (switches 1 and 2 are on in Figure 0.2). The techniques for identification in closed loop are presented in Chapter 9.

In many situations constraints on the complexity of the controller are imposed so Chapter 10 presents techniques for controller order reduction.

Appendix A reviews some basic concepts.

Appendix B offers an alternative time-domain approach to the design of RST digital controllers using one-step-ahead and long-range-predictive control strategies. Links and equivalence with the design methods presented in Chapter 3 are emphasized.

Appendix C presents a state space approach to the design of RST digital controllers. The equivalence with the design approach presented in Chapter 3 is emphasized. The linear quadratic control is also discussed.

Appendix D presents some important concepts in robustness.

Appendix E demonstrates the Youla–Kucera parametrization of digital controllers which is useful for a number of developments.

Appendix F describes a numerically robust algorithm for recursive identification.

Appendix G is dedicated to the presentation of suggested laboratory sessions that use data files and functions which can be downloaded from the book website.

Appendix H gives a list and a brief description of the MATLAB®- and Scilab-based functions and C^{++} programs implementing algorithms presented in the book. These functions and programs can also be downloaded from the book website.

The book website gives access, to the various functions and programs as well as to data files. It contains descriptions of additional laboratory sessions and slides for a number of chapters, tutorials and courses related to the material included in the book that can be downloaded; all the MATLAB® files used for generating the examples and figures in the text can also be found on the website.

How to Read the Book
The book can be read in different ways after the basic control concepts presented in Chapters 1 and 2 have been assimilated. If the reader is mainly interested in control algorithms, it would be useful for him/her to read Chapters 3 and 4 and then Chapters 5, 6, 7 and 8. If the reader is mainly interested in identification techniques, he or she can jump straight to Chapters 5, 6 and 7 and then return to Chapters 3, 4 and 8. Those who are familiar with the basics of computer-controlled systems can even start with Section 2.5. Chapters 9 and 10 follow dependently from Chapter 8.   Figure 0.3 shows the interdependence between the various chapters.

Course Configurations
A complete basic course on digital control should cover most of the material presented in Chapters 2, 3, 5, 8 and Section 4.1.   For an advanced course, all chapters might be included. For an introductory course in digital control one can use Chapters 2, 3 and 8. For an introductory course on system identification one can use Chapters 5, 6 and 7.

Why this Book?
The book reflects the first author's more than twenty-five years of experience in teaching, design and implementation of digital control systems. Involvement in many industrial projects and feedback from an industrial and academic audience from various countries in Europe, North and South America and Asia have played a major role in the selection, organization and presentation of the material. Experience from writing the book *System Identification and Control Design*, Prentice Hall, 1990[1] (Information and System Sciences Series) has been also very useful.

The present book is a revised translation of a book (*Commande des systèmes – conception, identification et mise en oeuvre*) published in 2002 by Hermes-Lavoisier, Paris.

---

[1] Revised taranslation of a book published by Hermes Paris, 1988 (second edition 1993).

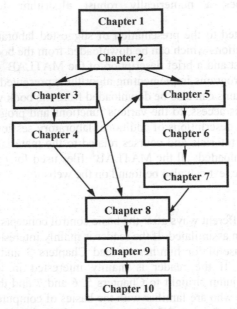

Figure 0.3. Logical dependence of the various chapters

The most recent academic courses based on the material in the present book include PhD courses delivered in 2004 at Universita Technologica de Valencia, Spain (robust discrete time controller design) and Escuela Superior de Ingenerios de Sevilla, Spain (system identification in open and closed loop).

Acknowledgments

We wish to acknowledge the large number of contributors on whose work our presentation is partly based as well as the discussions we have had with many of the experts in the field. In particular we wish to mention: G. Zames, V.M. Popov, B.D.O. Anderson, K.J. Aström, R. Bitmead, D. Clarke, G. Franklin, M. Gevers, G. Goodwin, T. Kailath, L. Ljung, M. Morari, M. Tomizuka and G. Vinnicombe.

We would like to thank M. M'Saad, N. M'Sirdi, M. Samaan, H.V. Duong, F. Rolland, A. Voda – Besançon, J. Langer, A. Franco and H. Bourlés for their collaboration in the development and implementation of various techniques for system identification and control design.

We would like to thank A. Karimi, H. Prochazka, A. Constantinescu, D. Rey, A. Rousset, P. Rival, F. Bouziani and G. Stroian who have contributed to this project.

The support of the Laboratoire d'Automatique de Grenoble (INPG/CNRS/UJF) and Adaptech for the experimental aspects has been essential.

We also would like to thank the participants in various courses. Their remarks and suggestion have been very valuable.

Grenoble,   *Ioan Doré Landau*
May 2005   *Gianluca Zito*

Contents

List of Principal Notation

f_s	sampling frequency
T_s	sampling period
ω	radian frequency (rad/s)
f	frequency (Hz) or normalized frequency(f/f_s)
t	continuous time or normalized discrete-time (with respect to the sampling period t/T_s)
k	normalized discrete time (t/T_s)
$u(t), y(t)$	plant input and output
$y^*(t+d+1)$	tracking reference
$r(t)$	reference or external excitation
$e(t)$	discrete time Gaussian white noise
q^{-1}	shift (delay) operator ($q^{-1}y(t) = y(t-1)$)
s, z	complex variables ($z = e^{sT_s}$)
$A(q^{-1}), B(q^{-1}), C(q^{-1})$	polynomials in the variable q^{-1}
d	delay of the discrete-time system (integer)
$\hat{A}(t,q^{-1})$, $\hat{B}(t,q^{-1})$, $\hat{C}(t,q^{-1})$	estimation of polynomials A, B, C at instant t
$\hat{a}_i(t)$, $\hat{b}_i(t)$, $\hat{c}_i(t)$	estimated coefficients of polynomials A, B, C
$H(q^{-1})$	pulse transfer operator (discrete time systems)
$H(z^{-1}), H(z)$	discrete-time transfer functions
τ	time delay of a continuous-time system
$R(q^{-1}), S(q^{-1}), T(q^{-1})$	pulse transfer operators used in a RST digital controller
$S_{xy}(s), S_{xy}(z^{-1})$	sensitivity functions
$P(z^{-1})$	closed loop characteristic polynomial
ΔM	modulus margin
$\Delta\tau$	delay margin
θ	parameter vector
$\hat{\theta}(t)$	estimated parameter vector
$\varphi(t), \Phi(\tau)$	measurement / observation vector
$F, F(t)$	adaptation gain
$\varepsilon°(t), \varepsilon(t)$	*a priori /a posteriori* prediction error

$\varepsilon^{\circ}_{CL}(t), \varepsilon_{CL}(t)$	closed loop *a priori* / *a posteriori* prediction error
$v^{\circ}(t), v(t)$	*a priori* / *a posteriori* adaptation error
A, F	matrices
$F > 0$	positive definite matrix
t_R	rise time
t_S	settling time
M	maximum overshoot
$\omega_o, \zeta,$	natural frequency and damping factor for a continuous-time second-order system
$E\{.\}$	expectation
MV	mean value
var.	variance
σ	standard deviation
$R(i)$	auto-correlation or cross-correlation
$RN(i)$	normalized auto-correlation or cross-correlation
t_{im}	maximum length of pulse in a PRBS
OL	open loop
CL	closed loop
BP	bandwidth
AF-CLOE	adaptive filtered closed loop output error
ARMAX	Auto-Regressive Moving Average with eXogenous input process
CLIM	closed loop input matching
CLOE	closed loop output error
CLOM	closed loop output matching
ELS	extended least squares
F-CLOE	filtered closed loop output error
GLS	generalized least squares
IVAM	instrumental variable with auxiliary model
OEAFO	output error with adaptive filtered observations
OEEPM	output error with extended prediction model
OEFC	output error with fixed compensator
OEFO	output error with filtered observations
PAA	parameter adaptation algorithm
PID	proportional + integral + derivative controller
PRBS	pseudo random binary sequence
RLS	recursive least squares
RML	recursive maximum likelihood
RST	two degrees of freedom digital controller
X-CLOE	extended closed loop output error

Warning
For sake of notation uniformity, we shall often use, in the case of linear systems with constant coefficients, q^{-1} notation both for the delay operator and the complex variable z^{-1}. The z^{-1} notation will be especially employed when an interpretation in the frequency domain is needed (in this case $z = e^{-j\omega T_s}$).

1

Continuous Control Systems: A Review

The aim of this chapter is to review briefly the main concepts of continuous control systems. The presentation is such that it will permit at a later stage an easy transition to digital control systems.

The subject matter handled relates to the description of continuous-time models in the time and frequency domains, the properties of closed-loop systems and PI and PID controllers.

1.1 Continuous-time Models

1.1.1 Time Domain

Equation 1.1.1 gives an example of a differential equation describing a simple dynamic system:

$$\frac{dy}{dt} = -\frac{1}{T}y(t) + \frac{G}{T}u(t) \tag{1.1.1}$$

In Equation 1.1.1 u represents the input (or the control) of the system and y the output. This equation may be simulated by continuous means as illustrated in Figure 1.1.

The step response illustrated in Figure 1.1 reveals the speed of the output variations, characterized by the time constant T, and the final value, characterized by the static gain G.

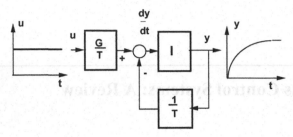

Figure 1.1. Simulation and time responses of the dynamic system described by Equation 1.1.1 (I - integrator)

Using the *differential operator p = d/dt*, Equation 1.1.1 is written as

$$(p+\frac{1}{T})\,y(t)=\frac{G}{T}\,u(t); \qquad p=\frac{d}{dt} \qquad\qquad (1.1.2)$$

For systems described by differential equations as in Equation 1.1.1 we distinguish three types of time response:

1. The "free" response: it corresponds to the system response starting with an initial condition $y(0)=y_0$ and for an identically zero input for all t ($u = 0$, $\forall$ t).
2. The "forced" response: it corresponds to the system response starting with an identically zero initial condition $y(0) = 0$ and for a non-zero input $u(t)$ for all $t \geq 0$ ($u(t) = 0$, $t < 0$; $u(t) \neq 0$, $t \geq 0$ and $y(t) = 0$ for $t \leq 0$).
3. The "total" response: it represents the sum of the "free" and "forced" responses (the system being linear, the superposition principle applies).

Nevertheless later we will consider separately the "free" response and the "forced" response.

1.1.2 Frequency Domain

The characteristics of the models in the form of Equation 1.1.1 can also be studied in the frequency domain. The idea is then to study the system behavior when the input u is a sinusoidal or a cosinusoidal input that varies over a given range of frequencies.

Remember that

$$e^{j\omega t} = \cos \omega t + j \sin \omega t \qquad\qquad (1.1.3)$$

and, consequently, it can be considered that the study of the dynamic system described by an equation of the type 1.1.1, in the frequency domain, corresponds to the study of the system output for inputs of the type $u(t) = e^{j\omega t}$.

Since the system is *linear*, the output will be a signal containing only the frequency ω, the input being amplified or attenuated (and possibly a phase lag will appear) according to ω; i.e. the output will be of the form

$$y(t) = H(j\omega)e^{j\omega t} \tag{1.1.4}$$

Figure 1.2 illustrates the behavior of a system for an input $u(t) = e^{j\omega t}$.

However there is nothing to stop us from considering that the input is formed by damped or undamped sinusoids and cosinusoids, which in this case are written as

$$u(t) = e^{\sigma t} e^{j\omega t} = e^{(\sigma + j\omega)t} = e^{st}; \qquad s = \sigma + j\omega \tag{1.1.5}$$

where s is interpreted as a complex frequency. As a result of the linearity of the system, the output will reproduce the input signal, amplified (or attenuated), with a phase lag or not, depending on the values of s; i.e. the output will have the form

$$y(t) = H(s)e^{st} \tag{1.1.6}$$

and it must satisfy Equation 1.1.1 for $u(t) = e^{st}$.[1]

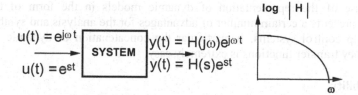

$$u(t) = e^{j\omega t} \qquad \boxed{\text{SYSTEM}} \qquad y(t) = H(j\omega)e^{j\omega t}$$
$$u(t) = e^{st} \qquad \qquad \qquad y(t) = H(s)e^{st}$$

Figure 1.2. Response of a dynamic system to periodic inputs

From Equation 1.1.6 one gets

$$\frac{dy(t)}{dt} = s\,H(s)e^{st} \tag{1.1.7}$$

and by introducing Equation 1.1.7 in Equation 1.1.1, while bearing in mind that $u(t) = e^{st}$, one obtains

$$(s + \frac{1}{T})H(s)e^{st} = \frac{G}{T}e^{st} \tag{1.1.8}$$

---

[1] e^{st} is an eigenfunction of the system because its functional properties are preserved when passing through the system (only the amplitude and the phase are modified).

$H(s)$, which gives the gain and phase deviation introduced by the system of Equation 1.1.1 at different complex frequencies, is known as the *transfer function*. The transfer function $H(s)$ is a function of only the complex variable s. It represents the ratio between the system output and input when the input is e^{st}. From Equation 1.1.8, it turns out that, for the system described by Equation 1.1.1, the transfer function is

$$H(s) = \frac{G}{1+sT} \qquad (1.1.9)$$

The transfer function $H(s)$ generally appears as a ratio of two polynomials in s ($H(s)=B(s)/A(s)$). The roots of the numerator polynomial ($B(s)$) define the "zeros" of the transfer function and the roots of the denominator polynomial ($A(s)$) define the "poles" of the transfer function. The "zeros" correspond to those complex frequencies for which the system gain is null and the "poles" correspond to those complex frequencies for which the system gain is infinite.

Note that the transfer function $H(s)$ can also be obtained by two other techniques:

- Replacing p by s in Equation 1.1.2 and algebraic computation of the y/u ratio.
- Using the Laplace transform (Ogata 1990).

The use of the representation of dynamic models in the form of transfer functions presents a certain number of advantages for the analysis and synthesis of closed-loop control systems. In particular the concatenation of dynamic models described by transfer functions is extremely easy.

1.1.3 Stability

The stability of a dynamic system is related to the asymptotic behavior of the system (when $t \rightarrow \infty$), starting from an initial condition and for an identically zero input.

For example, consider the first-order system described by the differential Equation 1.1.1 or by the transfer function given in Equation 1.1.9.

Consider the free response of the system given in Equation 1.1.1 for $u \equiv 0$ and from an initial condition $y(0) = y_0$:

$$\frac{dy}{dt} + \frac{1}{T} y(t) = 0 \; ; \; y(0) = y_0 \qquad (1.1.10)$$

A solution for y will be of the form

$$y(t) = Ke^{st} \qquad (1.1.11)$$

in which K and s are to be determined[2]. From Equation 1.1.11 one finds

$$\frac{dy}{dt} = s\,Ke^{st} \qquad (1.1.12)$$

and Equation 1.1.10 becomes

$$Ke^{st}\left(s + \frac{1}{T}\right) = 0 \qquad (1.1.13)$$

from which one obtains

$$s = -\frac{1}{T}\,;\; K = y_0 \qquad (1.1.14)$$

and respectively

$$y(t) = y_0 e^{-t/T} \qquad (1.1.15)$$

The response for $T > 0$ and $T < 0$ is illustrated in Figure 1.3.

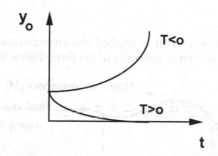

Figure 1.3. Free response of the first-order system

For $T > 0$, we have $s < 0$ and, when $t \rightarrow \infty$, the output will tend toward zero (asymptotic stability). For $T < 0$, we have $s > 0$ and, when $t \rightarrow \infty$, the output will diverge (instability). Note that $s = -1/T$ corresponds to the pole of the first-order transfer function of Equation 1.1.9.

We can generalize this result: it is the sign of the real part of the roots of the transfer function denominator that determines the stability or instability of the system.

In order that a system be *asymptotically stable*, all the roots of the transfer function denominator must be characterized by $Re\ s < 0$. If one or several roots of

---

[2] The structure of the solution for Equation 1.1.11 results from the theory of linear differential equations.

the transfer function denominator are characterized by $Re\ s > 0$, then the system is *unstable*.

For $Re\ s = 0$ we have a limit case of *stability* because the amplitude of $y(t)$ remains equal to the initial condition (e.g. pure integrator, $dy/dt = u(t)$; in this case $y(t)$ remains equal to the initial condition).

Figure 1.4 gives the stability and instability domains in the plane of the complex variable s.

Note that stability criteria have been developed, which allow determining the existence of unstable roots of a polynomial without explicitly computing its roots, (e.g. Routh-Hurwitz' criterion) (Ogata 1990).

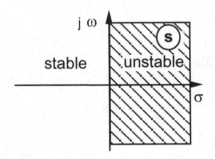

Figure 1.4. Stability and instability domains in the s-plane

1.1.4 Time Response

The response of a dynamic system is studied and characterized for a step input. The response of a stable system is generally of the form shown in Figure 1.5.

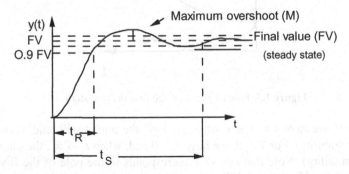

Figure 1.5. Step response

The step response is characterized by a certain number of parameters:

- t_R *(rise time):* generally defined as the time needed to reach 90% of the final value (or as the time needed for the output to pass from 10 to 90% of the final value). For systems that present an overshoot of the final value, or that have an oscillating behavior, we often define the rise time as the time needed to reach for the first time the final value. Subsequently we shall generally use the first definition of t_R.

- t_S (*settling time*): defined as the time needed for the output to reach and remain within a tolerance zone around the final value ($\pm 10\%$, $\pm 5\% \pm 2\%$).
- FV (*final value*): a fixed output value obtained for $t \to \infty$.
- M (*maximum overshoot*): expressed as a percentage of the final value.

For example, consider the first-order system

$$H(s) = \frac{G}{1 + sT}$$

The step response for a first-order system is given by

$$y(t) = G(1 - e^{-t/T})$$

Since the input is a unitary step one has
$FV = G$ (static gain); $t_R = 2.2\ T$

$t_S = 2.2\ T$ (for $\pm 10\%$ FV) ; $t_S = 3\ T$ (for $\pm 5\%$FV) ; $M = 0$
and the response of such a system is represented in Figure 1.6. Note that for $t = T$, the output reaches 63% of the final value.

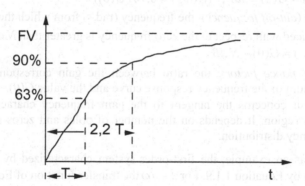

Figure 1.6. Step response for a first-order system

1.1.5 Frequency Response

The frequency response of a dynamic system is studied and characterized for periodic inputs of variable frequency but of constant magnitude. For continuous-time systems, the gain-frequency characteristic is represented on a double logarithmic scale and the phase frequency characteristic is represented on a logarithmic scale only for the frequency axis.

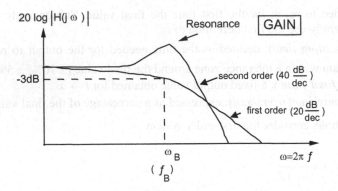

Figure 1.7. Frequency responses

The gain $G(\omega) = |H(j\omega)|$ is expressed in dB ($|H(j\omega)|$ $dB = 20 \log |H(j\omega)|$) on the vertical axis and the frequency ω, expressed in rad/s ($\omega = 2\pi f$ where f represents the frequency in Hz) is represented on the horizontal axis. Figure 1.7 gives some typical frequency response curves.

The characteristic elements of the frequency response are:

- $f_B(\omega_B)$ (*bandwidth*): the frequency (radian frequency) from which the zero-frequency (steady-state) gain $G(0)$ is attenuated more than 3 dB;
  $G(\omega_B) = G(0) - 3dB$; $(G(\omega_B) = 0.707\,G(0))$.
- $f_C(\omega_C)$ (*cut-off frequency*): the frequency (rad/s) from which the attenuation introduced with respect to the zero frequency is greater than N dB;
  $G(j\omega_C) = G(0) - N\,dB$.
- Q (*resonance factor*): the ratio between the gain corresponding to the maximum of the frequency response curve and the value $G(0)$.
- *Slope*: it concerns the tangent to the gain frequency characteristic in a certain region. It depends on the number of poles and zeros and on their frequency distribution.

Consider, as an example, the first-order system characterized by the transfer function given by Equation 1.1.9. For $s = j\omega$ the transfer function of Equation 1.1.9 is rewritten as

$$H(j\omega) = \frac{G}{1+j\omega T} = |H(j\omega)| e^{j\phi(\omega)} = |H(j\omega)| \angle\phi(\omega) \qquad (1.1.16)$$

where $|H(j\omega)|$ represents the modulus (gain) of the transfer function and $\angle\phi(\omega)$ the phase deviation introduced by the transfer function. We then have

$$G(\omega) = |H(j\omega)| = \frac{G}{\sqrt{(1+\omega^2 T^2)}} \qquad (1.1.17)$$

$$\angle\phi(\omega) = \tan^{-1}\left[\frac{\mathrm{Im}\,G(j\omega)}{\mathrm{Re}\,G(j\omega)}\right] = \tan^{-1}\left[-\omega T\right]$$ (1.1.18)

From Equation 1.1.17 and from the definition of the bandwidth ω_B, we obtain:

$$\omega_B = 1/T$$

Using Equation 1.1.18, we deduce that for $\omega = \omega_B$ the system introduces a phase deviation $\angle\phi(\omega_B) = -45°$. Also note that for $\omega = 0$, $G(0) = G$, $\angle\phi(0) = 0°$ and for $\omega \to \infty$, $G(\infty) = 0$, $\angle\phi(\infty) = -90°$.

Figure 1.8 gives the exact and asymptotic frequency characteristics for a first-order system (gain and phase).

As a general rule, each stable pole introduces an asymptotic slope of -20 dB/dec (or 6 dB/octave) and an asymptotic phase lag of -90°. On the other hand, each stable zero introduces an asymptotic slope of +20 dB/dec and an asymptotic phase shift of +90°.

It follows that the asymptotic slope of the gain-frequency characteristic in dB, for high frequencies, is given by

$$\frac{\Delta G}{\Delta\omega} = -(n - m)\times 20\,dB\,/\,dec$$ (1.1.19a)

where n is the number of poles and m is the number of zeros.

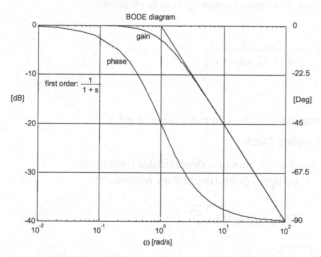

Figure 1.8. Frequency characteristic of a first-order system

The relation

$$\angle \phi(\infty) = -(n - m) \times 90° \tag{1.1.19b}$$

gives the asymptotic phase deviation.

Note that the rise time (t_R) for a system depends on its bandwidth (ω_B). We have the approximate relation

$$t_R \approx \frac{2\,to\,3}{\omega_B} \tag{1.1.20}$$

1.1.6 Study of the Second-order System

The normalized differential equation for a second-order system is given by:

$$\frac{d^2 y(t)}{dt^2} + 2\zeta\omega_0 \frac{dy(t)}{dt} + \omega_0^2 y(t) = \omega_0^2 u(t) \tag{1.1.21}$$

Using the operator $p = d/dt$, Equation 1.1.21 is rewritten as

$$(p^2 + 2\zeta\omega_0 p + \omega_0^2)\, y(t) = \omega_0^2 u(t) \tag{1.1.22}$$

Letting $u(t) = e^{st}$ in Equation 1.1.21, or $p = s$ in Equation 1.1.22, the normalized transfer function of a second-order system is obtained:

$$H(s) = \frac{\omega_0^2}{s^2 + 2\zeta\,\omega_0 s + \omega_0^2} \tag{1.1.23}$$

in which

- ω_0 : natural frequency in rad/s $(\omega_0 = 2\,\pi f_0)$
- ζ : damping factor

The roots of the transfer function denominator (poles) are
a) $|\zeta| < 1$, complex poles (oscillatory response) :

$$s_{1,2} = -\zeta\,\omega_0 \pm j\,\omega_0 \sqrt{1 - \zeta^2} \tag{1.1.24a}$$

$(\omega_0 \sqrt{1 - \zeta^2}$ is called "damped resonance frequency").
b) $|\zeta| \geq 1$, real poles (aperiodic response) :

$$s_{1,2} = -\zeta \omega_0 \pm \omega_0 \sqrt{\zeta^2 - 1} \tag{1.1.24b}$$

The following situations are thus obtained depending on the value of the damping factor ζ:

- $\zeta > 0$: asymptotically stable system
- $\zeta < 0$: unstable system

These different cases are summarized in Figure 1.9.

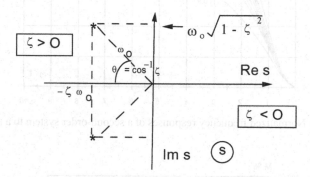

Figure 1.9. The roots of the second-order system as a function of ζ (for $|\zeta| \leq 1$)

The step response for the second-order system described by Equation 1.1.21 is given by the formula (for $|\zeta| \leq 1$)

$$y(t) = 1 - \frac{1}{\sqrt{1-\zeta^2}} e^{-\zeta\omega_0 t} (\sin \omega_0 \sqrt{1-\zeta^2}\, t + \theta) \tag{1.1.25}$$

in which

$$\theta = \cos^{-1} \zeta \tag{1.1.26}$$

Figure 1.10 gives the normalized step responses for the second-order system. This diagram makes it possible to determine both the response of a given second-order system and the values of ω_0 and ζ, in order to obtain a system having a given rise (or settling) time and overshoot.

To illustrate this, consider the problem of determining ω_0 and ζ so that the rise time (0 to 90% of the final value) is 2.75s with a maximum overshoot $\approx 5\%$. From Figure 1.10, it is seen that in order to ensure an overshoot $\approx 5\%$ we must choose $\zeta = 0.7$. The corresponding normalized rise time is: $\omega_0\, t_M \approx 2.75$. It can be concluded that to obtain a rise time of 2.75s, $\omega_0 = 1\ rad/sec$ must be taken.

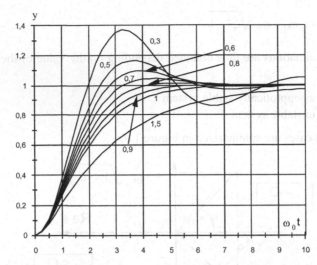

Figure 1.10. Normalized frequency responses of a second-order system to a step input

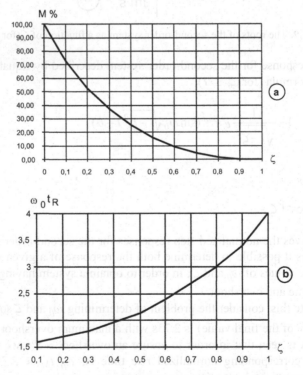

Figure 1.11. Second-order system: **a)** maximum overshoot M as a function of the damping factor ζ ; **b)** normalized rise time as a function of ζ

In order to make easier the determination of ω_0 and ζ for a given rise time t_R and a given maximum overshoot M, the graph of M as a function of ζ and the graph of $\omega_0 t_R$ as a function of ζ have been represented in Figure 1.11a, b.

The curve given in Figure 1.11a allows choosing the damping factor ζ for a given maximum overshoot M. Once the value of ζ chosen, the Figure 1.11b gives the corresponding value of $\omega_0 t_R$. This allows one to determine ω_0 for a given rise time t_R.

The functions *omega_damp.sci* (Scilab) and *omega_damp.m* (MATLAB®) allow one to obtain the values of ω_0 and ζ directly from the desired overshoot and rise time[3].

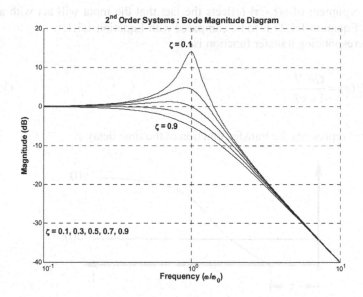

Figure 1.12. Normalized frequency responses of a second-order system (gain)

The settling time t_S, for different values of ζ and of the tolerance zone around the final value, can be determined from the normalized responses given in Figure 1.10.

Figure 1.12 gives the normalized frequency responses for a second-order system.

---

[3] To be downloaded from the book website.

1.1.7 Systems with Time Delay

Many industrial processes exhibit a step response of the form shown in Figure 1.13. The period of time during which the output does not react to the input is called *time delay* (denoted by τ).

A first-order dynamic system with a time delay τ is described by the following differential equation:

$$\frac{dy}{dt} = -\frac{1}{T}y(t) + \frac{G}{T}u(t - \tau)$$

(1.1.27)

where the argument of $u(t - \tau)$ reflects the fact that the input will act with a time delay of τ. Equation 1.1.27 is to be compared with Equation 1.1.1.

The corresponding transfer function is

$$H(s) = \frac{Ge^{-s\tau}}{1 + sT}$$

(1.1.28)

in which $e^{-s\tau}$ represents the transfer function of the time delay τ.

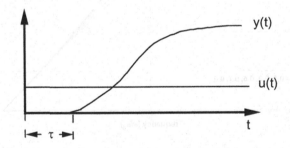

Figure 1.13. Step response of a system with time delay

Equations 1.1.27 and 1.1.28 can be straightforwardly extended to high-order systems with time delay.

Note that for the systems with time delay the rise time t_R is generally defined from $t = \tau$.

The frequency characteristics of the time delay are obtained by replacing $s = j\omega$ in $e^{-s\tau}$. We then obtain

$$H_{delay}(j\,\omega) = e^{-j\omega\tau} = |1| \cdot \angle\phi(\omega)$$

(1.1.29)

with

$$\angle\phi(\omega) = -\omega\tau \text{ (rad)}$$

(1.1.30)

Therefore a time delay does not modify the system gain, but it introduces a phase deviation proportional to the frequency.

1.1.8 Non-minimum Phase Systems

For continuous time systems (exclusively), non-minimum phase systems have one or more unstable zeros. In the continuous time case, the main effect of unstable zeros is the appearance of a negative overshoot at the beginning of the step response, as it is shown for example in Figure 1.14. The effect of the unstable zeros cannot be offset by the controller (one should use an unstable controller).

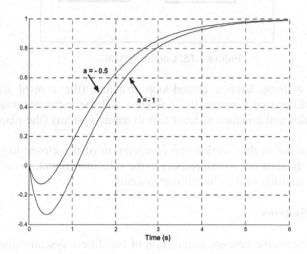

Figure 1.14. Step responses of a non-minimum phase system (H(s)=(1-sa)/(1+s)(1+0.5s), a=1,0.5)

As an example consider the system

$$H(s) = \frac{1-sa}{(1+s)(1+0.5s)}$$

with $a = 1$ and 0.5. Figure 1.14 represents the step response of the system.

1.2 Closed-loop Systems

Figure 1.15 shows a simple control system. $y(t)$ is the plant[4] output and represents the *controlled variable*, $u(t)$ is the input (control signal) applied to the plant by the controller (*manipulated variable*) and $r(t)$ is the reference signal.

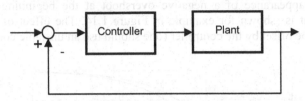

Figure 1.15. Control system

The control systems have a closed-loop structure (the control signal is a function of the difference between the reference and the measured value of the controlled variable) and contains at least two dynamic systems (the plant and the controller).

We shall examine in this section the computation of the closed-loop transfer function, the steady-state error with respect to the reference signal, the rejection of disturbances and stability of the closed-loop systems.

1.2.1 Cascaded Systems

Figure 1.16 represents the cascade connection of two linear systems characterized by the transfer functions $H_1(s)$ and $H_2(s)$.

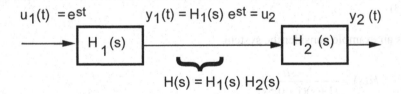

Figure 1.16. Cascade connection of two systems

If the input to $H_1(s)$ is $u_1(t) = e^{st}$ the following relations are found:

$$u_2(t) = y_1(t) = H_1(s)\, e^{st} \tag{1.2.1}$$
$$y_2(t) = H_2(s)\, u_2(t) = H_2(s)\, H_1(s)\, e^{st} = H(s)\, e^{st} \tag{1.2.2}$$

---

[4] the term "plant" defines the set : actuator, process to be controlled and sensor.

and we can conclude that the transfer function of two cascaded systems is

$$H(s) = H_2(s)\, H_1(s) \tag{1.2.3}$$

or in the general case of n cascaded systems

$$H(s) = H_n(s)\, H_{n-1}(s).....H_2(s)\, H_1(s) \tag{1.2.4}$$

1.2.2 Transfer Function of Closed-loop Systems

Consider the closed-loop system represented in Figure 1.17.

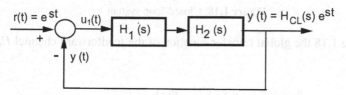

Figure 1.17. Closed-loop system

The output $y(t)$ of the closed-loop system in the case of an external reference $r(t) = e^{st}$ is written as

$$y(t) = H_{CL}(s)\, e^{st} = H_2(s)\, H_1(s)\, u_1(t) \tag{1.2.5}$$

But $u_1(t)$ is given by the relation

$$u_1(t) = r(t) - y(t) \tag{1.2.6}$$

Introducing this relation into Equation 1.2.5, one gets

$$[1 + H_2(s)\, H_1(s)]\, y(t) = H_2(s)\, H_1(s)\, r(t) \tag{1.2.7}$$

from which

$$H_{CL}(s) = \frac{H_2(s)\, H_1(s)}{1 + H_2(s)\, H_1(s)} \tag{1.2.8}$$

The stability of the closed-loop system will be determined by the real parts of the roots (poles) of the transfer function denominator $H_{CL}(s)$.

1.2.3 Steady-state Error

When carrying out the synthesis of a closed-loop system, our aim is to obtain an asymptotically stable system having a given response time, a specified overshoot and ensuring a zero steady-state error with respect to the reference signal. In Figure 1.18, it is desired that, in steady-state, $y(t)$ equals $r(t)$, i.e. the steady-state gain of the closed-loop system between y(t) and r(t) must be equal to 1.

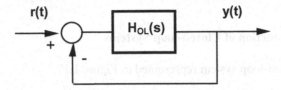

Figure 1.18. Closed-loop system

In Figure 1.18 the global transfer function of the feedforward channel $H_{OL}(s)$ is of the form

$$H_{OL}(s) = \frac{b_0 + b_1 s + \ldots + b_m s^m}{a_0 + a_1 s + \ldots + a_n s^n} = \frac{B(s)}{A(s)} \tag{1.2.9}$$

and the transfer function in closed-loop is given by

$$H_{CL}(s) = \frac{H_{OL}(s)}{1 + H_{OL}(s)} = \frac{B(s)}{A(s) + B(s)} \tag{1.2.10}$$

The steady-state corresponds to a zero frequency ($s = 0$). The steady-state gain is obtained by making $s = 0$ in the transfer function given by Equation 1.2.10.

$$y = H_{CL}(0)r = \frac{B(0)}{A(0) + B(0)} r = \frac{b_0}{a_0 + b_0} r \tag{1.2.11}$$

in which y and r represent the stationary values of the output and the reference.
    To obtain a unitary steady-state gain ($H_{CL}(0) = 1$), it is necessary that

$$\frac{b_0}{a_0 + b_0} = 1 \Rightarrow a_0 = 0 \tag{1.2.12}$$

This implies that the denominator of the transfer function $H(s)$ should be of the following form:

$$A(s) = s(a_1 s + a_2 s^2 + \ldots + a_{n-1} s^{n-1}) = s.A'(s) \tag{1.2.13}$$

and, respectively:

$$H_{OL}(s) = \frac{1}{s} \cdot \frac{B(s)}{A'(s)}$$ (1.2.14)

Thus to obtain a zero steady-state error in closed-loop when the reference is a step, the transfer function of the feedforward channel must contain an *integrator*.

This concept can be generalized for the case of time varying references as indicated below with the *internal model principle:* to obtain a zero steady-state error, $H_{OL}(s)$ must contain the *internal model* of the reference $r(t)$.

The internal model of the reference is the transfer function of the filter that generates $r(t)$ from the Dirac pulse. E.g., $step = (1/s) \cdot Dirac$, $ramp = (1/s^2) \cdot Dirac$). For more details see Appendix A.

Therefore, for a ramp reference, $H_{OL}(s)$ must contain a double integrator in order to obtain a zero steady-state error.

1.2.4 Rejection of Disturbances

Figure 1.19 represents the structure of a closed-loop system in the presence of a disturbance acting on the controlled output. $H_{OL}(s)$ is the open-loop global transfer function (controller + plant) and is given by Equation 1.2.9.

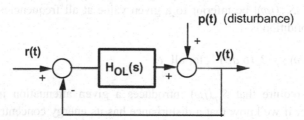

Figure 1.19. Closed-loop system in the presence of disturbances

Generally, we would prefer that the influence of the disturbance $p(t)$ on the system output be as weak as possible, at least in given frequency regions. In particular, we would prefer that the influence of a constant disturbance (step disturbance), often called "load disturbance", be zero during in steady-state regime $(t \to \infty, s \to 0)$.

The transfer function between the disturbance and the output is written as:

$$S_{yp}(s) = \frac{1}{1 + H_{OL}(s)} = \frac{A(s)}{A(s) + B(s)}$$ (1.2.15)

$S_{yp}(s)$ is called "output sensitivity function".

The steady-state regime corresponds to $s = 0$.

$$y = S_{yp}(0)p = \frac{A(0)}{A(0)+B(0)}p = \frac{a_0}{a_0+b_0}p \qquad (1.2.16)$$

in which y and p represent the steady-state values of the output and respectively of the disturbance.

$S_{yp}(0)$ must be zero for a perfect rejection of the disturbance in steady-state regime. It follows (as in Section 1.2.3) that, in order to obtain the desired property, we must have $a_0 = 0$. This implies the presence of an *integrator* in the direct path in order to have a perfect rejection of a step disturbance during steady-state regime (see previous section).

As a general rule, the direct path must contain the *internal model* of the disturbance in order to obtain a perfect rejection of a deterministic disturbance (see previous section).

Example: Sinusoidal Disturbance of Constant Frequency.

The internal model of the sinusoid is $1/(1+s^2/\omega_0^2)$ (the transfer function of the filter which, excited by a Dirac pulse, generates a sinusoid). For a perfect rejection (asymptotically) of this disturbance the controller must contain the transfer function $1/(1+s^2/\omega_0^2)$.

In general, we also have to check if there is not an amplification of the disturbance's effect in certain frequency regions. That is why we must require that the modulus of $|S_{yp}(j\omega)|$ be inferior to a given value at all frequencies. A typical value for this condition is

$$| S_{yp}(j\omega) | < 2 \ (6\ dB) \quad \text{for all } \omega \qquad (1.2.17)$$

We may also require that $S_{yp}(j\omega)$ introduces a given attenuation in a certain frequency range, if we know that a disturbance has its energy concentrated in this frequency range.

1.2.5 Analysis of Closed-loop Systems in the Frequency Domain: Nyquist Plot and Stability Criterion

The transfer function of the open-loop $H_{OL}(s)$ (Figure 1.19) can be represented in the complex plane when ω varies from 0 to ∞ as

$$H_{OL}(j\omega) = \text{Re}\,H_{OL}(j\omega) + j\,\text{Im}\,H_{OL}(j\omega) = |H_{OL}(j\omega)|.\angle\phi(\omega) \qquad (1.2.18)$$

The plot of the transfer function in this plane is graduated in frequencies (rad/s). This representation is often called a *Nyquist plot (or hodograph)*.

Figure 1.20 shows the Nyquist plot for $H_1(s) = 1/(1+s)$ and $H_2(s) = 1/[s(1+s)]$. Note that the plot of $H_2(s)$ corresponds to the typical case where an integrator is present in the loop (to ensure a zero steady-state error).

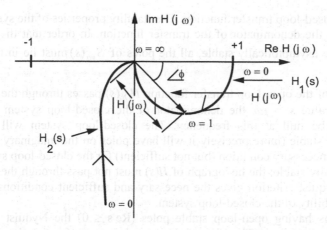

Figure 1.20. Nyquist plot for $H_1(s) = 1/(1+s)$ and $H_2(s) = 1/[s(1+s)]$

These curves show the gain and phase of the transfer function at different frequencies. The vector joining the origin to a point on the hodograph of the transfer function represents $H(j\omega)$ for a certain frequency ω.

In this diagram, the point $[-1, j0]$ plays a particularly important role (critical point). We can see in Figure 1.21 that the vector connecting the point $[-1, j0]$ to the plot of the open-loop transfer function $H_{OL}(j\omega)$ is given by

$$S_{yp}^{-1}(j\omega) = 1 + H_{OL}(j\omega) \tag{1.2.19}$$

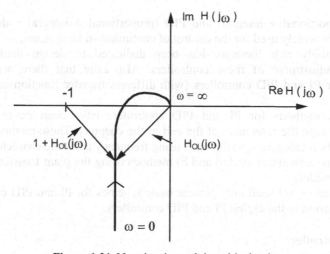

Figure 1.21. Nyquist plot and the critical point

This vector corresponds to the inverse of the output sensitivity function (see Equation 1.2.15). The *denominator* of the output sensitivity function defines the poles of the closed-loop transfer function. The stability properties of the system are determined by the denominator of the transfer function. In order that the closed-loop system be asymptotically stable, all the poles of $S_{yp}(s)$ must lie in the half-plane Re $s < 0$.

If the plot of the open-loop transfer function $H_{OL}(s)$ passes through the point [-1, j0], for a value $s = j\omega$, the denominator of the closed-loop system transfer function will be null at this frequency. The closed-loop system will not be asymptotically stable (more precisely it will have poles on the imaginary axis). It then follows a necessary condition (but not sufficient) for the closed-loop system to be asymptotically stable: the hodograph of $H(s)$ must not pass through the point [-1, j0]. The Nyquist criterion gives the necessary and sufficient conditions for the asymptotic stability of the closed-loop system.

For systems having open-loop stable poles (Re s ≤ 0) the Nyquist stability criterion is expressed as: *The plot of the open-loop transfer function H_{OL} (s) traversed in the sense of growing frequencies (from $\omega = 0$ to $\omega = \pi$) must leave the critical point [-1, j0] on the left.*

As a general rule, a controller will be computed for the nominal model of the plant so that the closed-loop system be asymptotically stable, i.e. H_{OL} (s) will leave the critical point on the left.

It is also obvious that the minimal distance to the critical point will characterize the "stability margin" or "robustness" of the closed-loop system in relation to variations of the system parameters (or uncertainties in parameter values).

1.3 PI and PID Controllers

The PI (proportional + integral) and PID (proportional + integral + derivative) controllers are widely used for the control of continuous-time systems.

An extremely rich literature has been dedicated to design methods and parameters adjustment of these controllers. Also note that there are several structures for PI and PID controllers (with different transfer function and tuning parameters).

Synthesis methods for PI and PID controllers have been developed and implemented (see the references at the end of the chapter). These methods can be divided into two categories: a) methods using frequency and time characteristics of the plant (non–parametric model) and b) methods using the plant transfer function (parametric model).

In this section, we shall only present basic schemes for PI and PID controllers as an introduction to the digital PI and PID controllers.

1.3.1 PI Controller

In general PI controllers have as input the difference between the reference and the measured output and as output the control signal delivered to the actuator (see Figure 1.15). A typical transfer function of a PI controller is

$$H_R(s) = K\left[1 + \frac{1}{T_i s}\right] = \frac{K(T_i s + 1)}{T_i s} \tag{1.3.1}$$

in which K is called the *proportional gain* and T_i the *integral action* of the PI controller. There also exist, however also PI controllers with independent actions, i.e.

$$H_R(s) = K_p + \frac{1}{T_i s}$$

In certain situations the proportional action may operate only on the measured output.

1.3.2 PID Controller

The transfer function of a typical PID controller is

$$H_{PID}(s) = K\left(1 + \frac{1}{T_i s} + \frac{T_d s}{1 + \frac{T_d}{N} s}\right) \tag{1.3.2}$$

in which K specifies the *proportional gain*, T_i characterizes the *integral action*, T_d characterizes the *derivative action* and $1 + (T_d / N) s$ introduces a filtering effect on the derivative action (low-pass filter).

By summing up the three terms, the transfer function given by Equation 1.3.2 can also be rewritten as

$$H_{PID}(s) = \frac{K\left[1 + s\left(T_i + \frac{T_d}{N}\right) + s^2\left(T_i T_d + \frac{T_i T_d}{N}\right)\right]}{T_i s\left(1 + \frac{T_d}{N} s\right)} \tag{1.3.3}$$

Several structures for PID controllers exist. In addition there are situations when the proportional and derivative actions act only on the measured output.

1.4 Concluding Remarks

The behavior of controlled plants around an operating point can in general be described by linear dynamic models. Linear dynamic models are characterized in the time domain by linear differential equations and in the frequency domain by transfer functions.

Control systems are closed-loop systems that contain the plant, the controller and the feedback connection. For these systems, the control applied to the plant is a function of the difference between the desired value and the measured value of the controlled variable. Control systems are characterized by a dynamic model that depends upon the structure and the coefficients of the plant and controller transfer functions.

The desired control performances can be expressed in terms of the desired characteristics of the dynamic model of the closed-loop system (ex.: transfer function with specified coefficients). This allows the synthesis of the controller if the plant model is known.

The plot in the complex plane of the transfer function of the open-loop system (controller + plant), also called *Nyquist plot,* plays an important role in the assessment of controller qualities. In particular, it allows studying of the stability and robustness properties of the closed-loop system.

1.5 Notes and References

Many books deal with the fundamentals of continuous-time control. Among the different titles we can mention:

Takahashi Y., Rabins M., Auslander D. (1970) Control. Addison Wesley, Readings, Mass

Franklin G., Powell J.D. (1986) Feedback Control of Dynamic Systems. Addison Wesley, Reading, Mass

Ogata K. (1990) Modern Control Engineering (second edition). Prentice Hall, N.J

Kuo B.C. (1991) Automatic Control Systems (sixth edition). Prentice Hall, N.J

PID controller adjustment techniques are discussed in:

Ziegler J.G., Nichols N.B. (1942) Optimum Settings for Automatic Controllers. Trans. ASME, vol. 64, pp. 759-768

Shinskey F.G. (1979) Process Control Systems. McGraw-Hill, N.Y.

Aström K.J., Hägglund I. (1995) PID Controllers Theory, Design and Tuning, 2nd edition. ISA, Research Triangle Park, N.C., U.S.A

Voda A., Landau I.D. (1995a) A method for the auto-calibration of PID controllers. Automatica, vol. 31, no. 1, pp. 45-53.

Voda A., Landau I.D. (1995b) The auto-calibration of PI controllers based on two frequency measurements. Int. J. of Adaptive Control and Signal Processing, vol. 9, no. 5, pp. 395-422

as well as in (Takahashi *et al.* 1970) and (Ogata 1990).

2

Computer Control Systems

In this chapter we present the elements and the basic concepts of computer-controlled systems. The discretization and choice of sampling frequency will be first examined, followed by a study of discrete-time models in the time and frequency domains, discrete-time systems in closed loop and basic principles for designing digital controllers.

2.1 Introduction to Computer Control

The first approach for introducing a digital computer or a microprocessor into a control loop is indicated in Figure 2.1. The measured error between the reference and the output of the plant is converted into digital form by an analog-to-digital converter (ADC), at sampling instants *k* defined by the synchronization clock. The computer interprets the converted signal *y(k)* as a sequence of numbers, which it processes using a **control algorithm** and generates a new sequence of numbers {*u(k)*} representing the control. By means of a digital-to-analog converter (DAC), this sequence is converted into an analog signal, which is maintained constant between the sampling instants by a zero-order hold (ZOH). The cascade: ADC-computer-DAC should behave in the same way as an analog controller (PID type), which implies the use of a high sampling frequency but the algorithm implemented on the computer is very simple (we just do not make use of the potentialities of the digital computer!).

A second and much more interesting approach for the introduction of a digital computer or microprocessor in a control loop is illustrated in Figure 2.2 which can be obtained from Figure 2.1 by moving the reference-output comparator after the analog-to-digital converter. The reference is now specified in a digital way as a sequence provided by a computer.

In Figure 2.2 the set DAC - plant - ADC is interpreted as a discretized system, whose control input is the sequence {*u(k)*} generated by the computer, the output being the sequence {*y(k)*} resulting from the A/D conversion of the system output *y(t)*. This discretized system is characterized by a **"discrete-time model"**, which describes the relation between the sequence of numbers $\{u(k)\}$ and the sequence of numbers $\{y(k)\}$. This model is related to the continuous-time model of the plant.

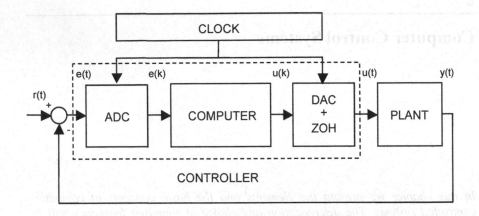

Figure 2.1. Digital realization of an « analog » type controller

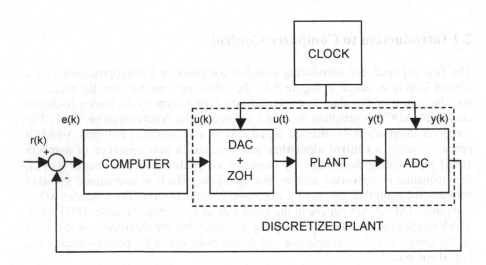

Figure 2.2. Digital control system

This approach offers several advantages. Among these advantages here we recall the following:

1. The sampling frequency is chosen in accordance with the "bandwidth" of the continuous-time system (it will be much lower than for the first approach).

2. Possibility of a direct design of the control algorithms tailored to the discretized plant models.
3. Efficient use of the computer since the increase of the sampling period permits the computation power to be used in order to implement algorithms which are more performant but more complex than a PID controller, and which require a longer computation time.

In fact, if one really wants to take advantage of the use of a digital computer in a control loop, the "language" must also be changed. This may be achieved by replacing the continuous-time system models by discrete-time system models, the continuous-time controllers by digital control algorithms, and by using dedicated control design techniques.

The changing over to this new "language" (discrete-time dynamic models) makes it possible to use various high performing control strategies which cannot be implemented by analog controllers.

The operating details of the ADC (analog-to-digital converter), the DAC (digital-to-analog converter) and the ZOH (zero-order hold) are illustrated in Figure 2.3.

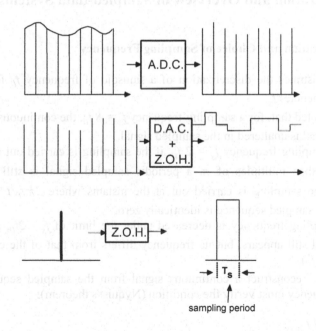

Figure 2.3. Operation of the analog-to-digital converter (ADC), the digital-to-analog converter (DAC) and the zero-order hold (ZOH)

The analog-to-digital converter implements two functions:
1. Analog signal sampling: this operation consists in the replacement of the continuous signal with a sequence of values equally spaced in the time domain (the temporal distance between two values is the sampling period), as these values correspond to the continuous signal amplitude at sampling instants.
2. Quantization: this is the operation by means of which the amplitude of a signal is represented with a discrete set of different values (quantized values of the signal), generally coded with a binary sequence.

The general use of high-resolution A/D converters (where the samples are coded with 12 bits or more) allows one to consider the quantification effects as negligible, and this assumption will hold in the following. Quantization effects will be taken into account in Chapter 8.

The digital-analog converter (DAC) converts at the sampling instants a discrete signal, digitally coded, in a continuous signal.

The zero-order hold (ZOH) keeps constant this continuous signal between two sampling instants (sampling period), in order to provide a continuous-time signal.

2.2 Discretization and Overview of Sampled-data Systems

2.2.1 Discretization and Choice of Sampling Frequency

Figure 2.4 illustrates the discretization of a sinusoid of frequency f_0 for several sampling frequencies f_s.

It can be noted that, for a sampling frequency $f_s = 8 f_0$, the continuous nature of the analog signal is unaltered in the sampled signal.

For the sampling frequency $f_s = 2 f_0$, if the sampling is carried out at instants $2\pi f_0 t$ other than multiples of π, a periodic sampled signal is still obtained. However if the sampling is carried out at the instants where $2\pi f_0 t = n\pi$, the corresponding sampled sequence is identically zero.

If the sampling frequency is decreased under the limit of $f_s = 2f_0$, a periodic sampled signal still appears, but its frequency differs from that of the continuous signal ($f = f_s - f_0$).

In order to reconstruct a continuous signal from the sampled sequence, the sampling frequency must verify the condition (Nyquist's theorem):

$$f_s > 2 f_{max} \tag{2.2.1}$$

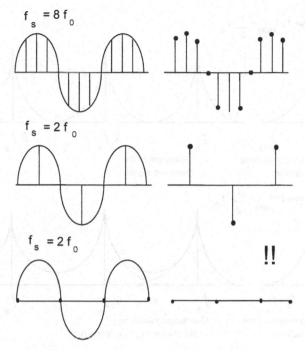

$f_s = 8 f_0$

$f_s = 2 f_0$

$f_s = 2 f_0$

!!

Figure 2.4. Sinusoidal signal discretization

in which f_{max} is the maximum frequency to be transmitted. The frequency $f_s = 2 f_{max}$ is a theoretical limit; in practice, a higher sampling frequency must be chosen.

The existence of a maximum limit for the frequency that may be converted without distortion, for a given sampling frequency, is also understandable when it is observed that the sampling of a continuous-time signal is a "magnitude modulation" of a "carrier" frequency f_s (analogy with the magnitude modulation in radio transmitters). The modulation effect may be observed in the replication of the spectrum of the modulating signal (in our case the continuous signal) around the sampling frequency and its multiples.

The spectrum of the sampled signal, if the maximum frequency of the continuous signal (f_{max}) is less than $(1/2) f_s$, is represented in the upper part of Figure 2.5.

The spectrum of the sampled signal, if $f_{max} > (1/2) f_s$, is represented in the lower part of Figure 2.5. The phenomenon of overlapping (aliasing) can be observed. This corresponds to the appearance of distortions. The frequency $(1/2) f_s$, which defines the maximum frequency (f_{max}) admitted for a sampling with no distortions, is known as "Nyquist frequency" (or Shannon frequency).

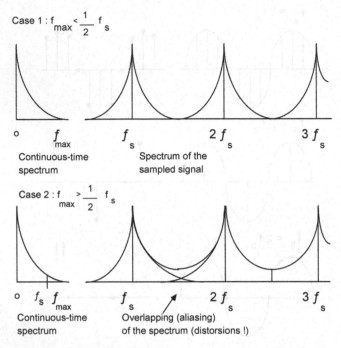

Figure 2.5. Spectrum of a sampled signal

For a given sampling frequency, in order to avoid the *folding* (aliasing) of the spectrum and thus of the distortions, the analog signals must be filtered prior to sampling to ensure that:

$$f_{max} < \frac{1}{2} f_s \qquad (2.2.2)$$

The filters used are known as "anti-aliasing filters". A good anti-aliasing filter must have a minimum of two cascaded second-order cells ($f_{max} << (1/2) f_s$). An example of an anti-aliasing filter of this type is given in Figure 2.6. These filters must introduce a large attenuation at frequencies higher than $(1/2) f_s$, but their bandwidth must be higher than the required bandwidth of the closed loop system (generally higher than open loop system bandwidth). Circuits of this type (or more complex) are currently available.

$$\rightarrow \boxed{\frac{\omega_0^2}{\omega_0^2 + 2\omega_0 \zeta s + s^2}} \rightarrow \boxed{\frac{\omega_0^2}{\omega_0^2 + 2\omega_0 \zeta s + s^2}} \rightarrow$$

Figure 2.6. Anti-aliasing filter

In the case of very low frequency sampling, first a sampling at a higher frequency is carried out (integer multiple of the desired frequency), using an appropriate analog anti-aliasing filter. The sampled signal thus obtained is passed through a digital anti-aliasing filter followed by a frequency divider (decimation) thereby giving a sampled signal having the required frequency. This procedure is shown in Figure 2.7. It is also employed every time the frequency of data acquisition is higher than the sampling frequency chosen for the loop that must be controlled (the sampling frequency should be an integer divider of the acquisition frequency).

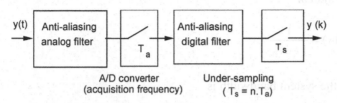

Figure 2.7. Anti-aliasing filtering with under-sampling

2.2.2 Choice of the Sampling Frequency for Control Systems

The sampling frequency for digital control systems is chosen according to the desired bandwidth of the closed loop system . Note that, no matter how the desired performances are specified, these can always be related to the closed loop system bandwidth.

Example: Let us consider the performances imposed in Section 1.1.6 on the step response (maximum overshoot 5%, rise time 2.75 s). The transfer function to be determined corresponds to the desired closed loop system transfer function. From the diagrams given in Figure 1.11 we have deduced that the closed loop transfer function must be a normalized second-order transfer function with $\zeta=0.7$ and $\omega_0=1$ *rad/s*. By immediately using the diagrams given in Figure 1.12, it can be observed that the bandwidth of the closed loop system is approximately equal to

$$f_B^{CL} = \frac{1}{2\pi} Hz$$

The rule used to choose the sampling frequency in control systems is the following:

$$f_s = (6\,to\,25)\,f_B^{CL} \tag{2.2.3}$$

where

f_s: sampling frequency, f_B^{CL} : closed loop system bandwidth

Rule of Equation 2.2.3 is equally used in open loop, when it is desired to choose the sampling frequency in order to identify the discrete-time model of a plant. In this case f_B^{CL} is replaced by an estimation of the bandwidth of the plant.

For information purposes, Table 2.1 gives the sampling periods ($T_s = 1/f_s$) used for the digital control of different types of plants.

The rule for choosing the sampling frequency given in Equation 2.2.3 can be connected to the transfer function parameters.

First- order system

$$H(s) = \frac{1}{1 + sT_0}$$

In this case the system bandwidth is

$$f_B = f_0 = \frac{1}{2\pi T_0}$$

(an attenuation greater than 3 db is introduced for frequencies higher than $\omega_0 = 1/T_0 = 2\pi f_0$).

Table 2.1. Choice of the sampling period for digital control systems (indicative values)

Type of variable (or plant))	Sampling period (s)
Flow rate	1 – 3
Level	5 – 10
Pressure	1 – 5
Temperature	10 - 180
Distillation	10 - 180
Servo-mechanisms	0.001 - 0.05
Catalytic reactors	10 - 45
Cement plants	20 - 45
Dryers	20 – 45

By applying the rule of Equation 2.2.3 the condition for choosing the sampling period is obtained ($T_s = 1/f_s$):

$$\frac{T_0}{4} < T_s < T_0 \tag{2.2.4}$$

This corresponds to the existence of two to nine samples on the rise time of a step response.

Second- order system

$$H(s) = \frac{\omega_0^2}{\omega_0^2 + 2\zeta\,\omega_0 s + s^2}$$

The bandwidth of the second-order system depends on ω_0 and on ζ (see Figure 1.12).

For example:

$$\zeta = 0.7 \Rightarrow f_B = \frac{\omega_0}{2\pi}$$

$$\zeta = 1 \Rightarrow f_B = \frac{0.6\,\omega_0}{2\pi}$$

By applying the rule of Equation 2.2.3, the following relations are obtained between the natural frequency ω_0 and the sampling period T_s:

$$0.25 \leq \omega_0 T_s \leq 1 \; ; \quad \zeta = 0.7 \tag{2.2.5}$$

and

$$0.4 \leq \omega_0 T_s \leq 1.75 \,; \quad \zeta = 1 \tag{2.2.6}$$

The lower values correspond to the choice of a high sampling frequency and the upper values to the choice of a low sampling frequency.

For simplicity's sake, given that in closed loop the behavior frequently chosen as the desired behavior is that of a second order having a damping factor ζ between 0.7 and 1, the following rule can be used (approximation of Equations 2.2.5 and 2.2.6):

$$0.25 \leq \omega_0 T_s \leq 1.5 \; ; \quad 0.7 \leq \zeta \leq 1 \tag{2.2.7}$$

2.3 Discrete-time Models

2.3.1 Time Domain

Figure 2.8 illustrates the response of a continuous-time system to a step input, a response that can be simulated by a first order system (an integrator with a feedback gain indicated in the figure).

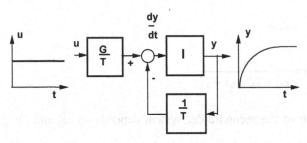

Figure 2.8. Continuous-time model

The corresponding model is described by the differential equation

$$\frac{dy}{dt} = -\frac{1}{T}y(t) + \frac{G}{T}u(t)$$ (2.3.1)

or by the transfer function

$$H(s) = \frac{G}{1 + sT}$$ (2.3.2)

where T is the time constant of the system and G is the gain.

If the input $u(t)$ and the output $y(t)$ are sampled with a specified sampling period, the representations of $u(t)$ and $y(t)$ are obtained as number sequences in which t (or k) is now the *normalized discrete-time* (real time divided by the sampling period, $t = t/T_s$). The relation between the input sequence $\{u(t)\}$ and the output sequence $\{y(t)\}$ can be simulated by the scheme given in Figure 2.9 by using a delay (backward shift) operator (symbolized by q^{-1}: $y(t-1) = q^{-1}y(t)$), instead of an integrator.

This relation is described in the time domain by the algorithm (known as recursive equation or difference equation)

$$y(t) = -a_1 y(t-1) + b_1 u(t-1)$$ (2.3.3)

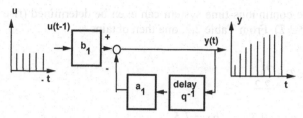

Figure 2.9. Discrete-time model

Let us now examine in greater detail the discrete-time model given by Equation 2.3.3 for a zero initial condition ($y(0) = 0$) and a discrete-time unit step input:

$$u(t) = \begin{cases} 0 & t < 0 \\ 1 & t \ge 0 \end{cases}$$

The response is directly computed by recursively using Equation 2.3.3 from $t = 0$ (in the case of discrete-time models there are no problem with the integration of the differential equations like in continuous time). We shall examine two cases.

Case 1. $a_1 = -0.5$; $b_1 = 0.5$

The output values for different instants are given in Table 2.2 and the corresponding sequence is represented in Figure 2.10.

Table 2.2. Step response of a first-order discrete-time model ($a_1 = -0.5$, $b_1 = 0.5$)

T	0	1	2	3	4	5
y(t)	0	0.5	0.75	0.875	0.937	0.969

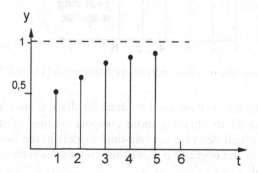

Figure 2.10. Step response of a first-order discrete- time model ($a_1 = -0.5$, $b_1 = 0.5$)

It is observed that the response obtained resembles the step response of a continuous-time first order system which has been sampled. An equivalent time constant for the continuous-time system can even be determined (rise time from 0 to 90 %: $t_R = 2.2\ T$). From Table 2.2, one then obtains

$$\frac{3T_s}{2.2} < T < \frac{4T_s}{2.2}$$

Case 2. $a_1 = 0.5\ ;\quad b_1 = 1.5$

Output values for different instants are given in Table 2.3 and the corresponding sequence is represented in Figure 2.11.

Table 2.3. Step response of a first-order discrete-time model (a_1=0.5; b_1=1.5)

T	0	1	2	3	4	5
y(t)	0	1.5	0.75	1.125	0.937	1.062

An oscillatory damped response is observed with a period equal to two sampling periods. This type of phenomenon cannot result from the discretization of a continuous-time first order system, since this latter is always a-periodic. It may thus be concluded that the first order discrete-time model corresponds to the discretization of a first order continuous-time system only if a_1 is negative[1].

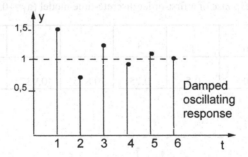

Figure 2.11. Step response of a first-order discrete-time model (a_1=0.5; b_1=1.5)

We now go back to the method used to describe discrete-time models. The delay operator q^{-1} is used to obtain a more compact writing of the recursive (difference) equations which describe discrete-time models in the time domain (it has the same function as the operator $p = d/dt$ for continuous-time systems). The following relations hold:

---

[1] For a positive a_1, this corresponds to the discretization of a 2nd order system, with a damped resonant frequency equal to $0.5f_s$ (see Section 2.3.2).

$$q^{-1}y(t) = y(t-1)$$
$$q^{-d}y(t) = y(t-d)$$

(2.3.4)

By using the operator q^{-1}, Equation 2.3.3 is rewritten as

$$(1 + a_1 q^{-1}) y(t) = b_1 q^{-1} u(t)$$

(2.3.5)

Discrete-time models may also be obtained by the discretization of the differential equations describing continuous-time models. This operation is used for the simulation of continuous-time models on a digital computer.

Let us consider Equation 2.3.1 and approximate the derivative by

$$\frac{dy}{dt} = \frac{y(t+T_s) - y(t)}{T_s}$$

(2.3.6)

Equation 2.3.1 will be rewritten as

$$\frac{y(t+T_s) - y(t)}{T_s} + \frac{1}{T} y(t) = \frac{G}{T} u(t)$$

(2.3.7)

By multiplying both sides of Equation 2.3.7 by T_s, and with the introduction of the normalized time $t (= t/T_s)$, it follows that

$$y(t+1) + \left(\frac{T_s}{T} - 1\right) y(t) = \frac{G}{T} T_s u(t)$$

(2.3.8)

which can be further rewritten as:

$$(1 + a_1 q^{-1}) y(t+1) = b_1 u(t)$$

(2.3.9)

where

$$a_1 = \frac{T_s}{T} - 1 \quad (<0) \; ; \quad b_1 = \frac{G}{T} T_s$$

Shifting Equation 2.3.9 by one step, Equation 2.3.3 is obtained.

We point out that, in order to represent a first-order continuous model with Equation 2.3.9, the condition $a_1 < 0$ must be verified. As a consequence, the sampling period T_s must be smaller than time constant T $(T_s < T)$. This result corresponds to the upper bound in Equation 2.2.4, introduced for sampling period selection of a first-order system as a function of the desired closed loop bandwidth.

If Equation 2.3.6 is the approximation of the "derivative", the digital integrator equation can be directly deduced. Thus, if normalized time is used, Equation 2.3.6 is written as

$$\frac{d}{dt}y = py \approx y(t) - y(t-1) = (1 - q^{-1})y(t) \qquad (2.3.10)$$

where $(1 - q^{-1})$ is now equivalent to p. As the integration is the opposite of the differentiation, one obtains:

$$s(t) = \int y\, dt = \frac{1}{p}y \approx \frac{1}{1 - q^{-1}}y(t) \qquad (2.3.11)$$

Multiplying both sides of Equation 2.3.11 by $(1 - q^{-1})$, it follows that

$$s(t)(1 - q^{-1}) = y(t) \qquad (2.3.12)$$

which we can rewrite as

$$s(t) = s(t\text{-}1) + 1 \cdot y(t) \qquad (2.3.13)$$

corresponding to the approximation of the integration operation by means of the rectangular rule, as illustrated in Figure 2.12 (if continuous-time is used, Equation 2.3.13 is written as $s(t) = s(t-T_s) + T_s y(t)$).

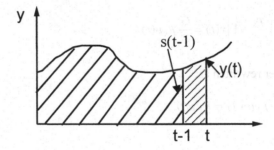

Figure 2.12. Numerical integration

2.3.2 Frequency Domain

The study of continuous-time models in the frequency domain has been carried out considering a periodic input of the complex exponential type

$$e^{j\omega t} = \cos \omega t + j \sin \omega t$$

or e^{st} with $s = \sigma + j\omega$.

For the study of discrete-time models in the frequency domain we shall consider complex (sampled) exponentials, i.e. sequences resulting from complex continuous-time exponentials evaluated at the sampling instants $t = k\,T_s$.

These sequences will thus be written as

$$e^{j\omega T_s k} \; ; \; e^{sT_s k} \; ; \; k=1,2,3...$$

Since the discrete-time models being considered are linear, if a signal of a certain frequency is applied to the input, a signal of the same frequency, but amplified or attenuated according to the frequency, will be found at the output. This is summarized in Figure 2.13. in which $H(s)$ is the "transfer function" of the system that expresses the dependence of the gain and the phase-deviation on the complex frequency s ($s = \sigma + j\omega$).

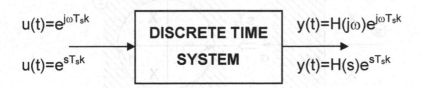

Figure 2.13. Frequency response of a discrete-time system

If the input of the system is in the form $e^{sT_s k}$, the output will be

$$y(t) = H(s)e^{sT_s k} \qquad (2.3.14)$$

and respectively

$$y(t-1) = H(s)e^{sT_s(k-1)} = e^{-sT_s}H(s)e^{sT_s k} = e^{-sT_s}y(t) \qquad (2.3.15)$$

It is thus observed that shifting backward by one step is equivalent to multiplying by e^{-sT_s}.

Let now determine the transfer function related to the recursive Equation 2.3.3. In this case $u(t) = e^{sT_s k}$ and the output will be in the form of Equation 2.3.14. By also using Equation 2.3.15 one obtains:

$$(1 + a_1 e^{-sT_s})H(s)e^{sT_s k} = b_1 e^{-sT_s}e^{sT_s k} \qquad (2.3.16)$$

from which results

$$H(s) = \frac{b_1 e^{-sT_s}}{1 + a_1 e^{-sT_s}} \tag{2.3.17}$$

We consider now the following change of variable:

$$z = e^{sT_s} \tag{2.3.18}$$

which corresponds to the transformation of the left half-plane of the s-plane into the interior of the unit circle centered at the origin in the z- plane, as illustrated by Figure 2.14.

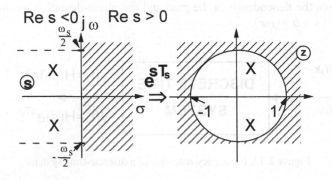

Figure 2.14. Effect of the transformation $z = e^{sT_s}$

With the transformation given by Equation 2.3.18 the transfer function given in Equation 2.3.17 becomes

$$H(z^{-1}) = \frac{b_1 z^{-1}}{1 + a_1 z^{-1}} \tag{2.3.19}$$

Note that the transfer function in z^{-1} can be directly obtained from the recursive Equation 2.3.3 by using the delay operator q^{-1} (see Equation 2.3.5), and afterwards by formally computing the ratio $y(t)/u(t)$ and replacing q^{-1} with z^{-1}. This procedure can obviously be applied to all models described by linear difference equations with constant coefficients, regardless of their complexity. The same result can be also derived by means of the z - *transform* (see Appendix A, Section A.2)

We also remark that the transfer functions of discrete-time models are often written in terms of q^{-1}. It is of course understood that the meaning of q^{-1} varies according to the context (delay operator or complex variable). When q^{-1} is considered as a delay operator, the expression $H(q^{-1})$ is named "transfer operator". It must be observed that the representation by transfer operators can also be used for models described by linear difference equations with time varying coefficients as well. In contrast, the interpretation of q^{-1} as a complex variable (z^{-1}) is only possible for linear difference equations with constant coefficients.

Properties of the Transformation $z = e^{sT_s}$

The transformation of Equation 2.3.18 is not bijective because several points in the s- plane are transformed at the same point in the z-plane. Nevertheless, we are interested in the s-plane being delimited between the two horizontal lines crossing the points $[0, +j\omega_s / 2]$ and $[0, -j\omega_s / 2]$ where $\omega_s = 2\pi f_s = 2\pi/T_s$. This region is called "primary strip".

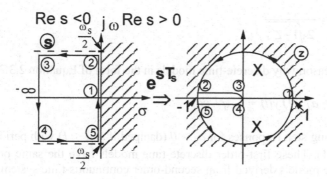

Figure 2.15. Effects of the transformation $z = e^{sT_s}$ on the points located in the "primary strip" in s-plane

The complementary bands are outside the frequency domain of interest if the conditions of the Shannon theorem (Section 2.2.1) have been satisfied.

Figure 2.15 gives a detailed image of the effects of the transformation $z = e^{sT_s}$ for the points that are inside the "primary strip".

Attention must be focused on an important aspect for continuous second-order systems in the form:

$$\frac{\omega_0^2}{\omega_0^2 + 2\zeta \omega_0 s + s^2} \quad (\zeta < 1)$$

for which the resonant damped frequency is equal to half the sampling frequency:

$$\omega_0 \sqrt{1 - \zeta^2} = \omega_s / 2$$

The image of their conjugates poles

$$s_{1,2} = -\zeta \omega_0 \pm j \frac{\omega_s}{2}$$

through the transformation $z = e^{sT_e}$ *corresponds to a single point placed on the real axis in the z- plane and with negative abscissa.*

One gets:

$$z_{1,2} = e^{s_{1,2}T_s} = e^{-\zeta\omega_0 T_s} \cdot e^{\pm j\frac{\omega_s}{2}T_s} = e^{-\zeta\omega_0 T_s} \cdot e^{\pm j\pi} = -e^{-\zeta\omega_0 T_s} = -e^{-\frac{\zeta}{\sqrt{1-\zeta^2}}\pi}$$

since:

$$\omega_0 = \frac{\omega_s}{2\sqrt{1-\zeta^2}}$$

This is the reason why discrete-time models in the form of Equation 2.3.3 such as

$$(1 + a_1 q^{-1}) y(t) = b_1 q^{-1} u(t)$$

give oscillating step responses for $a_1 > 0$ (damped if $|a_1| < 1$) with period $2T_s$ (see Section 2.3.1). These first-order discrete-time models have the same poles as the discrete-time models derived from second-order continuous-time systems having a damped resonant frequency equal to $\omega_s/2$.

2.3.3 General Forms of Linear Discrete-time Models

A linear discrete-time model is generally described as

$$y(t) = -\sum_{i=1}^{n_A} a_i\, y(t-i) + \sum_{i=1}^{n_B} b_i\, u(t-d-i) \tag{2.3.20}$$

in which d corresponds to a pure time delay which is an integer multiple of the sampling period.

Let us introduce the following notations:

$$1 + \sum_{i=1}^{n_A} a_i q^{-i} = A(q^{-1}) = 1 + q^{-1}A^*(q^{-1}) \tag{2.3.21}$$

$$A^*(q^{-1}) = a_1 + a_2 q^{-1} + \dots + a_{n_A} q^{-n_A+1} \tag{2.3.22}$$

$$\sum_{i=1}^{n_B} b_i q^{-i} = B(q^{-1}) = q^{-1}B^*(q^{-1}) \tag{2.3.23}$$

$$B^*(q^{-1}) = b_1 + b_2 q^{-1} + ... + b_{n_B} q^{-n_B+1}$$

(2.3.24)

By using the delay operator q^{-1} in Equation 2.3.20 and taking into account the notations of Equations 2.3.21 to 2.3.24, the Equation 2.3.20 describing the discrete-time system is written as

$$A(q^{-1}) y(t) = q^{-d} B(q^{-1}) u(t)$$

(2.3.25)

or in the predictive form (by multiplying both sides by q^d)

$$A(q^{-1}) y(t+d) = B(q^{-1}) u(t)$$

(2.3.26)

Equation 2.3.25 can also be written in a compact form using the *pulse transfer operator*

$$y(t) = H(q^{-1}) u(t)$$

(2.3.27)

where the pulse transfer operator is given by

$$H(q^{-1}) = \frac{q^{-d} B(q^{-1})}{A(q^{-1})}$$

(2.3.28)

The *pulse transfer function* characterizing the system described by Equation 2.3.20 is obtained from the pulse transfer operator given in Equation 2.3.28 by replacing q^{-1} with z^{-1} [2]

$$H(z^{-1}) = \frac{z^{-d} B(z^{-1})}{A(z^{-1})}$$

(2.3.29)

Pulse Transfer Function Order
To evaluate the order of a discrete time model represented by the pulse transfer function in the form of Equation 2.3.29, the representation in terms of positive power of z is needed. If d is the system pure time delay expressed as number of samples, n_A the degree of the polynomial $A(z^{-1})$ and n_B the degree of the polynomial $B(z^{-1})$, one must multiply both numerator and denominator of $H(z^{-1})$ by z^n in order to obtain a *proper*[3] pulse transfer function $H(z)$ on the positive powers of z, where

---

[2] The pulse transfer operator $H(q^{-1})$ can be used for a compact representation of the input-output relationship even in the case of $A(q^{-1})$ and $B(q^{-1})$ have time depending coefficients. The pulse transfer function $H(z^{-1})$ is only defined for the case of $A(q^{-1})$ and $B(q^{-1})$ are with constant coefficients.
[3] This means that the denominator degree is greater than (or equal to) the numerator degree.

$$n = max \ (n_A, \ n_B + d)$$

n represents the *discrete-time system order* (the higher power of a term in z in the pulse transfer function denominator).

Example 1:

$$H(z^{-1}) = \frac{z^{-3}(b_1 z^{-1} + b_2 z^{-2})}{1 + a_1 z^{-1}}$$

$n = max \ (1, \ 5) = 5$

$$H(z) = \frac{b_1 z + b_2}{z^5 + a_1 z^4}$$

Example 2:

$$H(z^{-1}) = \frac{b_1 z^{-1} + b_2 z^{-2}}{1 + a_1 z^{-1} + a_2 z^{-2}}$$

$n = max \ (2, \ 2) = 2$

$$H(z) = \frac{b_1 z + b_2}{z^2 + a_1 z + a_2}$$

One notes that the order n of an irreducible pulse transfer function also corresponds to the number of states for a minimal state space system representation associated to the transfer function (See Appendix C).

2.3.4 Stability of Discrete-time Systems

The stability of discrete-time systems can be studied either from the recursive (differences) equation describing the discrete-time system in the time domain, or from the interpretation of difference equations solutions as sums of discretized exponentials. We shall use examples to illustrate both these approaches.

Let us assume that the recursive equation is

$$y(t) = -a_1 y(t-1) \ ; \quad y(0) = y_0 \tag{2.3.30}$$

which is obtained from Equation 2.3.3 when the input $u(t)$ is identically zero. The free response of the system is written as

$$y(1) = -a_1 y_0 \ ; \quad y(2) = (-a_1)^2 y_0 \ ; \quad y(t) = (-a_1)^t y_0 \tag{2.3.31}$$

The asymptotic stability of the system implies

$$\lim_{t\to\infty} y(t) = 0 \tag{2.3.32}$$

The condition of asymptotic stability thus results from Equation 2.3.31. It is necessary and sufficient that

$$|a_1| < 1 \tag{2.3.33}$$

On the other hand, it is known that the solution of the recursive (difference) equations is of the form (for a first-order system):

$$y(t) = K\, e^{sT_s t} = Kz^t \tag{2.3.34}$$

By introducing this solution into Equation 2.3.30, and taking into account Equation 2.3.15, one obtains

$$(1 + a_1 e^{-sT_s})Ke^{sT_s t} = (1 + a_1 z^{-1})Kz^t = 0 \tag{2.3.35}$$

from which it follows that

$$z = e^{sT_s} = e^{(\sigma + j\omega)T_s} = e^{\sigma T_s} e^{j\omega T_s} = -a_1 \tag{2.3.36}$$

For this solution to be asymptotically stable, it is necessary that $\sigma = Re\, s < 0$ which implies that $e^{\sigma T_s} < 1$ and respectively $|z| < 1$ (or $|a_1| < 1$).

However, the term $(1 + a_1 z^{-1})$ is nothing more than the denominator of the pulse transfer function related to the system described by Equation 2.3.3 (see Equation 2.3.19).

The result obtained can be generalized. For a discrete-time system to be *asymptotically stable,* all the roots of the transfer function denominator must be inside the unit circle (see Figure 2.14):

$$1 + a_1 z^{-1} + \dots\dots + a_n z^{-n} = 0 \implies |z| < 1 \tag{2.3.37}$$

In contrast, if one or several roots of the transfer function denominator are in the region defined by $|z| > 1$ (outside the unit circle), this implies that $Re\, s > 0$ and thus the discrete-time system will be unstable.

As for the continuous-time case, some stability criteria are available (Jury criterion, Routh-Hurwitz criterion applied after the change of variable $w = (z + 1)/(z-1)$) for establishing the existence of unstable roots for a polynomial in the variable z with no explicit calculation of the roots (Åström and Wittenmark 1997).

A helpful tool to test z-polynomial stability is derived from a *necessary condition* for the stability of a z^{-1}-polynomial. This condition states: the evaluations of the polynomial $A(z^{-1})$ given by Equation 2.3.37 in $z = 1$, $(A(1))$ and in $z = -1$ $(A(-1))$ must be positive (the coefficient of $A(q^{-1})$ corresponding to z^0 is supposed to be positive).

Example:

$$A(z^{-1}) = 1 - 0.5 \, z^{-1} \quad \text{(stable system)}$$

$$A(1) = 1 - 0.5 = 0.5 > 0 \; ; \quad A(-1) = 1 + 0.5 = 1.5 > 0$$

$$A(z^{-1}) = 1 - 1.5 \, z^{-1}; \quad \text{(unstable system)}$$

$$A(1) = -0.5 < 0 \; ; \quad A(-1) = 2.5 > 0$$

2.3.5 Steady-state Gain

In the case of continuous-time systems, the steady-state gain is obtained by making $s = 0$ (zero frequency) in the transfer function. In the discrete case, $s = 0$ corresponds to

$$s = 0 \Rightarrow z = e^{sT} = 1 \tag{2.3.38}$$

and thus the steady-state gain $G(0)$ is obtained by making $z = 1$ in the pulse transfer function. Therefore for the first-order system one obtains:

$$G(0) = \left. \frac{b_1 z^{-1}}{1 + a_1 z^{-1}} \right|_{z=1} = \frac{b_1}{1 + a_1}$$

Generally speaking, the steady-state gain is given by the formula

$$G(0) = H(1) = \left. H(z^{-1}) \right|_{z=1} = \left. \frac{z^{-d} B(z^{-1})}{A(z^{-1})} \right|_{z=1} = \frac{\sum\limits_{i=1}^{n_B} b_i}{1 + \sum\limits_{i=1}^{n_A} a_i} \tag{2.3.39}$$

In other words, the steady-state gain is obtained as the ratio between the sum of the numerator coefficients and the sum of the denominator coefficients. This formula is quite different from the continuous-time systems, where the steady-state gain appears as a common factor of the numerator (if the denominator begins with 1).

The steady-state gain may also be obtained from the recursive equation describing the discrete-time models, the steady-state being characterized by $u(t) = const.$ and $y(t) = y(t-1) = y(t-2)....$

From Equation 2.3.3, it follows that

$$(1 + a_1)\, y(t) = b_1\, u(t)$$

and respectively

$$y(t) = \frac{b_1}{1 + a_1} u(t) = G(0)u(t)$$

2.3.6 Models for Sampled-data Systems with Hold

Up to this point we have been concerned with sampled-data systems models corresponding to the discretization of inputs and outputs of a continuous-time system. However, in a computer controlled system, the control applied to the plant is not continuous. It is constant between the sampling instants (effect of the zero-order hold) and varies discontinuously at the sampling instants, as is illustrated in Figure 2.16.

It is important to be able to relate the model of the discretized system, which gives the relation between the control sequence (produced by the digital controller) and the output sequence (obtained after the analog-to-digital converter), to the transfer function H(s) of the continuous-time system. The zero-order hold, whose operation is reviewed in Figure 2.17 introduces a transfer function in cascade with H(s).

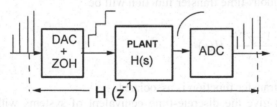

Figure 2.16. Control system using an analog-to-digital converter followed by a zero-order hold

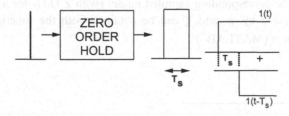

Figure 2.17. Operation of the zero-order hold

The hold converts a Dirac pulse given by the digital-to-analog converter at the sampling instant into a rectangular pulse of duration T_s, which can be interpreted as the difference between a step and the same step shifted by T_s. As the step is the integral of the Dirac pulse, it follows that the zero-order hold transfer function is

$$H_{ZOH}(s) = \frac{1 - e^{-sT_s}}{s} \qquad (2.3.40)$$

Equation 2.3.40 allows one to consider the zero-order hold as a filter having a frequency response given by

$$H_{ZOH}(j\omega) = \frac{1 - e^{-j\omega T_s}}{j\omega} = T_s \frac{\sin(\omega T_s / 2)}{\omega T_s / 2} e^{-j\omega \frac{T_s}{2}}$$

From the study of this response in the frequency region $0 \le f \le f_s / 2$ ($0 \le \omega \le \omega_s / 2$), one can conclude:

1. The ZOH gain at the zero frequency is equal to: $G_{ZOH}(0) = T_s$.
2. The ZOH introduces an attenuation at high frequencies. For $f = f_s / 2$ one gets $G(f_s / 2) = \frac{2}{\pi} T_s = 0.637 \, T_s \, (-3. \, 92 \, dB)$.
3. The ZOH introduces a phase lag which grows with the frequency. This phase lag is between 0 (for $f = 0$) and $-\pi/2$ (for $f = f_s / 2$) and should be added to the phase lag due to $H(s)$.

The global continuous-time transfer function will be

$$H'(s) = \frac{1 - e^{-sT_s}}{s} H(s) \qquad (2.3.41)$$

to which a pulse transfer function is associated.

Tables which give the discrete-time equivalent of systems with a zero-order hold are available. Some typical situations are summarized in Table 2.4.

The computation of ZOH sampled models for transfer functions of different orders can be done by means of the functions: cont2disc.sci (Scilab) or cont2disc.m (MATLAB®). The corresponding sampled model (with Z.O.H) for a second-order system characterized by ω_0 and ζ can be obtained with the functions ft2pol.sci (Scilab) or ft2pol.m (MATLAB®)[4].

---

[4] To be downloaded from the book website.

2.3.7 Analysis of First-order Systems with Time Delay

The continuous-time model is characterized by the transfer function

$$H(s) = \frac{G e^{-s\tau}}{1 + T_s}$$

(2.3.42)

where G is the gain, T is the time constant and τ is the pure time delay. If T_s is the sampling period, then τ is expressed as

$$\tau = d T_s + L \; ; \; 0 < L < T_s$$

(2.3.43)

where L is the fractional time delay and d is the integer number of sampling periods included in the delay and corresponding to a sampled delay of d-periods. From Table 2.4, one derives the transfer function of the corresponding sampled model (when a zero-order hold is used)

$$H(z^{-1}) = \frac{z^{-d}(b_1 z^{-1} + b_2 z^{-2})}{1 + a_1 z^{-1}} = \frac{z^{-d-1}(b_1 + b_2 z^{-1})}{1 + a_1 z^{-1}}$$

(2.3.44)

with

$$a_1 = -e^{-\frac{T_s}{T}} \; ; \; b_1 = G(1 - e^{\frac{L-T_s}{T}}) \; ; \; b_2 = G e^{-\frac{T_s}{T}} (e^{\frac{L}{T}} - 1)$$

The effect of the fractional time delay can be seen in the appearance of the coefficient b_2 in the transfer function. For $L = 0$, one gets $b_2 = 0$. On the other hand, if $L = T_s$, it follows that $b_1 = 0$, which correspond to an additional delay of one sampling period. For $L < 0.5 T_s$ one has $b_2 < b_1$, and for $L > 0.5 T_s$ one has $b_2 > b_1$. For $L = 0.5 T_s$ $b_2 \approx b_1$. Therefore, a fractional delay introduces a zero in the pulse transfer function. For $L > 0.5 T_s$ the relation $|b_2| > |b_1|$ holds and the zero is outside the unit circle (unstable zero)[5].

The pole-zero configuration in the z plane for the first-order system with ZOH is represented in Figure 2.18. The term z^{-d-1} introduces $d+1$ poles at the origin $[H(z) = (b_1 z + b_2) / z^{d+1} (z + a_1)]$.

---

[5] The presence of unstable zeros has no influence on the system stability, but it imposes constraints on the use of controller design techniques based on the cancellation of model zeros by controller poles.

Table 2.4. Pulse transfer functions for continuous-time systems with zero-order hold

$H(s)$	$H(z^{-1})$
$\dfrac{1}{s}$	$\dfrac{T_s z^{-1}}{1-z^{-1}}$
$\dfrac{G}{1+sT}$	$\dfrac{b_1 z^{-1}}{1+a_1 z^{-1}}$; $b_1 = G(1 - e^{-T_s/T})$; $a_1 = -e^{-T_s/T}$
$\dfrac{Ge^{-sL}}{1+sT}$; $L < T_s$	$\dfrac{b_1 z^{-1} + b_2 z^{-2}}{1+a_1 z^{-1}}$; $b_1 = G(1 - e^{(L-T_s)/T})$; $b_2 = Ge^{-T_s/T}(e^{L/T} - 1)$; $a_1 = -e^{-T_s/T}$
$\dfrac{\omega_0^2}{\omega_0^2 + 2\zeta\omega_0 s + s^2}$ $\omega = \omega_0\sqrt{1-\zeta^2}$ $\zeta < 1$	$\dfrac{b_1 z^{-1} + b_2 z^{-2}}{1+a_1 z^{-1} + a_2 z^{-2}}$ $b_1 = 1 - \alpha\left(\beta + \dfrac{\zeta\omega_0}{\omega}\partial\right)$; $b_2 = \alpha^2 + \alpha\left(\dfrac{\zeta\omega_0}{\omega}\partial - \beta\right)$ $a_1 = -2\alpha\beta$; $a_2 = \alpha^2$ $\alpha = e^{-\zeta\omega_0 T_s}$; $\beta = \cos(\omega T_s)$; $\partial = \sin(\omega T_s)$

Figure 2.19 represents the step responses for a system characterized by a pulse transfer function

$$H(z^{-1}) = \frac{b_1 z^{-1}}{1+a_1 z^{-1}} \qquad (2.3.45)$$

with $\dfrac{b_1}{1+a_1} = 1$ (steady-state gain = 1) for different values of the parameter a_1:

$$a_1 = -0.2 \; ; \; -0.3 \; ; \; -0.4 \; ; \; -0.5 \; ; \; -0.6 \; ; \; -0.7 \; ; \; -0.8 \; ; \; -0.9$$

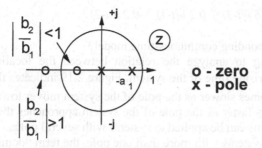

Figure 2.18. Pole-zero configuration of the sampled-data system described by Equation 2.3.44 (first order system with ZOH)

On the basis of these responses, it is easy to derive the time constant of the corresponding continuous-time system, expressed in terms of the sampling period (the time constant is equal to the time required to reach 63% of the final value).

The presence of a time delay equal to an integer multiple of the sampling period only causes a time shift in the responses given in Figure 2.19.

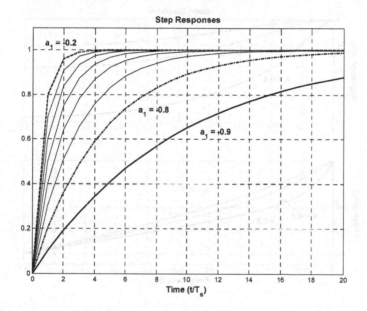

Figure 2.19. Step responses of the discrete-time system $b_1 z^{-1}/(1+a_1 z^{-1})$ for different values of a_1 and $[b_1/(1+a_1)]=1$

The presence of a fractional time delay has as a main consequence a modification at the beginning of the step response, if compared to the case with no fractional time delay.

Exercise. Assuming that the sampled-data system model is

$$y(t) = -0.6\,y(t-1) + 0.2\,u(t-1) + 0.2\,u(t-2)$$

What is the corresponding continuous-time model?

It is interesting to analyze the relation between the location of the pole ($z = -a_1$) and the rising time of the system. Figure 2.19 indicates that the response of the system becomes slower as the pole of the system moves toward the point [1, j0], and it becomes faster as the pole of the system approaches the origin ($z = 0$). These considerations can be applied to systems with several poles.

In the case of systems with more than one pole, the term "dominant pole(s)" is introduced to characterize the pole (or the poles) that is (are) the closest to the point [$1,j0$], i.e. which is the slowest pole(s).

Figure 2.20 shows the frequency responses (magnitude and phase) of the first-order discrete system given by 2.3.45 for $a_1 = -0.8; -0.5; -0.3$. It can be observed that the bandwidth increases when the system pole is approaching the origin (faster pole). We can also remark that the phase lag at the frequency $0.5f_s$ is $-180°$ due to the presence of the ZOH (see Section 2.3.6).

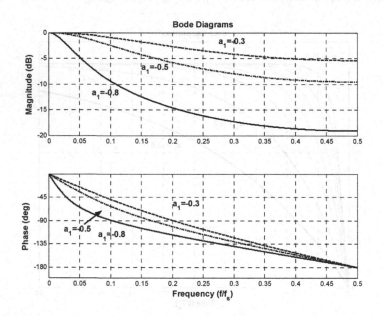

Figure 2.20. Frequency responses (magnitude and phase) of the discrete-time model $b_1\,z^{-1}$ / $(1 + a_1\,z^{-1})$ for different values of a_1 and b_1

2.3.8 Analysis of Second-order Systems

The pulse transfer function corresponding to the discretization with a zero-order hold of a normalized second-order continuous-time system, characterized by a natural frequency ω_0 and a damping ζ, is given by

$$H(z^{-1}) = \frac{z^{-d}(b_1 z^{-1} + b_2 z^{-2})}{1 + a_1 z^{-1} + a_2 z^{-2}} \tag{2.3.46}$$

where d represents the integer number of sampling periods contained in the delay. The values of a_1, a_2, b_1, b_2 as a function of ω_0 and ζ for a pure time delay $\tau = d \cdot T_s$ are given in Table 2.4.

It is interesting to express the poles of the discretized system as a function of ω_0, ζ and the sampling period T_s (or the sampling frequency f_s).

From Table 2.4 the following relations are easily found (for $\zeta < 1$):

$$a_1 = -2 e^{-\zeta \omega_0 T_s} \cos \omega_0 \sqrt{1 - \zeta^2}\, T_s = -e^{-\zeta \omega_0 T_s}(e^{j\omega_0 \sqrt{1-\zeta^2} T_s} + e^{-j\omega_0 \sqrt{1-\zeta^2} T_s})$$

$$a_2 = -e^{-2\zeta \omega_0 T_s}$$

The poles of the pulse transfer function (roots of the denominator) are found by solving the equation

$$z^2 + a_1 z + a_2 = (z - z_1)(z - z_2) = 0$$

From the expressions of a_1 and a_2 the solutions are directly derived:

$$z_{1,2} = e^{-\zeta \omega_0 T_s \pm j\omega_0 \sqrt{1-\zeta^2} T_s}$$

Note that the poles of the discretized system correspond to the poles of the continuous-time system $s_{1,2} = \zeta \omega_0 \pm j\omega_0 \sqrt{1-\zeta^2}$ by applying the transformation $z = e^{sT_s}$.

For $\zeta < 1$, the poles of the discretized system are complex conjugate and, consequently, symmetric with respect to the real axis. They are characterized by a module and a phase given by

$$|z_{1,2}| = e^{-\zeta \omega_0 T_s} = e^{-2\pi\zeta \frac{f_0}{f_s}} = e^{-2\pi\zeta \frac{\omega_0}{\omega_s}}$$

$$\angle z_{1,2} = \pm\sqrt{1 - \zeta^2}\, \omega_0 T_s = \pm 2\pi\sqrt{1 - \zeta^2}\, \frac{f_0}{f_s} = \pm 2\pi\sqrt{1 - \zeta^2}\, \frac{\omega_0}{\omega_s}$$

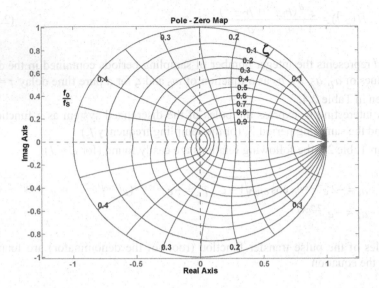

Figure 2.21. The curves ζ = constant and $\omega_0 T_s / 2\pi = f_0 / f_s$ = constant in the z-plane for a second-order discrete-time system

Note that the poles location depends upon ζ and $\omega_0 T_s$ (or $\omega_0/\omega_s = f_0/f_s$).

That is:

$$z = f(\zeta, \, \omega_0 T_s / 2\pi) = f(\zeta, f_0/f_s)$$

and in the z-plane the following curves can be drawn:

$$z = f(\omega_0 T_s/2\pi) = f(f_0/f_s) \qquad \text{for } \zeta = \text{constant}$$

and

$$z = f(\zeta) \qquad \text{for } \omega_0 T_s/2\pi = f(f_0/f_s) = \text{constant}$$

We must remember (see Figure 1.9) that in the s-plane (continuous system) the curves $\zeta = constant$ are straight lines forming an angle $\theta = cos^{-1}\zeta$ with the real axis and the curves $\omega_0 = constant$ are circles with radius ω_0 (these two sets of curves are orthogonal). In the z-plane the curves $z = f(\omega_0 T_s)$ for $\zeta = constant$ are logarithmic spirals that are orthogonal in each point to the curves $z: f(\zeta)$ for $\omega_0 T_s = constant$.

Figure 2.21 shows the set of curves $z = f(\zeta)$ for $\omega_0 T_s/2\pi = constant$ and $z = f(\omega_0 T_s/2\pi)$ for $\zeta = constant$ corresponding to different values of ζ and $\omega_0 T_s/2\pi$ (respectively f_0/f_s).

We should also remember (see Section 2.3.2) that for $f_0/f_s = \omega_0/\omega_s = 1/\left(2\sqrt{1-\zeta^2}\right)$, the corresponding poles in the z- plane are confounded ($\angle z_{1,2} = \pm\pi$), and they are located on the segment of the real axis (-1,0) having an abscissa coordinate equal to $-e^{-\dfrac{\pi\zeta}{\sqrt{1-\zeta^2}}}$.

The stability domain of the second-order discrete-time system in the plane of the parameters $a_1 - a_2$ is a triangle (see Figure 2.22). For values of a_1, a_2 placed inside of the triangle, the roots of the denominator of the pulse transfer function are inside the unit circle.

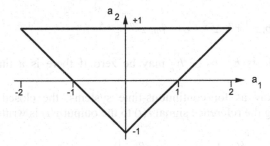

Figure 2.22. Stability domain for the second-order discrete-time system

2.4 Closed Loop Discrete-time Systems

2.4.1 Closed Loop System Transfer Function

Figure 2.23 gives the diagram of a closed loop discrete-time system. The transfer function on the feedforward channel can result from the cascade of a digital controller and of the group DAC+ ZOH + continuous-time system + ADC (discretized system).

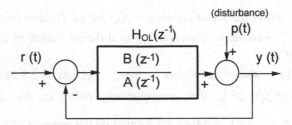

Figure 2.23. Closed loop discrete-time system

Let

$$H_{OL}(z^{-1}) = \frac{B(z^{-1})}{A(z^{-1})} \qquad (2.4.1)$$

be the feed-forward channel transfer function with

$$B(z^{-1}) = b_1 z^{-1} + b_2 z^{-2} + \ldots + b_{n_B} z^{-n_B} \qquad (2.4.2)$$

where the coefficients b_1, b_2 ... b_d may be zero if there is a time delay of d sampling periods.

In the same way as for continuous-time systems, the closed loop transfer function connecting the reference signal $r(t)$ to the output $y(t)$ is written as

$$H_{CL}(z^{-1}) = \frac{H_{OL}(z^{-1})}{1 + H_{OL}(z^{-1})} = \frac{B(z^{-1})}{A(z^{-1}) + B(z^{-1})} \qquad (2.4.3)$$

The denominator of the closed loop transfer function, whose roots correspond to the closed loop system poles, is also called *characteristic polynomial of the closed loop*.

2.4.2 Steady-state Error

The steady-state is obtained for $r(t) = constant$ by making $z = 1$, corresponding to the zero frequency ($z = e^{sTs} = 1$ for $s = 0$).

It follows from Equation 2.4.3 that

$$y = H_{CL}(1)r = \frac{\displaystyle\sum_{i=1}^{n_B} b_i}{A(1) + \displaystyle\sum_{i=1}^{n_B} b_i} r \qquad (2.4.4)$$

where $H_{CL}(1)$ is the steady-state gain (static gain) of the closed loop system. In order to obtain a zero steady-state error between the reference signal r and the output y, it is necessary that

$$H_{CL}(1) = 1 \tag{2.4.5}$$

From Equation 2.4.4 the following conditions are derived:

$$\sum_{i=1}^{n_B} b_i \neq 0 \text{ and } A(1) = 0 \tag{2.4.6}$$

In order to obtain $A(1) = 0$, $A(z^{-1})$ must have the following structure:

$$A(z^{-1}) = (1 - z^{-1}) \cdot A'(z^{-1}) \tag{2.4.7}$$

where

$$A'(z^{-1}) = 1 + a_1 z^{-1} + ... + a_{n_{A'}} z^{-n_{A'}} \tag{2.4.8}$$

and thus the global transfer function of the feedforward channel must be of the type

$$H_{OL}(z^{-1}) = \frac{1}{1 - z^{-1}} \cdot \frac{B(z^{-1})}{A'(z^{-1})} \tag{2.4.9}$$

It is thus observed that *the feedforward channel must contain a digital integrator* in order to obtain a *zero steady-state error* in closed loop. This situation is similar to the continuous case (see Section 1.2.3) and *internal model principle* is also applicable to discrete-time systems.

2.4.3 Rejection of Disturbances

In the presence of a disturbance $p(t)$ acting on the controlled output (see Figure 2.23), the objective is to reduce its effect as much as possible, at least in some frequency regions.

In particular, the constant disturbance effect (a step), often called "load disturbance", is expected to be zero in steady-state $(t \rightarrow \infty, z \rightarrow 1)$.

The pulse transfer function, which links the disturbance to the output, is

$$S_{yp}(z^{-1}) = \frac{1}{1 + H_{OL}(z^{-1})} = \frac{A(z^{-1})}{A(z^{-1}) + B(z^{-1})}$$

As for the continuous-time case, $S_{yp}(z^{-1})$ is called "output sensitivity function". The steady-state is obtained for $z = 1$. It follows that

$$y = S_{yp}(1)\, p = \frac{A(1)}{A(1) + B(1)}\, p$$

(where p is the stationary value of the disturbance).

In order to achieve a perfect steady-state disturbance rejection, it is necessary that $S_{yp}(1) = 0$ and thus $A(1) = 0$. This implies that $A(z^{-1})$ must have the form given in Equation 2.4.7, corresponding to the *integrator* insertion in the feedforward channel.

Similarly, to the continuous case, a perfect steady-state disturbance rejection implies that the feedforward channel must contain *the internal model* of the disturbance (the transfer function that produces $p(t)$ from a Dirac pulse).

As in the continuous-time case, it should be avoided that an amplification of the disturbance effect occurs in certain frequency regions. This is the reason why $|S_{yp} (e^{-j\omega})|$ must be lower than a specified value for all frequencies $f = \omega/2\pi \leq f_s / 2$.

A typical value used as upper bound is

$$|S_{yp}\, (e^{-j\omega})| \leq 2 \qquad 0 \leq \omega \leq \pi f_s$$

Furthermore, if it is known that a disturbance has its energy concentrated in a particular frequency region, $|S_{yp} (e^{j\omega})|$ may be constrained to introduce a desired attenuation in this frequency region.

2.5 Basic Principles of Modern Methods for Design of Digital Controllers

2.5.1 Structure of Digital Controllers

Figure 2.24 gives the diagram of a PI type analog controller. The controller contains two channels (a proportional channel and an integral channel) that process the error between the reference signal and the output.

In the case of sampled-data systems the controller is digital, and the only operations it can carry out are additions, multiplications, storage and shift. All the digital control algorithms have the same structure. Only "the memory" of the controller is different, that is the number of coefficients.

Figure 2.25 illustrates the computation structure of the control $u(t)$ applied to the plant at the instant t by the digital controller. This control is a weighted average of the measured output at instants $t, t-1,....,$ $t-n_A$ .., of the previous control values at instants $t-1, t-1 ..., t-n_B...$ and of the reference signal at instants $t, t-1,...,$ the weights being the coefficients of the controller.

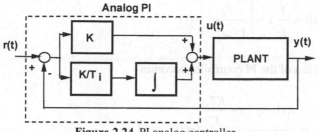

Figure 2.24. PI analog controller

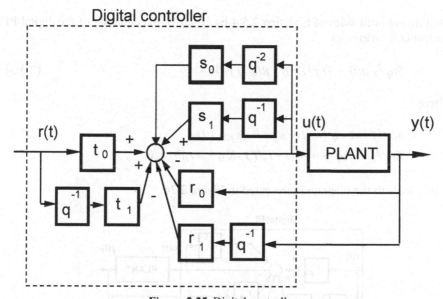

Figure 2.25. Digital controller

This type of control law can even be obtained by the discretization of a PI or PID analog controller. We shall consider, as an example, the discretization of a PI controller. The control law for an analog PI controller is given by

$$u(t) = K\left[1 + \frac{1}{pT_i}\right][r(t) - y(t)] \qquad (2.5.1)$$

For the discretization of the PI controller, p (the differentiation operator) is approximated by $(1 - q^{-1})/T_s$ (see Section 2.3.1, Equation 2.3.6). This yields

$$\frac{dx}{dt} = px \approx \frac{x(t) - x(t-1)}{T_s} = \frac{1 - q^{-1}}{T_s}x(t) \qquad (2.5.2)$$

$$\int x dt = \frac{1}{p} x \approx \left[\frac{T_s}{1-q^{-1}} \right] x(t) \tag{2.5.3}$$

and the equation of the PI controller becomes

$$u(t) = \frac{K(1-q^{-1}) + \dfrac{KT_s}{T_i}}{1-q^{-1}} [r(t) - y(t)] \tag{2.5.4}$$

Multiplying both sides of Equation 2.5.4 by $(1 - q^{-1})$, the equation of the digital PI controller is written as

$$S(q^{-1}) u(t) = T(q^{-1}) r(t) - R(q^{-1}) y(t) \tag{2.5.5}$$

where

$$S(q^{-1}) = 1 - q^{-1} = 1 + s_1 q^{-1} \quad (s_1 = -1) \tag{2.5.6}$$
$$R(q^{-1}) = T(q^{-1}) = K (1+T_s/T_i) - K q^{-1} = r_0 + r_1 q^{-1} \tag{2.5.7}$$

which leads to the diagram represented in Figure 2.26.

Digital PI

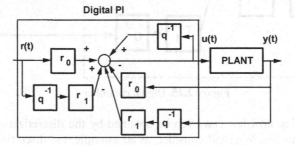

Figure 2.26. Digital PI controller

Taking into account the expression of $S(q^{-1})$, the control signal $u(t)$ is computed on the basis of Equation 2.5.5, by means of the formula

$$u(t) = u(t-1) - R(q^{-1}) y(t) + T(q^{-1}) r(t)$$

$$= u(t-1) - r_0 y(t) - r_1 y(t-1) + r_0 r(t) + r_1 r(t-1) \tag{2.5.8}$$

which corresponds to the diagram given in Figure 2.26.

2.5.2 Digital Controller Canonical Structure

Dividing by $S(q^{-1})$ both sides of Equation 2.5.5, one obtains

$$u(t) = -\frac{R(q^{-1})}{S(q^{-1})}y(t) + \frac{T(q^{-1})}{S(q^{-1})}r(t) \tag{2.5.9}$$

from which we derive the digital controller canonical structure presented in Figure 2.27 (three branched RST structure).

In general, $T(q^{-1})$ in Figure 2.27 is different from $R(q^{-1})$.

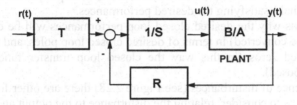

Figure 2.27. Digital controller canonical structure

Consider

$$H(z^{-1}) = \frac{B(z^{-1})}{A(z^{-1})} \tag{2.5.10}$$

as the pulse transfer function of the cascade DAC + ZOH + continuous-time system + ADC, then the transfer function of the open loop system is written as

$$H_{OL}(z^{-1}) = \frac{B(z^{-1})R(z^{-1})}{A(z^{-1})S(z^{-1})} \tag{2.5.11}$$

and the closed loop transfer function between the reference signal $r(t)$ and the output $y(t)$, using a digital controller canonical structure, has the expression

$$H_{CL}(z^{-1}) = \frac{B(z^{-1})T(z^{-1})}{A(z^{-1})S(z^{-1}) + B(z^{-1})R(z^{-1})} = \frac{B(z^{-1})T(z^{-1})}{P(z^{-1})} \tag{2.5.12}$$

where

$$P(z^{-1}) = A(z^{-1})S(z^{-1}) + B(z^{-1})R(z^{-1}) = 1 + p_1 z^{-1} + p_2 z^{-2} + ... \tag{2.5.13}$$

is the denominator of the closed loop transfer function that defines the closed loop system poles. Note that $T(q^{-1})$ introduces one more degree of freedom, which allows one to establish a distinction between tracking and regulation performances specifications.

We also remark that $r(t)$ is often replaced by a "desired trajectory" $y^*(t)$, obtained either by filtering the reference signal $r(t)$ with the so-called shaping filter or tracking reference model, or saving in the memory of the digital computer the sequence of the desired trajectory values.

The digital controller represented in Figure 2.27 is also defined as "RST digital controller". It is a *two degrees of freedom* controller, which allows one to impose different specifications in terms of desired dynamics for the tracking and regulation problems.

The goal of the digital controller design is to find the polynomials R, S, and T in order to obtain the closed loop transfer functions, with respect to the reference and disturbance signals, satisfying the desired performances.

This explains why the desired closed loop performances will be expressed, (if not, they will be converted) in terms of desired closed loop poles, and eventually in terms of desired zeros (in this way the closed loop transfer function will be completely imposed).

In the presence of disturbances (see Figure 2.28) there are other four important transfer functions to consider, relating the disturbance to the output and the input of the plant.

The transfer function between the disturbance $p(t)$ and the output $y(t)$ (*output sensitivity function*) is given by

$$S_{yp}(z^{-1}) = \frac{A(z^{-1})S(z^{-1})}{A(z^{-1})S(z^{-1}) + B(z^{-1})R(z^{-1})} \qquad (2.5.14)$$

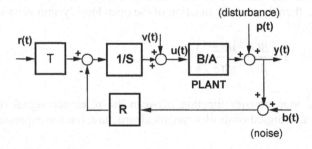

Figure 2.28. Digital control system in presence of disturbances and noise

This function allows the characterization of the system performances from the point of view of disturbances rejection. In addition, certain components of $S(z^{-1})$ can be pre-specified in order to obtain satisfactory disturbance rejection properties.

Thus, if a perfect disturbance rejection is required at a specified frequency, $S(z^{-1})$ must include a zero corresponding to this frequency. In particular, if a perfect load disturbance rejection in steady-state (i.e. zero frequency) is desired, $S_{yp}(z^{-1})$ must include a term $(1 - z^{-1})$ in the numerator, which leads to a value of the gain equal to zero for $z = 1$. This is coherent with the result given in Section 2.4.3., because a zero of $S_{yp}(z^{-1})$ corresponds to a pole of the open loop system.

The transfer function between the disturbance $p(t)$ and the input of the plant $u(t)$ (*input sensitivity function*) is given by

$$S_{up}(z^{-1}) = \frac{-A(z^{-1})R(z^{-1})}{A(z^{-1})S(z^{-1}) + B(z^{-1})R(z^{-1})} \qquad (2.5.15)$$

The analysis of this function allows one to evaluate the influence of a disturbance upon the plant input, and to specify a factor of the polynomial $R(z^{-1})$ if the controller must not react to disturbances concentrated in a particular frequency region.

When noise is added to the measured output (see Figure 2.28), important information can be retrieved by the transfer function that relates the noise $b(t)$ to the plant output $y(t)$ (*noise-output sensitivity function*).

$$S_{yb}(z^{-1}) = \frac{-B(z^{-1})R(z^{-1})}{A(z^{-1})S(z^{-1}) + B(z^{-1})R(z^{-1})} \qquad (2.5.16)$$

As the noise energy is often concentrated at high frequency, attention should be paid in order to obtain a low gain of the transfer function $\left| S_{yb}(e^{-j\omega}) \right|$ in this frequency region.

For $T=R$, the sensitivity function between r and y (also called *complementary sensitivity function*) is defined as

$$S_{yr}(z^{-1}) = \frac{B(z^{-1})R(z^{-1})}{A(z^{-1})S(z^{-1}) + B(z^{-1})R(z^{-1})} = -S_{yb}(z^{-1}) \qquad (2.5.17)$$

Note that

$$S_{yp}(z^{-1}) - S_{yb}(z^{-1}) = S_{yp}(z^{-1}) + S_{yr}(z^{-1}) = 1$$

which implies an interdependence between these sensitivity functions.

Notice that $S_{ub}(z^{-1})$, the transfer function between the noise and the plant input, is equal to $S_{up}(z^{-1})$.

Another important transfer function describes the influence on the output of a disturbance $v(t)$ on the plant input. This sensitivity function (*input disturbance-output sensitivity function*) is given by

$$S_{yv}(z^{-1}) = \frac{B(z^{-1})S(z^{-1})}{A(z^{-1})S(z^{-1}) + B(z^{-1})R(z^{-1})} \qquad (2.5.18)$$

The importance of this sensitivity function is that it enhances the possible simplification of unstable plant poles by the zeros of $R(z^{-1})$.

In order to clarify this point, let us consider the assumption $R(z^{-1})=A(z^{-1})$ (plant poles compensation by controller zeros) and suppose that the plant to be controlled is unstable ($A(z^{-1})$ has roots outside the unit circle). In this case

$$S_{yp}(z^{-1}) = \frac{A(z^{-1})S(z^{-1})}{A(z^{-1})S(z^{-1})+B(z^{-1})A(z^{-1})} = \frac{S(z^{-1})}{S(z^{-1})+B(z^{-1})}$$

$$S_{up}(z^{-1}) = -\frac{A(z^{-1})A(z^{-1})}{A(z^{-1})S(z^{-1})+B(z^{-1})A(z^{-1})} = -\frac{A(z^{-1})}{S(z^{-1})+B(z^{-1})}$$

$$S_{yb}(z^{-1}) = -\frac{B(z^{-1})A(z^{-1})}{A(z^{-1})S(z^{-1})+B(z^{-1})A(z^{-1})} = -\frac{B(z^{-1})}{S(z^{-1})+B(z^{-1})}$$

$$S_{yv}(z^{-1}) = \frac{B(z^{-1})S(z^{-1})}{A(z^{-1})S(z^{-1})+B(z^{-1})A(z^{-1})} = \frac{B(z^{-1})S(z^{-1})}{A(z^{-1})[S(z^{-1})+B(z^{-1})]}$$

Note that S_{yp}, S_{up}, S_{yb} are stable transfer functions if $S(z^{-1})$ is chosen in order to have $S(z^{-1})+B(z^{-1})$ stable, that is

$$S(z^{-1})+B(z^{-1}) = 0 \Rightarrow |z| < 1$$

while the sensitivity function $S_{yv}(z^{-1})$ is unstable.

This remark yields to the following general statement:

The feedback system presented in Figure 2.28 is asymptotically stable if and only if all the four sensitivity functions Syp, Sup, Syb (or Syr) and Syv (describing the relations between disturbances on one hand and plant input or output on the other hand) are asymptotically stable.

The set of five transfer functions $H_{OL}(z^{-1})$, $S_{yp}(z^{-1})$, $S_{up}(z^{-1})$, $S_{yb}(z^{-1})$ (or $S_{yr}(z^{-1})$) and $S_{yv}(z^{-1})$ also play an important role in the closed loop system robustness analysis.

2.5.3 Control System with PI Digital Controller

In this section the design of digital PI controllers will be illustrated. The transfer (function) operator of the discretized plant with zero-order hold is given by

$$H(q^{-1}) = \frac{B(z^{-1})}{A(z^{-1})} = \frac{B(q^{-1})}{A(q^{-1})} = \frac{b_1 q^{-1}}{1+a_1 q^{-1}} \qquad (2.5.19)$$

For the sake of notation uniformity, we shall often use, in the case of constant coefficients, q^{-1} notation both for the delay operator and the complex variable z^{-1}.

The z^{-1} notation will be specially employed when an interpretation in the frequency domain is needed (in this case $z = e^{j\omega T_s}$).

The digital PI controller is characterized by the polynomials (see Equations 2.5.6 and 2.5.7):

$$R(q^{-1}) = T(q^{-1}) = r_0 + r_1 q^{-1} \qquad (2.5.20)$$
$$S(q^{-1}) = 1 - q^{-1} \qquad (2.5.21)$$

The closed loop system transfer function (with respect to the reference $r(t)$) in the general form is given by Equation 2.5.12.

The characteristic polynomial $P(q^{-1})$, whose roots are the desired closed loop system poles, essentially defines the performances. As a general rule, it is chosen as a second-order polynomial corresponding to the discretization of a second-order continuous-time system with a specified natural frequency ω_0 and damping ζ (ω_0 and ζ, for example, and can be obtained on the basis of the diagrams given in Figures 1.10 or 1.11) starting from specifications in the time domain. The coefficients corresponding to the polynomial $P(q^{-1})$ are obtained either by conversion tables mentioned in Table 2.4, or by Scilab and MATLAB® functions given in Section 2.3. In this case, sampling period T_s, natural frequency ω_0 and damping ζ must be specified.

We recall that the relation between ω_0 and T_s must be respected (see Section 2.2.2, Equation 2.2.7):

$$0.25 \leq \omega_0 T_s \leq 1.5 \; ; \; 0.7 \leq \zeta \leq 1 \qquad (2.5.22)$$

For a plant having an equivalent discrete-time transfer operator (function) given by Equation 2.5.19, and the use of a digital PI controller, the closed loop system poles are given by Equation 2.5.13, and they are

$$(1 + a_1 q^{-1})(1 - q^{-1}) + b_1 q^{-1}(r_0 + r_1 q^{-1}) = 1 + p_1 q^{-1} + p_2 q^{-2} \qquad (2.5.23)$$

By rearranging the terms in Equation 2.5.23 in ascending q^{-1} powers, we get

$$1 + (a_1 - 1 + r_0 b_1) q^{-1} + (b_1 r_1 - a_1) q^{-2} = 1 + p_1 q^{-1} + p_2 q^{-2} \qquad (2.5.24)$$

For the polynomial Equation 2.5.24 to be verified, it is necessary that the coefficients of the same q^{-1} powers must be equal on both sides. Thus the following system is obtained:

$$\begin{cases} a_1 - 1 + r_0 b_1 &= p_1 \\ b_1 r_1 - a_1 &= p_2 \end{cases} \qquad (2.5.25)$$

which gives for r_0 and r_1 the results

$$r_1 = \frac{p_2 + a_1}{b_1} \quad ; \quad r_0 = \frac{p_1 - a_1 + 1}{b_1} \tag{2.5.26}$$

One can see that the parameters of the controller depend upon the performance specifications (the desired closed loop poles) and the plant model parameters.

By using Equation 2.5.7, one can obtain the parameters of the continuous-time PI controller:

$$K = -r_1 \quad ; \quad T_i = -\frac{T_s r_1}{r_1 + r_0}$$

2.6 Analysis of the Closed Loop Sampled-Data Systems in the Frequency Domain

2.6.1 Closed Loop Systems Stability

In the case of continuous-time systems, it was shown in Chapter 1, Section 1.2.5, how to use the open loop transfer function representation in the complex plane (the Nyquist plot) in order to analyze the closed loop system stability and the robustness with respect to the parameters variations (or uncertainties on the parameters value). The same approach can be applied to the case of sampled-data systems. The Nyquist plot for sampled-data systems can be drawn using the functions *Nyquist-ol.sci* (Scilab) and *Nyquist-ol.m* (MATLAB®)[6].

Figure 2.29 shows the Nyquist plot of an open loop sampled-data system including a plant (represented by the corresponding transfer function $H(z^{-1}) = B(z^{-1})/A(z^{-1})$) and a RST controller.

In this case, the open loop transfer function is given by

$$H_{OL}(e^{-j\omega}) = \frac{B(e^{-j\omega})R(e^{-j\omega})}{A(e^{-j\omega})S(e^{-j\omega})} \tag{2.6.1}$$

The vector linking the plane origin to a point belonging to the Nyquist plot of the transfer function represents H_{OL} ($e^{j\omega}$) for a specified normalized radian frequency $\omega = \omega T_s = 2 \pi f/f_s$. The considered range of variation of the radian natural

---

[6] To be downloaded from the book website.

frequency ω is between 0 and π (corresponding to an unnormalized frequency variation between 0 and $0.5 f_s$).

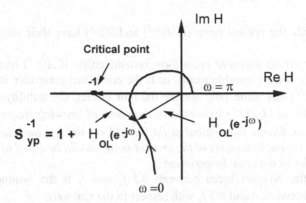

Figure 2.29. Nyquist plot for a sampled-data system transfer function and the critical point

In this diagram the point $[-1, j0]$ is the "critical point". As Figure 2.29 clearly shows, the vector linking the point $[- 1, j0]$ to the Nyquist plot of $H_{OL} (e^{j\omega})$ has the expression

$$S_{yp}^{-1}(z^{-1}) = 1 + H_{OL}(z^{-1}) = \frac{A(z^{-1})S(z^{-1}) + B(z^{-1})R(z^{-1})}{A(z^{-1})S(z^{-1})} \qquad (2.6.2)$$

This vector represents the inverse of the output sensitivity function S_{yp} (z^{-1}) (see Equation 2.5.14) and the zeros of S^{-1}_{yp} (z^{-1}) correspond to the closed loop system poles (see Equation 2.5.13). In order to have an asymptotically stable closed loop system, it is necessary that all the zeros of S^{-1}_{yp} (z^{-1}) (that are the poles of S_{yp} (z^{-1})) be inside the unit circle ($|z| < 1$). The necessary and sufficient conditions for the asymptotic stability of the closed loop system are given by the Nyquist criterion.

For systems having stable poles in open loop (in this case $A(z^{-1}) = 0$ and $S(z^{-1})$ $= 0 \rightarrow |z| \leq 1$) the Nyquist stability criterion states (stable open loop system): *The Nyquist plot of $H_{OL}(z^{-1})$ traversed in the sense of growing frequencies (from $\omega = 0$ to $\omega = \pi$), leaves the critical point $[-1, j0]$ on the left.*

As a general rule, for the given nominal plant model $B(z^{-1})/A(z^{-1})$, polynomials $R(q^{-1})$ and $S(q^{-1})$ are computed in order to have

$$A(z^{-1}) S(z^{-1}) + B(z^{-1}) R(z^{-1}) = P(z^{-1}) \qquad (2.6.3)$$

where $P(z^{-1})$ is a polynomial with asymptotically stable roots. As a consequence, for the nominal values of $A(z^{-1})$ and $B(z^{-1})$, since the closed loop system is stable, the open loop transfer function:

$$H_{OL}(z^{-1}) = \frac{B(z^{-1})R(z^{-1})}{A(z^{-1})S(z^{-1})}$$

does not encircle the critical point (if $A(z^{-1})$ and $S(z^{-1})$ have their roots inside the unit circle).

In the case of an *unstable open loop system*, either if $A(z^{-1})$ has some pole outside the unit circle (unstable plant), or if the computed controller is unstable in open loop ($S(z^{-1})$ has some pole outside the unit circle), the stability criterion is: *The Nyquist plot of $H_{OL}(z^{-1})$ traversed in the sense of growing frequencies (from $\omega = 0$ to $\omega = \pi$), leaves the critical point $[-1, j0]$ on the left and the number of counter clockwise encirclements of the critical point should be equal to the number of unstable poles of the open loop system[7].*

Note that the Nyquist locus between $0.5\,f_s$ and f_s is the symmetric of the Nyquist locus between 0 and $0.5\,f_s$ with respect to the real axis.

The general Nyquist criterion formula that gives the number of encirclements around the critical point is

$$N = P_{CL}^i - P_{OL}^i$$

where P_{CL}^i is the number of closed loop unstable poles and P_{OL}^i is the number of open loop unstable poles. Positive values of N correspond to clockwise encirclements around the critical point. In order that the closed loop system be asymptotically stable it is necessary that $N = -P_{OL}^i$. Figure 2.30 shows two interesting Nyquist loci.

If the plant is stable in open loop and the controller is computed on the basis of Equation 2.6.3 to obtain a desired stable closed loop polynomial $P(z^{-1})$ (this means that the nominal closed loop system is stable too), then, if a Nyquist plot of the form of Figure 2.30a is obtained, one concludes that the controller is unstable in open loop. This situation must be generally avoided[8], and this can be achieved by reducing the desired closed loop dynamic performances (by modifying $P(z^{-1})$).

---

[7] The criterion holds even if an unstable pole-zero cancellation occurs. The number of encirclements should be equal to the number of unstable poles without taking into account the possible cancellations.

[8] Note that there exist some « pathological » transfer functions $B(z^{-1})/A(z^{-1})$ with unstable poles and/or zeros that can be only stabilized by controllers that are unstable in open loop.

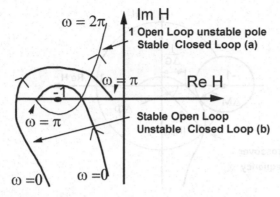

Figure 2.30. Nyquist plots: **a**) unstable system in open loop but stable in closed loop; **b**) stable system in open loop but unstable in closed loop

2.6.2 Closed Loop System Robustness

When designing a control system, one has to take into account the plant model uncertainties (uncertainties of the parameter values or of the frequency characteristics, variations of the parameters, etc.). It is therefore extremely important to assess if the stability of the closed loop is guaranteed in the presence of the plant model uncertainties. The closed loop will be termed "robust" if the stability is guaranteed for a given set of model uncertainties.

   The *robustness* of the closed loop is related to the minimal distance between the Nyquist plot for the nominal plant model and the "critical point" as well as to the frequency characteristics of the modulus of the sensitivity functions.

   The following elements help to evaluate how far is the critical point $[-1, j0]$ (see Figure 2.31):

- Gain margin;
- Phase margin;
- Delay margin;
- Modulus margin.

Gain Margin

The gain margin (ΔG) equals the inverse of $H_{OL}(e^{-j\omega})$ gain for the frequency corresponding to a phase shift $\angle\phi = -180°$ (see Figure 2.31).

   The gain margin is often expressed in dB. In other words, the gain margin gives the maximum admissible increase of the open loop gain for the frequency corresponding to $\angle\phi(\omega) = -180°$.

$$\Delta G = \frac{1}{\left|H_{OL}(e^{-j\omega_{180}})\right|} \qquad \text{for } \angle\phi(\omega_{180}) = -180°$$

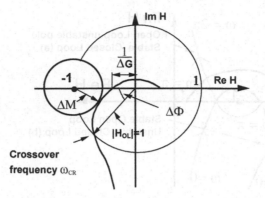

Figure 2.31. Gain, phase and modulus margins

Typical values for a good gain margin are

$$\Delta G \geq 2 \quad (6\ dB) \quad [min:\ 1.6\ (4\ dB)]$$

If the Nyquist plot crosses the real axis at several frequencies $\omega_{i\pi}$ characterized by a phase lag

$$\angle\phi(\omega_{i\pi}) = -i\ 180°;\ i = 1, 3, 5 \ldots$$

and the corresponding gains of the open loop system are denoted by $\left|H_{OL}(e^{-j\omega_{i\pi}})\right|$, then the *gain margin* is defined by[9]

$$\Delta G = \min_{i} \frac{1}{\left|H_{OL}(e^{-j\omega_{i\pi}})\right|}$$

Phase Margin
The phase margin ($\Delta\phi$) is the additional phase that we must add at the crossover frequency, for which the gain of the open-loop system equals *1*, in order to obtain a total phase shift $\angle\phi = -180°$ (see Figure 2.31).

$$\Delta\phi = 180^0 - \angle\phi(\omega_{cr}) \quad for \quad \left|H_{OL}(e^{-j\omega_{cr}})\right| = 1$$

in which ω_{cr} is called *crossover frequency* and it corresponds to the frequency for which the Nyquist plot crosses the unit circle (see Figure 2.31).

---

[9] Note that if the Nyquist plot crosses the real axis for values less than −*1* and leaves the critical point to the left, there is a minimal value of the gain margin under which the system becomes unstable.

Typical values for a good phase margin are

$$30° \leq \Delta\phi \leq 60°$$

If the Nyquist plot crosses the unit circle at several frequencies ω_{cr}^i characterized by the corresponding phase margins:

$$\Delta\phi_i = 180^0 - \angle\phi(\omega_{cr}^i)$$

then the system *phase margin* is defined as

$$\Delta\phi = \min_i \Delta\phi_i$$

Delay Margin
A time-delay introduces a phase shift proportional to the frequency ω. For a certain frequency ω_0, the phase shift introduced by a time-delay τ is

$$\angle\phi(\omega_0) = -\omega_0\tau$$

We can therefore convert the phase margin in a "time-delay margin", i.e. to compute the maximum admissible increase of the delay of the open-loop system without making the closed-loop system unstable. It then follows that:

$$\Delta\tau = \frac{\Delta\phi}{\omega_{cr}}$$

If the Nyquist plot intersects the unit circle at several frequencies ω_{cr}^i, characterized by the corresponding phase margins $\Delta\phi_i$, the *delay margin* is defined as

$$\Delta\tau = \min_i \frac{\Delta\phi_i}{\omega_{cr}^i}$$

Note that *a good phase margin does not guarantee a good delay margin* (if the frequency ω_{cr} is high, then the delay margin is low even if the phase margin is important).

The typical value of the delay margin is $\Delta\tau \geq T_S$ [min: $0.75T_S$]

Modulus Margin
This concerns a more global measure of the distance between the critical point $[-1, j0]$ and the plot of $H_{OL}(z^{-1})$. The *modulus margin* (ΔM) is defined as the radius

of the circle centered in [-1, j0] and tangent to the plot of $H_{OL}(z^{-1})$ (see Figure 2.31).

From the definition of Equation 2.6.2 of the vector connecting the critical point [-1, j0] to the plot of $H_{OL}(e^{-j\omega})$ it follows immediately that

$$\Delta M = \left|1 + H_{OL}(z^{-1})\right|_{min} = \left|S_{yp}^{-1}(z^{-1})\right|_{min} = \left(\left|S_{yp}(z^{-1})\right|_{max}\right)^{-1} =$$

$$\left(\left|\frac{A(z^{-1})S(z^{-1})}{A(z^{-1})S(z^{-1}) + B(z^{-1})R(z^{-1})}\right|_{max}\right)^{-1} \quad for \quad z^{-1} = e^{-j2\pi f} \tag{2.6.4}$$

In other words, the *modulus margin* ΔM is equal to the inverse of the maximum value of the sensitivity function $S_{yp}(z^{-1})$ magnitude. By plotting $S_{yp}(z^{-1})$ magnitude in dB scale, the following condition is immediately derived:

$$\left|S_{yp}(e^{-j\omega})\right|_{max} dB = \Delta M^{-1} dB = -\Delta M \ dB \tag{2.6.5}$$

Figure 2.32 shows the relation between the sensitivity function S_{yp} and the modulus margin.

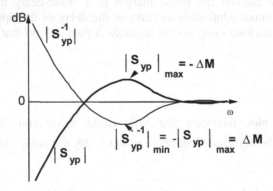

Figure 2.32. Relation between the output sensitivity function and the modulus margin

Therefore, the reduction (or the minimization) of $|S_{yp}(j\omega)|_{max}$ will lead to an increase (or maximization) of the modulus margin[10].

Typical values for a good modulus margin are

$\Delta M \geq 0.5 \ (-6 \ dB) \quad [min: \ 0.4 \ (-8 \ dB)]$

---

[10] $|S_{yp}(j\omega)|_{max}$ corresponds to the H_{∞} norm of the output sensitivity function.

Note that $\Delta M \geq 0.5$ implies a gain margin $\Delta G \geq 2$ (6 dB) and a phase margin $\Delta\phi >$ 29°. As a general rule, a good modulus margin guarantees satisfactory values for the gain and phase margins[11].

To summarize, typical values for the stability margins in a robust design are:

- gain margin: $\Delta G \geq 2$ (6dB)   [*min.: 1.6 (4dB)*]
- phase margin: $30° \leq \Delta\phi \leq 60°$

- delay margin: $\Delta\tau = \dfrac{\Delta\phi}{\omega_{cr}} \geq T_s$ [*min.: 0.75 T_s*]

- modulus margin: $\Delta M \geq 0.5$ (-6dB),   [*min.: 0.4 (-8dB)*]

If the plant model is known with a very good precision for a certain region of operation, the imposed robustness margins can eventually be less restrictive.

The *modulus margin* is very important because:

- It defines the maximum admissible value for the modulus of the *output sensitivity function* and therefore the low limits of the performance in disturbance rejection;
- It defines the tolerance with respect to nonlinear or time varying elements that may belong to the system (the circle criterion - see below).

Tolerance with Respect to Nonlinear Elements

In control systems we frequently have components with static nonlinear or time-varying characteristics (often in the actuators).

The characteristics of these components, without being accurately known, generally lie inside the conic region defined by a minimum linear gain (α) and a maximum linear gain (β) – see Figure 2.33.

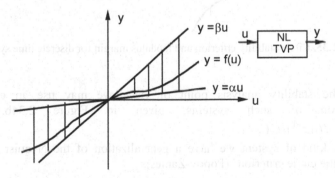

Figure 2.33. Nonlinear or time-varying characteristics, contained in the conic domain (α, β)

The closed-loop system looks, for example, like those in Figure 2.34a.

---

[11] The converse is not true. Systems having satisfactory gain and phase margins may have a very low modulus margin.

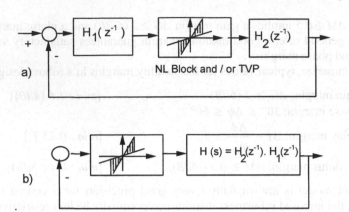

a)

NL Block and / or TVP

b)

Figure 2.34. Closed-loop systems containing a nonlinear block (NL) and / or time-varying parameters (TVP): **a)** block diagram **b)** equivalent representation

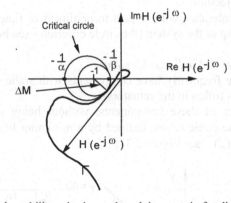

Figure 2.35. Circle stability criterion and modulus margin for discrete time systems

From the stability analysis point of view, we may use an equivalent representation of such systems, given in Figure 2.34b, where $H_{OL}(z^{-1}) = H_1(z^{-1})H_2(z^{-1})$.

For this kind of system we have a generalization of the Nyquist criterion, known as "the circle criterion" (Popov-Zames).

Circle (Stability) Criterion
The feedback system represented in Figure 2.34b is asymptotically stable for the set of nonlinear and/or time-varying characteristics lying in the conic domain $[\alpha,\beta]$ (with $\alpha, \beta > 0$) if the plot of $H_{OL}(z^{-1})$, traversed in the sense of growing frequencies, leaves on the left, without crossing it, the circle centered on the real axis and passes through the points $[-1/\beta, j0]$ and $[-1/\alpha, j0]$.

The modulus margin ΔM defines a circle of radius ΔM centered in $[-1, j0]$ that is outside the Nyquist plot of the open loop transfer function.

Thus, the closed loop system can tolerate non-linear blocks or time-variable parameters described by input-output characteristics lying in a conic sector delimited by a minimum linear gain $(1/(1+\Delta M))$ and a maximum linear gain $(1/(1-\Delta M))$ (see Figure 2.35).

Tolerances to Plant Transfer Function Uncertainties and/or Parameters Variations.

Figure 2.36 shows the effect of the plant nominal model uncertainties and parameters variations on the Nyquist plot of the open loop transfer function. As a general rule, the Nyquist plot of the plant nominal model lies inside a "tube" corresponding to the accepted tolerances of the parameters variations (or the uncertainties) of the plant model transfer function.

In order to ensure the stability of the closed loop system for an open loop transfer function $H'_{OL}(z^{-1})$ that is different from the nominal one $H_{OL}(z^{-1})$ (but having the same number of unstable poles as $H_{OL}(z^{-1})$), it is necessary that the Nyquist plot of the open loop transfer function $H'_{OL}(z^{-1})$ leaves the critical point $[-1, j0]$ on the left when traversed in the sense of growing frequencies from 0 to 0.5 f_s. This condition is satisfied if the difference between the real open loop transfer function $H'_{OL}(z^{-1})$ and the nominal one $H_{OL}(z^{-1})$ is smaller than the distance between the Nyquist plot of the open loop nominal transfer function and the critical point for all frequencies (see Figure 2.36). This robust stability condition is expressed by the inequality

$$\left| H'_{OL}(z^{-1}) - H_{OL}(z^{-1}) \right| < \left| 1 + H_{OL}(z^{-1}) \right| = \left| S_{yp}^{-1}(z^{-1}) \right| -$$

$$\left| \frac{A(z^{-1})S(z^{-1}) + B(z^{-1})R(z^{-1})}{A(z^{-1})S(z^{-1})} \right| = \left| \frac{P(z^{-1})}{A(z^{-1})S(z^{-1})} \right| \quad ; \quad z^{-1} = e^{-j\omega} \tag{2.6.6}$$

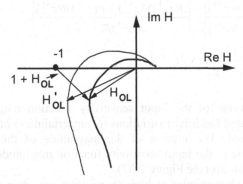

Figure 2.36. Nyquist plot for the nominal open loop transfer function and the real open loop transfer function in presence of uncertainties and parameters variations (H_{OL} and H'_{OL} are stable)

where $S(z^{-1})$ and $R(z^{-1})$ are computed on the basis of Equation 2.6.3 for the nominal values of $A(z^{-1})$ and $B(z^{-1})$.

In other words, the magnitude of S^{-1}_{yp} $(e^{j\omega})$ function (evaluated in dB units), obtained by symmetry from S_{yp} $(e^{j\omega})$ (see Figure 2.32), gives, at each frequency, a sufficient condition for the accepted difference (computed as the Euclidian distance) between the real open loop transfer function and the nominal open loop transfer function, in order to guarantee the stability of the closed loop.

This tolerance is higher at low frequencies (see Figure 2.32) where the gain of the open loop system is high (especially when an integrator is included), and it has a minimum value at the frequency (or frequencies) where S^{-1}_{yp} $(e^{j\omega})$ reaches its minimum value ($= \Delta M$), that is the frequency where S_{yp} $(e^{j\omega})$ has the maximum value.

It is necessary to ensure that at these frequencies the plant model variations are compatible with the obtained modulus margin. If this is not the case, the solution is to provide a more accurate model, or to modify the specifications in order to maintain the closed loop stability.

Equation 2.6.6 expresses a robustness condition in terms of open loop transfer function variations (controller + plant). It is interesting to express this robustness condition in terms of the plant model variations only. Note that Equation 2.6.6 can be further expressed as

$$\left| \frac{B'(z^{-1})R(z^{-1})}{A'(z^{-1})S(z^{-1})} - \frac{B(z^{-1})R(z^{-1})}{A(z^{-1})S(z^{-1})} \right| = \left| \frac{R(z^{-1})}{S(z^{-1})} \right| \cdot \left| \frac{B'(z^{-1})}{A'(z^{-1})} - \frac{B(z^{-1})}{A(z^{-1})} \right|$$
$$< \left| \frac{A(z^{-1})S(z^{-1}) + B(z^{-1})R(z^{-1})}{A(z^{-1})S(z^{-1})} \right| \qquad (2.6.7)$$

where $B(z^{-1})/A(z^{-1})$ corresponds to the nominal plant transfer function.

Multiplying by $|S(z^{-1})/R(z^{-1})|$ both sides of Equation 2.6.7 one gets the condition

$$\left| \frac{B'(z^{-1})}{A'(z^{-1})} - \frac{B(z^{-1})}{A(z^{-1})} \right| < \left| \frac{A(z^{-1})S(z^{-1}) + B(z^{-1})R(z^{-1})}{A(z^{-1})R(z^{-1})} \right|$$
$$= \left| \frac{P(z^{-1})}{A(z^{-1})R(z^{-1})} \right| = \left| S^{-1}_{up}(z^{-1}) \right| \qquad (2.6.8)$$

By plotting the inverse of the input sensitivity function magnitude, sufficient conditions for tolerated (additive) variations (or uncertainties) of the plant transfer function are obtained. The inverse of the magnitude of the input sensitivity function is symmetric to the input sensitivity function magnitude in dB units with respect to the axis at 0 dB (see Figure 2.37).

As plant model uncertainties at high frequencies are often present, one must verify that the maximum of $|S_{up}$ $(e^{j\omega})|$ at high frequencies is small. On the other hand, the input sensitivity function S_{up} is an effective image of the actuator stress in the frequency domain when disturbances act on the system. The physical characteristics of the actuator often impose a bound on actuator stress at high frequencies, and an upper bound of the maximum of $|S_{up} (e^{j\omega})|$ at these frequencies should be imposed.

Notice that (from Equation 2.6.8) the admitted tolerances (neglecting the term $1/|R(z^{-1})|$) depend to a large extent upon the relation between the open loop system poles (defined by $A(z^{-1})$) and the desired closed loop poles (defined by $P(z^{-1})$).

In order to understand this phenomenon in greater detail, Figure 2.38 shows the $S_{up}(z^{-1})$ magnitude functions for a plant model characterized by $A(z^{-1})= 1 - 0.8 \, z^{-1}$; $B(z^{-1})= z^{-1}$ and for two different desired closed loop system characteristic polynomials: $P_1(z^{-1})=1-0.6 \, z^{-1}$ and $P_2(z^{-1})=1-0.3 \, z^{-1}$ (the controller includes an integrator). Note that $P_2(z^{-1})$ corresponds to a closed loop system faster than the one specified by $P_1(z^{-1})$, and both closed loop systems are faster than the plant (open loop system).

The $|S_{up}(z^{-1})|$ maximum for $P_2(z^{-1})$ is greater than for $P_1(z^{-1})$, and then the inverse of $|S_{up}(z^{-1})|$ will be smaller. As a consequence, the accepted tolerances for the frequency response variations (especially at high frequencies) are smaller in the case of $P_2(z^{-1})$ with respect to the case of desired closed loop performances imposed by $P_1(z^{-1})$.

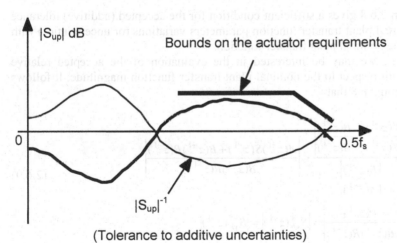

Figure 2.37. The input sensitivity function and its inverse

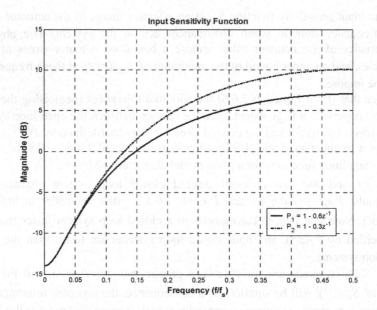

Figure 2.38. Input sensitivity function for the plant model $q^{-1}/(1 - 0.8\,q^{-1})$ as a function of the desired closed loop performances

Equation 2.6.8 gives a sufficient condition for the accepted (additive) tolerance in terms of real plant transfer function parameters variations (or uncertainties) with respect to the nominal plant transfer function.

Moreover, we may be interested in the evaluation of the accepted relative tolerance with respect to the nominal plant transfer function magnitude. It follows from Equation 2.6.8 that

$$\frac{\left|\dfrac{B'(z^{-1})}{A'(z^{-1})} - \dfrac{B(z^{-1})}{A(z^{-1})}\right|}{\left|\dfrac{B(z^{-1})}{A(z^{-1})}\right|} < \left|\frac{A(z^{-1})S(z^{-1}) + B(z^{-1})R(z^{-1})}{B(z^{-1})R(z^{-1})}\right| \qquad (2.6.9)$$

$$= \left|\frac{P(z^{-1})}{B(z^{-1})R(z^{-1})}\right| = \left|S_{yb}^{-1}(z^{-1})\right| = \left|S_{yr}^{-1}(z^{-1})\right|$$

where S_{yb} is the "noise-output sensitivity function" and S_{yr} is the *complementary sensitivity function*.

The noise-output sensitivity function S_{yb} allows the definition of a frequency "template" to ensure that the "delay margin" constraint is fulfilled. Let consider the case of a delay margin $\Delta\tau = 1 \cdot T_s$.

It follows that

$$H(z^{-1}) = \frac{z^{-d}B(z^{-1})}{A(z^{-1})} \quad ; \quad H'(z^{-1}) = \frac{z^{-d-1}B(z^{-1})}{A(z^{-1})} \qquad (2.6.10)$$

and consequently

$$\frac{H'(z^{-1}) - H(z^{-1})}{H(z^{-1})} = z^{-1} - 1 \qquad (2.6.11)$$

Equation 2.6.9 becomes

$$\left| S_{yb}(z^{-1}) \right| < \frac{1}{\left| z^{-1} - 1 \right|} \quad ; \quad z = e^{j\omega} \qquad 0 \le \omega \le \pi \qquad (2.6.12)$$

or in dB units

$$\left| S_{yb}^{-1}(z^{-1}) \right| dB < -20 \log \left| 1 - z^{-1} \right| \quad ; \quad z = e^{j\omega} \qquad 0 \le \omega \le \pi \qquad (2.6.13)$$

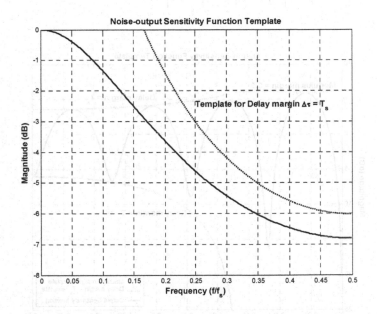

Figure 2.39. Frequency template for the noise-output sensitivity function and $\Delta\tau = T_S$

This expression defines a frequency robustness template for the sensitivity function S_{yb}. This template corresponds to the frequency response of a digital integrator and is represented in Figure 2.39.

As the modulus margin introduces a frequency template on the output sensitivity function ($\left|S_{yp}(e^{-j\omega})\right| \leq -\Delta M$; $0 \leq \omega \leq \pi$), we are interested in finding what template is introduced by the delay margin on $|S_{yp}|$.

From Equations 2.5.14 and 2.5.17 it results that

$$S_{yp}(z^{-1}) = 1 + S_{yb}(z^{-1}) \tag{2.6.14}$$

and by means of the triangle inequality it follows that

$$1 - \left|S_{yb}(z^{-1})\right| \leq \left|S_{yp}(z^{-1})\right| \leq 1 + \left|S_{yb}(z^{-1})\right| \tag{2.6.15}$$

Taking into account the frequency bound on S_{yb} given by Equation 2.6.12, the following condition is obtained :

$$1 - \left|1 - z^{-1}\right|^{-1} \leq \left|S_{yp}(z^{-1})\right| \leq 1 + \left|1 - z^{-1}\right|^{-1} \tag{2.6.16}$$

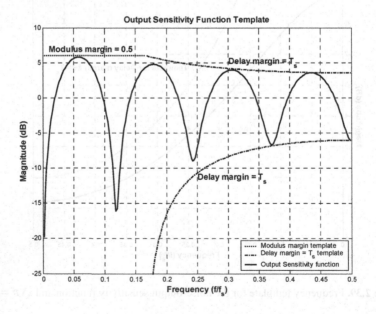

Figure 2.40. Frequency template on the output sensitivity function for $\Delta M = 0.5$ (-6dB) and $\Delta \tau = T_S$

that leads to the robustness template on $|S_{yp}|$ represented in Figure 2.40.

Notice that, from the point corresponding to $0.17 f_s$, $|S_{yp}|$ must lie inside a region delimited by an upper and a lower bound and that, for frequencies below $0.17 f_s$, the frequency template for the modulus margin also assures the delay margin constraint to be respected.

It is important to note that the template on S_{yp} will not always guarantee the desired delay margin (it is an approximation). If the condition on $|S_{yb}|$ is satisfied, then the condition on $|S_{yp}|$ will also be satisfied. However, if the condition on $|S_{yb}|$ is violated, this will not imply necessarily that the condition on $|S_{yp}|$ will also be violated. In practice, the results obtained by using the template on the $|S_{yp}|$ are very reliable.

The following remark is important: *the closed loop system robustness will be, in general, reduced when the closed loop system bandwidth is increased with respect to the open loop system bandwidth. Conversely, for a relevant reduction of the rise time for the closed loop system, with respect to the open loop system rise time, a good estimation of the plant model is required (especially in the frequency regions where $|S_{yp}(z^{-1})|$ is high).*

As a consequence, robustness constraints can imply either a small reduction of the closed loop system rise time (with respect to the open loop system rise time), or a controller design which takes into account the bounds on the sensitivity functions.

An important challenge in control system design is the maximization of the controller robustness for given performances. This is obtained by minimizing the sensitivity functions maximum in the critical frequency regions.

2.7 Concluding Remarks

Recursive (differences) equations of the form

$$y(t) = -\sum_{i=1}^{n_A} a_i y(t-i) + \sum_{i=1}^{n_B} b_i u(t-d-i) \qquad (2.7.1)$$

where u is the input, y is the output and d is the discrete-time delay, are used to describe discrete-time dynamic models.

The delay operator q^{-1} [$q^{-1} y(t) = y(t-1)$] is a simple tool to handle recursive equations. If the operator q^{-1} is used, the recursive Equation 2.7.1 takes the form

$$A(q^{-1})y(t) = q^{-d} B(q^{-1})u(t)$$

where

$$A(q^{-1}) = 1 + \sum_{i=1}^{n_A} a_i q^{-i} \qquad B(q^{-1}) = 1 + \sum_{i=1}^{n_B} b_i q^{-i}$$

The input-output relation for a discrete-time model is also conveniently described by the pulse transfer operator $H(q^{-1})$:

$$y(t) = H(q^{-1})u(t)$$

where

$$H(q^{-1}) = \frac{q^{-d} B(q^{-1})}{A(q^{-1})}$$

The pulse transfer function of a discrete-time linear system is expressed as function of the complex variable $z = e^{sT_s}$ (T_s = sampling period). The pulse transfer function can be derived from the pulse transfer operator $H(q^{-1})$ by replacing q^{-1} with z^{-1}.

The asymptotical stability of a discrete-time model is ensured if, and only if, all pulse transfer function poles (in z) lie inside the unit circle.

The order of a pulse transfer function is

$$n = max \ (n_A, \ n_B + d)$$

In computer controlled systems, the input signal applied to the plant is held constant between two sampling instants by means of a zero-order hold (ZOH). The zero-order hold is characterized by the following transfer function:

$$H_{ZOH}(s) = \frac{1 - e^{-sT_s}}{s}$$

Therefore, the continuous-time part of the system (between digital-to-analog converter and the analog-to-digital converter) is characterized by the continuous-time transfer function

$$H'(s) = H_{ZOH}(s) \cdot H(s)$$

where $H(s)$ is the plant transfer function.

In computer controlled systems, the input signal applied to the plant at time t is a weighted average of the plant output at times t, $t-1$, ..., $t-n_A+1$, of the previous input signal values at instants $t-1$, $t-2$, ..., $t-n_B-d$, and of the reference signal at instants t, $t-1,...,$ the weights being the coefficients of the controller. The corresponding control law (controller RST) is written as

$$S(q^{-1})\, u(t) = - R(q^{-1})\, y(t) + T(q^{-1})\, r(t) \tag{2.7.2}$$

where $u(t)$ is the control (input) signal to the plant, $y(t)$ is the plant output and $r(t)$ is the reference.

The transfer function of the closed loop system (between the reference signal and the plant output) that includes the digital controller of Equation 2.7.2 is given by

$$H_{CL}(z^{-1}) = \frac{B(z^{-1})T(z^{-1})}{A(z^{-1})S(z^{-1}) + B(z^{-1})R(z^{-1})}$$

where $H(z^{-1}) = B(z^{-1})/A(z^{-1})$ is the pulse transfer function of the discretized plant (in this case $B(z^{-1})$ may include possible delays).

The characteristic polynomial defining the closed loop system poles is given by

$$P(z^{-1}) = A(z^{-1})\, S(z^{-1}) + B(z^{-1})\, R(z^{-1})$$

The disturbance rejection properties on the output result from the output sensitivity function frequency response

$$S_{yp}(z^{-1}) = \frac{A(z^{-1})S(z^{-1})}{A(z^{-1})S(z^{-1}) + B(z^{-1})R(z^{-1})}$$

Robust stability of the closed loop system, with respect to the plant transfer function uncertainties or parameters variations, is essentially characterized by the *modulus margin and the delay margin.*

The *modulus margin* and the *delay margin* introduce frequency constraints on the magnitude of the sensitivity functions. These constraints lead to the definition of frequency robustness templates that must be respected.

The robust stability (or performance) of the closed loop system robustness, with respect to the plant transfer function uncertainties or parameters variations, depends upon the choice of the desired closed loop system performances (bandwidth, rise time) with respect to the open loop system dynamics. A significant reduction of the closed loop system rise time (or a significant augmentation of the bandwidth of the closed loop system), compared to the open loop system rise time (or bandwidth), requires a good estimation of the plant model.

In order to ensure closed loop system robustness, when a good estimation of the plant model is not available, or when large system parameters variations occur, the closed loop system rise time acceleration, compared to the open loop system rise time, must be moderate. However, some methods exist for maximizing the controller robustness with respect to plant model uncertainties (or parameters variations), for given nominal performance.

2.8 Notes and References

Some basic books on computer control system are:

Kuo B. (1980) Digital Control Systems, Holt Saunders, Tokyo.

Åström K.J., Wittenmark B. (1997) Computer Controlled Systems - Theory and Design, 3rd edition, Prentice-Hall, Englewood Cliffs, N.J.

Ogata K. (1987) Discrete-Time Control Systems, Prentice Hall, N.J.

Franklin G.F., Powell J.D., Workman M.L. (1998) Digital Control of Dynamic Systems , 3rd edition, Addison Wesley, Reading, Mass.

Wirk G.S. (1991) Digital Computer Systems, MacMillan, London.

Phillips, C.L., Nagle, H.T. (1995) Digital Control Systems Analysis and Design, 3rd edition, Prenctice Hall, N.J.

For an introduction to robust control theory see:

Doyle J.C., Francis B.A., Tanenbaum A.R. (1992) Feedback Control Theory, Mac Millan, N.Y.

Kwakernaak H. (1993) Robust control and H_{inf} optimization – a tutorial Automatica, vol.29, pp.255-273.

Morari M., Zafiriou E. (1989) Robust Process Control, Prentice Hall International, Englewood Cliffs, N.J.

The delay margin was introduced by:

Anderson B.D.O, Moore J.B. (1971) Linear Optimal Control, Prentice Hall, Englewood Cliffs, N.J.

For a detailed discussion on the delay margin:

Bourlès H., Irving E. (1991) La méthode LQG/LTR: une interprétation polynomiale, temps continu / temps discret, RAIRO-APII, Vol. 25, pp. 545-568.

Landau I.D. (1995) Robust digital control systems with time delay (the Smith predictor revisited), Int. J. of Control, Vol. 62, no. 2, pp. 325-347.

The *circle criterion* is presented in:

Zames G. (1966) On the input-output stability of time-varying nonlinear feedback systems, IEEE-TAC, vol. AC-11, April, pp. 228-238, July pp. 445-476.

Narendra K.S., Taylor J.H. (1973) Frequency Domain Criteria for Absolute Stability, Academic Press, New York.

For a generalization of the concepts of robustness margins and robust stability introduced in this chapter, see Appendix D.

3

Robust Digital Controller Design Methods

In this chapter the design of model based robust digital controllers is discussed. The design of digital PID controllers is first presented, emphasizing the general structure of digital controllers (three branched structure known as RST), the special features of the digital approach and the limitations of the digital PID. The following design methods are then presented: pole placement, tracking and regulation with independent objectives and tracking and regulation with internal model control. The presentation is done from the perspective of robust control. These methods permit the control of systems of any order with or without time delay. The last section of the chapter presents a general methodology for the design of robust digital controllers by means of sensitivity functions shaping.

3.1 Introduction

The use of a digital computer or microprocessor in control loops offers several advantages. These include:

- Considerable choice of strategies for controller design
- Possibility of using algorithms which are both more complex and more efficient than the PID
- Technique perfectly suited for the control of systems *with time dela;*
- Technique well suited for the control of systems characterized by linear dynamic models of high order (including systems with multiple low damped vibration modes)

Moreover, by combining the controller design methods with the system model identification techniques, a rigorous, high performance controller design procedure can be implemented. These aspects are covered in Chapters 7, 8 and 9.

The digital controller design methods that are presented in this chapter relate to single input-single output control in the presence of deterministic disturbances.

These methods are:

- Digital PID

85

- Pole placement
- Tracking and regulation with independent objectives
- Tracking and regulation with internal model control
- Pole placement with shaping of sensitivity functions

All the controllers, irrespective of their design method, will have the same RST three-branched structure (see Figure 2.27). Only the *memory* of the controller (number of coefficients) will vary depending on the complexity of the system.

Note that in the case of simple systems (at most second-order and small time delay), the various controllers designed by means of the pole placement, tracking and regulation with independent objectives and internal model control, correspond to digital PID controllers having differently tuned parameters.

The *tracking and regulation with independent objectives* and *tracking and regulation with internal model control* can be considered as particular cases of the *pole placement*. They result from a particular choice of the desired closed loop poles.

The robustness of the designed control system with respect to plant model uncertainties is a very important issue. The *pole placement with shaping of sensitivity functions* is a general methodology of digital control design that allows one to take into account simultaneously robustness and performances specifications for the closed loop.

The digital control design by *pole placement* (and the various particular cases) is a *predictive control* (the controller implicitly contains a predictor of the plant output). This will be illustrated in Sections 3.4.4 and 3.5.3 and in Appendix B.

The *design and tuning of digital controllers require the knowledge of the discrete-time model of the plant to be controlled (model based control)*. It is not possible to implement effectively a high performance control loop without identifying the plant model. Fortunately the system identification methodology is a mature subject and toolboxes or dedicated software are available. System identification is discussed in Chapters 5, 6, 7 and 9.

If the continuous-time model of the plant to be controlled is available, the discrete-time model of the sampled-data system can be obtained by using appropriate discretization techniques. The functions *cont2disc.sci* (Scilab) and *cont2disc.m* (MATLAB®), available on the book website, can be used for this purpose.

3.2 Digital PID Controller

The basic version of the digital PID controller considered in this book results from the discretization of the continuous-time PID controller presented in Section 1.3.2. Another version, which provides some advantages, will also be presented.

The methodology for the digital PID controllers design to be presented is rigorously applicable only to[1]:

- Plants that can be modeled by a continuous-time system characterized by a transfer function of maximum degree equal to 2, with or without time delay
- Plants having a time delay which is less than one sampling period

Although continuous-time PID parameters can be recovered from the digital PID controller design, in some cases, the discrete-time methodologies have not been developed for the tuning of continuous-time (or pseudo-digital) PID controllers. Specific methodologies exist and should be used for the tuning of continuous time PID controllers (see Chapter 1, Section 1.3). Moreover, it should be pointed out that certain tunings of digital PID controller parameters offering excellent performances have no counterpart in terms of continuous-time PID controller parameters.

3.2.1 Structure of the Digital PID 1 Controller

Consider the transfer function of the continuous-time PID controller (Equation 1.3.2):

$$H_{PID}(s) = K\left[1 + \frac{1}{T_i s} + \frac{T_d s}{1 + \frac{T_d}{N} s}\right] \qquad (3.2.1)$$

This controller is characterized by four tuning parameters:

- K — proportional gain
- T_i — integral action
- T_d — derivative action
- T_d/N — filtering of the derivative action

Several discretization methods may be used to derive the structure of the digital PID controller. The relations between the continuous-time and discrete-time parameters will depend on the method used, but the structure of the digital controller will remain the same.

Since in our case the design and implementation of the controller will be in the discrete-time, the discretization method is not essential. For this reason we will use the *backward difference approximation*. It follows that s (derivative) will be approximated by $(1 - q^{-1})/T_s$ and $1/s$ (integration) will be approximated by $T_s/(1 - q^{-1})$ (see Section 2.3, Equations 2.3.6 to 2.3.11).

---

[1] A method for the design of digital PID controllers for plants characterized by high order models is presented in Chapter 10, Section 10.5. It is based on the complexity reduction of a model based controller.

This produces

$$\frac{1}{T_i s} = \frac{T_s}{T_i} \cdot \frac{1}{1-q^{-1}} \tag{3.2.2}$$

$$T_d s = \frac{T_d}{T_s} \cdot (1-q^{-1}) \tag{3.2.3}$$

$$\frac{1}{1+\dfrac{T_d}{N}s} = \frac{1}{1+\dfrac{T_d}{NT_s}(1-q^{-1})} = \frac{\dfrac{NT_s}{T_d+NT_s}}{1-\dfrac{T_d}{T_d+NT_s}q^{-1}} \tag{3.2.4}$$

By introducing these expressions in Equation (3.2.1), the pulse transfer function (operator) of the digital PID 1 controller is obtained[2]:

$$H_{PID1}(q^{-1}) = \frac{R(q^{-1})}{S(q^{-1})} = K\left[1 + \frac{T_s}{T_i} \cdot \frac{1}{1-q^{-1}} + \frac{\dfrac{NT_s}{T_d+NT_s}(1-q^{-1})}{1-\dfrac{T_d}{T_d+NT_s}q^{-1}}\right] \tag{3.2.5}$$

The expression in terms of the ratio of two polynomials is obtained by summing up the three terms. Polynomials $R(q^{-1})$ and $S(q^{-1})$ have the form

$$R(q^{-1}) = r_0 + r_1 q^{-1} + r_2 q^{-2} \tag{3.2.6}$$
$$S(q^{-1}) = (1 - q^{-1})(1 + s'_1 q^{-1}) = (1 + s_1 q^{-1} + s_2 q^{-2}) \tag{3.2.7}$$

where

$$s'_1 = -\frac{T_d}{T_d + NT_s} \quad ; \quad r_0 = K\left(1 + \frac{T_s}{T_i} - N\frac{T_s}{T_d}s'_1\right)$$

$$r_1 = K\left[s_1\left(1 + \frac{T_s}{T_i} + 2N\frac{T_s}{T_d}\right) - 1\right] \quad ; \quad r_2 = -Ks_1\left(1 + N\frac{T_s}{T_d}\right)$$

---

[2] As indicated in Section 2.5.3, for systems with constant coefficients the notation q^{-1} will be used both for the backward delay operator and the complex variable z^{-1}, with the exception of cases that require an interpretation in the frequency domain.

The digital PID 1 controller has four parameters (r_0, r_1, r_2, s'_1) as the continuous-time PID controller.

Note that the pulse transfer function (operator) of the digital PID 1 controller contains as a common factor of the denominator the term $(1-q^{-1})$, which assures the behavior of the numerical integration. The denominator also contains the term $(1+s'_1 q^{-1})$, which is a digital filter that plays the role of the filter $[1 + (T_d/N) s]$ in the continuous-time PID controller.

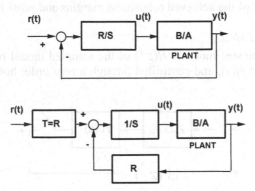

Figure 3.1. Equivalent block diagrams of a digital control loop using the digital PID 1 controller

The equivalent block diagram is given in the upper part of Figure 3.1. By taking $T(q^{-1}) = R(q^{-1})$, the digital PID 1 controller can take the standard three branched structure of the RST controller, as shown in the lower part of Figure 3.1.

The pulse transfer function (operator) of the closed loop relating the reference $r(t)$ and the output $y(t)$ is

$$H_{CL}(q^{-1}) = \frac{B(q^{-1})R(q^{-1})}{A(q^{-1})S(q^{-1}) + B(q^{-1})R(q^{-1})} = \frac{B(q^{-1})R(q^{-1})}{P(q^{-1})} \qquad (3.2.8)$$

in which the polynomial $P(q^{-1})$ defines the desired closed loop poles (directly linked to the desired regulation performances).

The product $B(q^{-1})R(q^{-1})$ defines the closed loop zeros. The digital PID 1 controller, in general, does not simplify the plant zeros (unless $B(q^{-1})$ is chosen as a factor of $P(q^{-1})$) and thus can be used for the regulation of plants having a discrete-time model with unstable zeros (a situation occurring, for example, if there is a fractional time delay greater than half of a sampling period, see Section 2.3.7).

Furthermore, the digital PID 1 controller introduces additional zeros defined by $R(q^{-1})$ that will depend on $A(q^{-1})$, $B(q^{-1})$ and $P(q^{-1})$ and thus which cannot be specified a priori. In certain situations, these zeros may produce undesirable overshoots during the transient (see the examples given later on in Section 3.2.3).

3.2.2 Design of the Digital PID 1 Controller

The computation of the parameters involves several stages:

1. Determination of discrete time plant model
2. Specification of the performances
3. Computation of the controller parameters (the coefficients of the polynomial $R(q^{-1})$ and $S(q^{-1})$)
4. Verification of the achieved robustness margins and sensitivity functions

Discrete-Time Plant Model
This is the pulse transfer function $H(z^{-1})$ of the sampled model of a plant having the transfer function $H(s)$, and controlled through a zero order hold as is indicated in Figure 3.2.

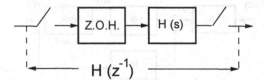

Figure 3.2. Pulse transfer function of a plant

Continuous-time transfer functions of the following form may be considered:

$$H(s) = \frac{Ge^{-s\tau}}{1+sT} \qquad (3.2.9)$$

or

$$H(s) = \frac{\omega_0^2 e^{-s\tau}}{\omega_0^2 + 2\zeta\omega_0 s + s^2} \qquad (3.2.10)$$

with the restriction

$$\tau < T_s \qquad (3.2.11)$$

Note moreover that for the first-order system the sampling period T_s must be smaller than T ($T_s < T$ – see Equation 2.2.4). It thus means that the digital PID controller design can be correctly applied only to the first-order system with a time delay verifying the condition $\tau < T$.

For these two types of continuous-time models, discretization with hold results in a pulse transfer function (operator) of the form

$$H(q^{-1}) = \frac{b_1 q^{-1} + b_2 q^{-2}}{1 + a_1 q^{-1} + a_2 q^{-2}} = \frac{B(q^{-1})}{A(q^{-1})} \tag{3.2.12}$$

The transfer function $H(q^{-1})$ may be obtained:

- Directly by identification of the discrete-time plant model
- From knowledge of $H(s)$ and T_s using discretization routines (as for example *cont2disc.sci* (Scilab) and *cont2disc.m* (MATLAB®)) or transformation tables

Specification of the Performances
As a general rule, the desired closed loop system performances can be expressed in terms of the parameters of a pulse transfer function. This may be expressed by the condition

$$H_{CL}(q^{-1}) = \frac{B(q^{-1})R(q^{-1})}{A(q^{-1})S(q^{-1}) + B(q^{-1})R(q^{-1})} = \frac{B_M(q^{-1})}{P(q^{-1})} \tag{3.2.13}$$

However, $B_M(q^{-1})$ cannot be specified *a priori* since, in general, $B(q^{-1})$ is not simplified (unless $B(q^{-1})$ is stable, which corresponds to cases where the time delay is negligible); moreover, the controller itself introduces zeros by means of $R(q^{-1})$. The closed loop polynomial remains to be specified. The polynomial $P(q^{-1})$ is chosen of the form

$$P(q^{-1}) = 1 + p'_1 q^{-1} + p'_2 q^{-2} \tag{3.2.14}$$

A recommended method for defining p'_1 and p'_2 consists first in considering a second-order normalized continuous-time model (see Section 1.1.6), enabling a rise time (t_R) or a settling time (t_S) and a maximum overshoot (M) to be obtained in accordance with the specifications. This choice may be done using the diagrams given in Figures 1.10 and 1.11 or using the functions *omega_dmp.sci* (Scilab) and *omega_dmp.m* (MATLAB®)[3]. That allows determining the parameters ω_0 and ζ of the second-order system. The sampling period T_s and the natural frequency ω_0 should verify the condition

$$0.25 \leq \omega_0 T_s \leq 1.5 \quad ; \quad 0.7 \leq \zeta \leq 1 \tag{3.2.15}$$

The discretized model with hold can be computed either by means of the functions *cont2disc.sci* (Scilab) and *cont2disc.m* (MATLAB®), or by means of

---

[3] Available from the book website.

transformation tables (see Section 2.3.6). The denominator of the pulse transfer function thus obtained will represent the polynomial $P(q^{-1})$.

Computation of the Coefficients of the Digital Controller
From Equation 3.2.13 it results that the following polynomial equation must be solved:

$$P(q^{-1}) = A(q^{-1})\, S(q^{-1}) + B(q^{-1})\, R(q^{-1}) \tag{3.2.16}$$

in the unknown polynomials $S(q^{-1})$ and $R(q^{-1})$. In Equation 3.2.16 $P(q^{-1})$ is given by Equation 3.2.14 and $A(q^{-1})$ and $B(q^{-1})$ are given by Equation 3.2.12 . The structures of $R(q^{-1})$ and $S(q^{-1})$ are given respectively by Equations 3.2.6 and 3.2.7. Equation 3.2.16 is a Bezout polynomial equation.

The detailed version of Equation 3.2.16 is

$$
\begin{aligned}
P(q^{-1}) &= 1 + p_1' q^{-1} + p_2' q^{-2} = A(q^{-1})S(q^{-1}) + B(q^{-1})R(q^{-1}) \\
&= (1 + a_1 q^{-1} + a_2 q^{-2})(1 - q^{-1})(1 + s_1' q^{-1}) \\
&\quad + (b_1 q^{-1} + b_2 q^{-2})(r_0 + r_1 q^{-1} + r_2 q^{-2}) \\
&= A'(q^{-1})S'(q^{-1}) + B(q^{-1})R(q^{-1})
\end{aligned}
\tag{3.2.17}
$$

where

$$
\begin{aligned}
A'(q^{-1}) &= A(q^{-1})(1 - q^{-1}) = (1 + a_1' q^{-1} + a_2' q^{-2} + a_3' q^{-3}) \\
&= (1 + (a_1 - 1)q^{-1} + (a_2 - a_1)q^{-2} - a_2 q^{-3})
\end{aligned}
\tag{3.2.18}
$$

$$S'(q^{-1}) = 1 + s_1' q^{-1} \tag{3.2.19}$$

Solving a polynomial equation implies that coefficients related to the powers of q should be equal on both sides. One observes that the maximum order on the right hand side is q^{-4} (it is a system of four equations in four variables that must be solved). The higher order of the left side of the equation is q^{-2}, that corresponds to a zero value for the coefficients of q^{-3} and q^{-4} powers in the polynomial $P(q^{-1})$. In fact, it is possible to impose non-zero values for these coefficients.

In general one imposes

$$
\begin{aligned}
P(q^{-1}) &= (1 + p_1 q^{-1} + p_2 q^{-2} + p_3 q^{-3} + p_4 q^{-4}) = \\
&= (1 + p_1' q^{-1} + p_2' q^{-2})(1 + \alpha_1 q^{-1})(1 + \alpha_2 q^{-2})
\end{aligned}
\tag{3.2.20}
$$

where coefficients p_1' and p_2' result from the discretization of a second-order continuous-time system with ω_0 and ζ corresponding to the specified nominal performances, and with α_1, α_2 corresponding to aperiodic "auxiliary poles" $-\alpha_1$

and $-\alpha_2$ located on the real axis (inside the unit circle) and corresponding to a frequency higher than $\omega_0 / 2\pi$ (see Figure 3.3).

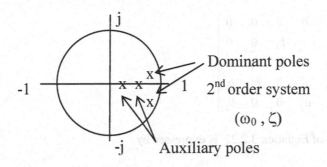

Figure 3.3. Dominant and auxiliary poles for the design of a digital PID controller

One observes that $-\alpha_1$ and $-\alpha_2$ are smaller than the real part of the dominant poles (then they are faster poles). The introduction of auxiliary poles allows improving the robustness of the controller. In practice, a typical choice is either $\alpha_1 = \alpha_2$, or $\alpha_2 = 0$ and the ranges for their values are

$$-0.05 \le \alpha_1, \alpha_2 \le -0.5 \tag{3.2.21}$$

The system of four equations to solve is

$$p_1 = b_1 r_0 + s_1' + a_1'$$

$$p_2 = b_2 r_0 + b_1 r_1 + s_1' a_1' + a_2'$$

$$p_3 = b_2 r_1 + b_1 r_2 + s_1' a_2' + a_3'$$

$$p_4 = b_2 r_2 + s_1' a_3'$$

The equivalent matrix form of this system of equations is

$$Mx = p \tag{3.2.22}$$

where

$$x^T = [1, s_1', r_0, r_1, r_2]$$ (3.2.23)

$$p^T = [1, p_1, p_2, p_3, p_4]$$ (3.2.24)

and the matrix M is of the form

$$
\begin{bmatrix}
1 & 0 & 0 & 0 & 0 \\
a_1' & 1 & b_1 & 0 & 0 \\
a_2' & a_1' & b_2 & b_1 & 0 \\
a_3' & a_2' & 0 & b_2 & b_1 \\
0 & a_3' & 0 & 0 & b_2
\end{bmatrix}
$$

The solution of Equation 3.2.22 is expressed by

$$x = M^{-1}p$$ (3.2.25)

where M^{-1} is the inverse matrix of M. In order to assure the existence of this inverse, it is necessary that the determinant of M is non-null. It can be shown that this condition is verified if, and only if, $A(q^{-1})$ and $B(q^{-1})$ are coprime polynomials (no simplification between zeros and poles).

Exercise
Let

$$B(q^{-1}) = b_1 q^{-1} + b_2 q^{-2} = b_1 q^{-1}(1 + \frac{b_2}{b_1} q^{-1})$$

$$A(q^{-1}) = (1 + \frac{b_2}{b_1} q^{-1})(1 + c q^{-1})$$

Show that in this case $\det M = 0$.

In order to solve the Bezout equation, one can use the function *bezoutd.sci* (Scilab) or the function *bezoutd.m* (MATLAB®) available on the book website[4].

The parameters of the continuous-time PID controller, which by discretization with a sampling period equal to T_s gives the digital PID controller with polynomials $R(q^{-1})$ and $S(q^{-1})$, are computed by means of the following relations:

$$K = \frac{r_0 s_1' - r_1 - (2 + s_1') r_2}{(1 + s_1')^2}$$ (3.2.26)

---

[4] In Scilab and MATLAB® environments one must specify $A'(q^{-1})$ and $B(q^{-1})$ to obtain $S'(q^{-1})$ and $R(q^{-1})$. $S(q^{-1})$ is obtained by using Equation 3.2.7.

$$T_i = T_s \cdot \frac{K(1 + s_1')}{r_0 + r_1 + r_2} \tag{3.2.27}$$

$$T_d = T_s \cdot \frac{s_1' r_0 - s_1' r_1 + r_2}{K(1 + s_1')^3} \tag{3.2.28}$$

$$\frac{T_d}{N} = \frac{-s_1' T_s}{1 + s_1'} \tag{3.2.29}$$

For the digital PID controller to be equivalent to a continuous-time PID controller, it is necessary that the coefficient s_1' verify the condition

$$-1 < s_1' \leq 0 \tag{3.2.30}$$

In the opposite case ($0 < s_1' < 1$), the digital filter $1/(1 + s_1' q^{-1})$ is stable but without a continuous-time equivalent (as a first-order filter). The digital PID controller in this case may provide very good performances but an equivalent continuous-time PID controller cannot be obtained (see examples given in Section 3.2.3).

3.2.3 Digital PID 1 Controller: Examples

We consider the case of the regulation of a first-order plant with time delay having the following characteristics:

- Gain $(G) = 1$;   Time Constant $(T) = 10$ s;   Pure time delay $(\tau) = 3$ s
- A sampling period $T_s = 5$ s is chosen in order to verify the conditions:
  $\tau < T_s$ and $T_s < T$

The objective of the design is to obtain the best closed loop performances without overshoot. The damping factor of the model that specifies the performances is fixed at $\zeta = 0.8$, and the natural frequency ω_0 will be chosen in the range: $0.25 \leq \omega_0 T_s \leq 1.5$.

For $\omega_0 = 0.05$ ($\omega_0 T_s = 0.25$), the results obtained are summed up in Table 3.1.

The model output $y(t)$ and the control signal $u(t)$ are displayed in Figure 3.4. It can be observed that the closed loop step response (about $13Ts = 65$ s) is slower than the open loop step response (about $2.2T + \tau = 25$ s) since the control $u(t)$ sent

Table 3.1. Digital PID 1 controller, $\omega_0 = 0.05$ rad/s

Plant:

- $B(q^{-1}) = 0.1813\ q^{-1} + 0.2122\ q^{-2}$
- $A(q^{-1}) = 1 - 0.6065\ q^{-1}$
- $T_s = 5s,\ G = 1,\ T = 10s,\ \tau = 3$

Performances $\rightarrow$ $T_s = 5s$, $\omega_0 = 0.05$ rad/s, $\zeta = 0.8$

*** CONTROL LAW ***

$$S(q^{-1}) \cdot u(t) + R(q^{-1}) \cdot y(t) = T(q^{-1}) \cdot r(t)$$

Controller:

- $R(q^{-1}) = 0.0621 + 0.0681\ q^{-1}$
- $S(q^{-1}) = (1 - q^{-1}) \cdot (1 - 0.0238\ q^{-1})$
- $T(q^{-1}) = R(q^{-1})$

Gain margin: 7.712	Phase margin: 67.2 deg
Modulus margin: 0.751 (-2.49dB)	Delay margin: 45.4 s
Cont. time PID: k = -0.073, Ti = -2.735, Td = - 0.122, Td/N = 0.122	

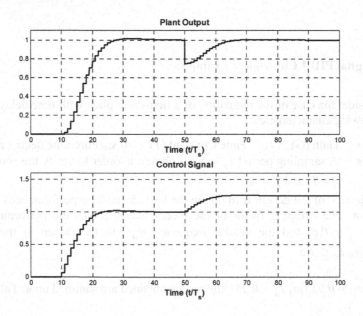

Figure 3.4. Performances of the digital PID 1 controller, $\omega_0 = 0.05$ rad/s for tracking and regulation

by the controller represents a filtered step. This situation is well known in practice. For time delays greater than 20 or 25% of the time constant, the continuous time PID controller slows down the closed loop system with respect to the open loop system. Can a digital PID controller therefore provide improved performances?

Take $\omega_0 = 0.1$ rad/s (twice faster desired closed loop response). The computation results are summed up in Table 3.2 and the evolutions of the plant output and of the control applied to the plant are represented in Figure 3.5.

By examining the results given in Table 3.2, first one can see that s'_1 is positive, and therefore an equivalent continuous-time PID controller does not exist.

An acceleration in the plant response is observed, but this is accompanied by the appearance of a slight overshoot (larger than the one corresponding to $\zeta = 0.8$).

The explanation of this overshoot is given by the coefficients of $T(q^{-1})=R(q^{-1})$. The polynomial $T(q^{-1})=R(q^{-1})$, which introduces zeros in the pulse transfer function, has now its second coefficient with a negative sign. As the zeros have influence on the first instants of the response, the contribution of $R(q^{-1})$ is a positive jump followed by a negative one, and the step on the reference filtered, by $R(q^{-1})$ is characterized by a peak related to the difference $r_0 - r_1$, that is not completely attenuated by the remaining components of the pulse transfer function.

In order to avoid an excessive overshoot with this structure of the PID, one needs (for $\zeta > 0.7$) to have all coefficients of $R(q^{-1})$ positive.

Taking $\omega_0 = 0.15$ rad/s (Table 3.3 and Figure 3.6), a significant acceleration of the step response is obtained, but the overshoot becomes larger. As in the previous case, there is no equivalent continuous-time PID controller. It is also observed that the difference $r_0 - r_1$ is greater, thus explaining the increased overshoot.

Table 3.2. Digital PID 1 controller, $\omega_0 = 0.1$ rad/s

Plant:
• $B(q^{-1}) = 0.1813\ q^{-1} + 0.2122\ q^{-2}$
• $A(q^{-1}) = 1 - 0.6065\ q^{-1}$
• $T_S = 5s,\ G = 1,\ T = 10s,\ \tau = 3$
Performances → $T_S = 5s$, $\omega_0 = 0.1$ rad/s, $\zeta = 0.8$
*** CONTROL LAW ***
$S(q^{-1}) . u(t) + R(q^{-1}) . y(t) = T(q^{-1}) . r(t)$
Controller:
• $R(q^{-1}) = 0.8954 - 0.4671\ q^{-1}$
• $S(q^{-1}) = (1 - q^{-1}) . (1 + 0.16343\ q^{-1})$
• $T(q^{-1}) = R(q^{-1})$
Gain margin: 6.046   Phase margin: 65.9 deg
Modulus margin: 0.759 (- 2, 39 dB)   Delay margin: 16.8 s
Cont. time PID: (no equivalent)

Table 3.3. Digital PID 1 controller, $\omega_0 = 0.15$ rad/s

Plant:
- $B(q^{-1}) = 0.1813\ q^{-1} + 0.2122\ q^{-2}$
- $A(q^{-1}) = 1 - 0.6065\ q^{-2}$
- $T_s = 5s,\ \ G = 1,\ \ T = 10s,\ \ \tau = 3$

Performances   → $T_s = 5s$, $\omega_0 = 0.15\ rad/s$, $\zeta = 0.8$

*** CONTROL LAW ***

$$S(q^{-1})\ .\ u(t) + R(q^{-1})\ .\ y(t) = T(q^{-1})\ .\ r(t)$$

Controller:
- $R(q^{-1}) = 1.6874 - 0.8924\ q^{-1}$
- $S(q^{-1}) = (1 - q^{-1})\ .\ (1 + 0.3122\ q^{-1})$
- $T(q^{-1}) = R(q^{-1})$

Gain margin: 3.681   Phase margin: 58.4 deg
Modulus margin: 0.664 (- 3.56 dB)   Delay margin: 9.4 s
Cont. time PID: (no equivalent)

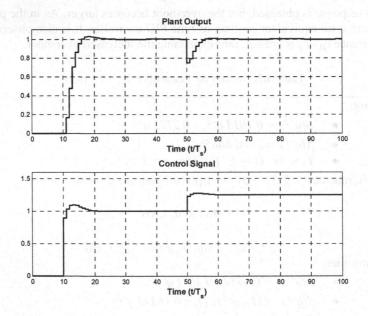

Figure 3.5. Performances of the digital PID 1 controller, $\omega_0 = 0.1$ rad/s for tracking and regulation

In order to eliminate the undesirable effect of the zeros introduced by the digital PID 1 controller, another structure must therefore be chosen for $T(q^{-1})$, which does not introduce additional zeros in the closed loop transfer function. That leads to the digital PID 2 controller.

3.2.4 Digital PID 2 Controller

This is a digital PID that does not introduce additional zeros.

The desired closed loop transfer function (between the reference and the output) will be of the form

$$H_{CL}(q^{-1}) = \frac{P(1)}{B(1)} \cdot \frac{B(q^{-1})}{P(q^{-1})}$$ (3.2.31)

in which $B(q^{-1})$ contains the plant zeros that will remain unchanged. $P(q^{-1})$ defines the desired closed loop poles and the term $P(1)/B(1)$ is introduced in order to ensure a unit gain between the reference and the output in steady state.

The controller will have the general structure

$$S(q^{-1})\, u(t) + R(q^{-1})\, y(t) = T(q^{-1})\, r(t)$$ (3.2.32)

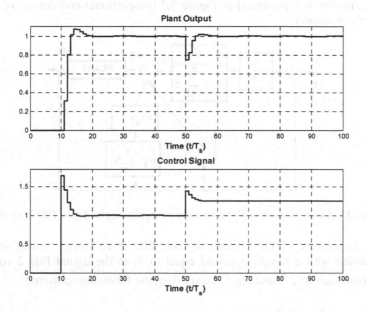

Figure 3.6. Performances of the digital PID 1 controller, $\omega_0 = 0.15$ rad/s for tracking and regulation

in which $S(q^{-1})$ and $R(q^{-1})$ are given by Equations 3.2.6 and 3.2.7 respectively.

The closed loop transfer function using the controller specified by Equation 3.2.32 will be

$$H_{CL}(q^{-1}) = \frac{T(q^{-1})B(q^{-1})}{A(q^{-1})S(q^{-1}) + B(q^{-1})R(q^{-1})} = \frac{[P(1)/B(1)]B(q^{-1})}{P(q^{-1})} \qquad (3.2.33)$$

In the same way as the PID1, the coefficients of the polynomials $S(q^{-1})$ and $R(q^{-1})$ will be obtained by solving Equation 3.2.16 . It thus results from Equation 3.2.33 that

$$T(q^{-1}) = \frac{P(1)}{B(1)} = \frac{B(1)R(1)}{B(1)} = R(1) \qquad (3.2.34)$$

since $S(1) = 0$ (which implies $P(1) = B(1) R(1)$). Then $T(q^{-1})$ will be a gain equal to the sum of the coefficients of $R(q^{-1})$.

To conclude: the digital PID 2 controller has the same polynomials $S(q^{-1})$ and $R(q^{-1})$ as the digital PID 1 controller; the only difference is that now $T(q^{-1}) = R(1)$ instead of $R(q^{-1})$, thereby preserving the unitary gain of the closed loop system in steady state without however introducing the effect of the zeros of $R(q^{-1})$.

The continuous-time PID (structure 2) that leads by discretization to the digital PID 2 controller is represented in Figure 3.7 (proportional and derivative actions only on the measure).

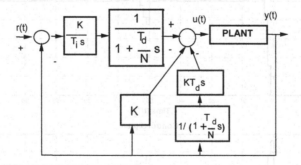

Figure 3.7. Continuous-time PID controller corresponding to digital PID 2 controller

The parameters of the continuous-time PID 2 controller, which leads by discretization with a sampling period equal to T_s to the digital PID 2 controller with polynomials $R(q^{-1})$ and $S(q^{-1})$, result from the following relations:

$$K = \frac{-(r_1 + 2r_2)}{1 + s_1'} \qquad (3.2.35)$$

$$T_i = T_s \cdot \frac{-(r_1 + 2r_2)}{r_0 + r_1 + r_2} \tag{3.2.36}$$

$$T_d = T_s \cdot \frac{s_1' r_1 + (s_1' - 1) r_2}{(r_1 + 2r_2)(1 + s_1')} \tag{3.2.37}$$

$$\frac{T_d}{N} = \frac{-s_1' T_s}{1 + s_1'} \tag{3.2.38}$$

Just like the PID 1 digital controller, for the PID 2 digital controller to be equivalent to a continuous-time PID controller, the condition of Equation 3.2.30 on the coefficient s'_1 must be satisfied (that is $-1 < s'_1 \le 0$).

Table 3.4 gives the computation results of the digital PID 2 controller for the same plant which was considered in the case of the digital PID 1 controller, and with a desired closed loop natural frequency $\omega_0 = 0.15$ rad/s (the results are to be compared with those given in Table 3.3; the values of the coefficients of $R(q^{-1})$ and $S(q^{-1})$ are the same).

Table 3.4. Digital PID controller (structure 2) $\omega_0 = 0.15$ rad/s

Plant:
• $B(q^{-1}) = 0.1813 \, q^{-1} + 0.2122 \, q^{-2}$
• $A(q^{-1}) = 1 - 0.6065 \, q^{-1}$
• $T_S = 5s, \ G = 1, \ T = 10s, \ \tau = 3$
Performances → $T_S = 5s$, $\omega_0 = 0.15$ rad/s, $\zeta = 0.8$
*** CONTROL LAW ***
$S(q^{-1}) \, u(t) + R(q^{-1}) \, y(t) = T(q^{-1}) \, r(t)$
Controller:
• $R(q^{-1}) = 1.6874 - 0.8924 \, q^{-1}$
• $S(q^{-1}) = (1 - q^{-1}) (1 + 0.3122 \, q^{-1})$
• $T(q^{-1}) = 0.795$
Gain margin: 3.681   Phase margin: 58.4 deg
Modulus margin: 0.664 (- 3.56 dB)   Delay margin: 9.4 s
Cont. time PID: (no equivalent)

The performances obtained are illustrated in Figure 3.8, which must be compared with the curves of Figure 3.6. It can be observed that for the same values of the polynomials $R(q^{-1})$ and $S(q^{-1})$, the overshoot during the transient disappears when the digital PID 2 controller is used (for steps on the reference). Moreover, the response to a disturbance is the same for PID1 and PID2.

Note also that the values obtained for the required robustness margins (gain, phase, modulus, delay) are without doubt satisfactory.

3.2.5 Effect of Auxiliary Poles

Figures 3.9 and 3.10 show the frequency characteristics of $|S_{yp}|$ and $|S_{up}|$ for three PID controllers designed for the same plant model used in the previous sections with the following performances specifications:

- $\omega_0 = 0.25$ rad/s ; $\zeta = 0.8$;
- $\omega_0 = 0.20$ rad/s ; $\zeta = 0.8$;
- $\omega_0 = 0.25$ rad/s ; $\zeta = 0.8$ and two auxiliary poles $-\alpha_1 = -\alpha_2 = 0.15$.

In Figure 3.9 the frequency region (low frequencies) where $|S_{yp}| < 0\,dB$ corresponds to the attenuation band for the disturbances. The frequency regions where $|S_{yp}| > 0\,dB$ correspond to an amplification of the disturbances. At the frequencies where $|S_{yp}| = 0dB$ the system behaves "in open loop" since the disturbances, at these frequencies, are neither amplified nor attenuated.

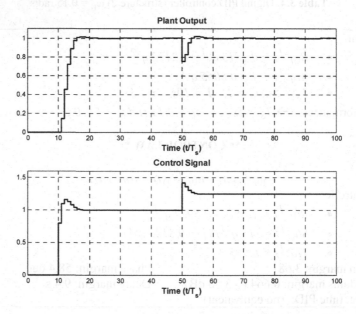

Figure 3.8. Performances of the digital PID 2 controller, $\omega_0 = 0.15$ rad/s for tracking and regulation

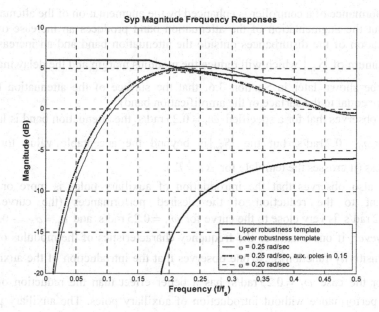

Figure 3.9. Frequency characteristics of the modulus of the output sensitivity function S_{yp} for different PID controllers

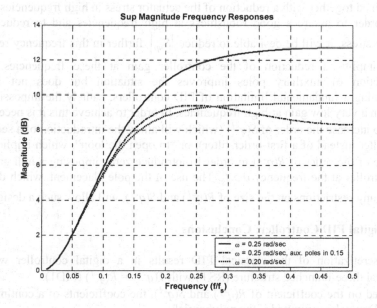

Figure 3.10. Frequency characteristics of the modulus of the input sensitivity function S_{up} for different PID controllers

The performance of a controller is enhanced by the augmentation of the attenuation band. But the augmentation of the attenuation band produces an increase of the amplification of the disturbances outside the attenuation band and an increase of the maximum of $|S_{yp}|$ which will reduce the modulus margin and the delay margin (it will be shown later in Section 3.6. that the surface of the attenuation band should be equal to the surface of the amplification band).

One observes that for a specified $\omega_0 = 0.25$ rad/s the attenuation band is larger than for $\omega_0 = 0.2$ rad/s but the $|S_{yp}|$ is beyond the acceptable value for the robustness (it crosses the template for $\Delta \tau = T_s$).

One also observes that the introduction of auxiliary poles is more or less equivalent to the reduction of the desired performances (the curve for $\omega_0 = 0.2$ rad/s is very close to the curve for $\omega_0 = 0.25$ rad/s and $\alpha_1 = \alpha_2 = -0.15$).

However, if one examines the frequency characteristics of the modulus of the input sensitivity function $|S_{up}|$, one observes that the introduction of the auxiliary poles for the case $\omega_0 = 0.25$ rad/s has a better effect than the reduction of the desired performance without introduction of auxiliary poles. The auxiliary poles allow reducing the value of $|S_{up}|$ in the high frequency region without affecting the regulation performances (see the frequency characteristics of $|S_{yp}|$). This means that a better robustness with respect to model uncertainties in high frequencies will be obtained together with a reduction of the actuator stress in high frequencies.

In order to assure a good robustness at high frequencies and to reduce the actuator stress, it will be desirable to reduce $|S_{up}|$ further in this frequency region, which implies a reduction of the controller gain at these frequencies. The introduction of auxiliary poles improves the situation but does not allow overcoming a fundamental limitation of PID controllers, namely the impossibility to obtain a very low gain at high frequencies. In order to achieve this it is necessary either to increase the order of $S(q^{-1})$ (which will allow the introduction of a second-order filter instead of a first-order filter), or "to open the loop", which implies the increase of the order of $R(q^{-1})$ in order to introduce zeros imposing a null gain of the controller at the frequency $0.5 f_s$. The use of the pole placement, which do not impose any restrictions on the size of $R(q^{-1})$ and $S(q^{-1})$, will allow such a design.

3.2.6 Digital PID Controller: Conclusions

The discretization of the classical PID results in a digital controller with a canonical three-branched structure (RST) with $T(q^{-1}) = R(q^{-1})$ (PID1).

Based on the coefficient of $R(q^{-1})$ and $S(q^{-1})$, the coefficients of a continuous-time PID may be computed if the polynomial

$$S(q^{-1}) = 1 + s'_1 q^{-1}$$

has $s'_1 \in] - 1,0]$.

The model based digital PID controller considered can deal with first-or second-order systems with time delay, if this latter is less than T_s (sampling period).

For time delays $\tau \geq 0.25\ T$ (time constant of the system of the plant), the continuous-time PID leads to responses that are slower in closed loop than the ones in open loop!

For systems with time delay, the closed loop performances may be significantly improved by choosing coefficients of R, S, T that do not result in an equivalent continuous-time PID controller.

The overshoots that may appear in closed loop can be eliminated by replacing

$$T(q^{-1}) = R(q^{-1})\ \text{[PID 1]} \quad \text{with} \quad T(q^{-1}) = R(1)\ \text{[PID 2]}$$

Digital PID controller design can be carried on the basis of the discrete-time plant model and of the desired closed loop performances.

Every PID design should be concluded by an analysis of the robustness margins and of the frequency response of the input sensitivity function at high frequencies.

3.3 Pole Placement

The computation of the digital PID controller parameters is a special case of the "pole placement" strategy.

The *pole placement* strategy allows the design of a RST digital controller both for stable and unstable systems:

- Without restriction on the degrees of the polynomials $A(q^{-1})$ and $B(q^{-1})$ of the discrete-time plant model (provided that they do not have common factors)
- Without restriction on the time delay
- Without restriction on the plant zeros (stable or unstable)

This method does not simplify the system zeros (this is why they can be unstable). The only restriction concerns the possible common factors of $A(q^{-1})$ and $B(q^{-1})$, which must be simplified before the computations are carried out.

3.3.1 Structure

The structure of the closed loop system is given in Figure 3.11.

The plant to be controlled is characterized by the pulse transfer function (irreducible)

$$H(q^{-1}) = \frac{q^{-d}B(q^{-1})}{A(q^{-1})} \tag{3.3.1}$$

in which d is the integer number of sampling periods contained in the time delay and

$$A(q^{-1}) = 1 + a_1 q^{-1} + \ldots + a_{n_A} q^{-n_A} \qquad (3.3.2)$$

$$B(q^{-1}) = b_1 q^{-1} + b_2 q^{-2} + \ldots + b_{n_B} q^{-n_B} = q^{-1} B^*(q^{-1}) \qquad (3.3.3)$$

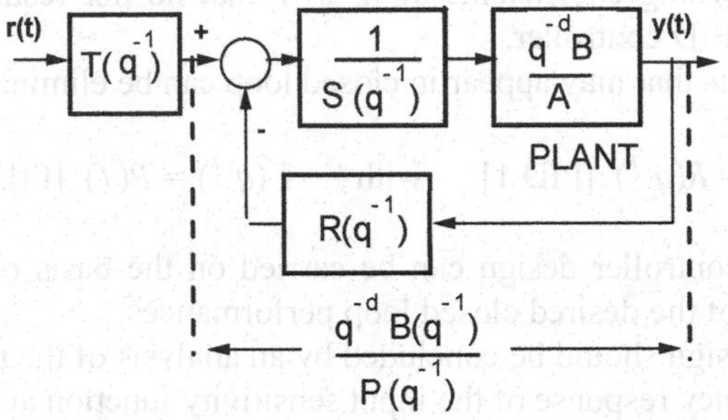

Figure 3.11. Pole placement with RST controller

The closed loop transfer function is given by

$$H_{CL}(q^{-1}) = \frac{q^{-d} T(q^{-1}) B(q^{-1})}{A(q^{-1}) S(q^{-1}) + q^{-d} B(q^{-1}) R(q^{-1})} = \frac{q^{-d} T(q^{-1}) B(q^{-1})}{P(q^{-1})} \qquad (3.3.4)$$

in which

$$P(q^{-1}) = A(q^{-1}) S(q^{-1}) + q^{-d} B(q^{-1}) R(q^{-1}) = 1 + p_1 q^{-1} + p_2 q^{-2} + \ldots \quad (3.3.5)$$

defines the closed loop poles that play an essential role for the regulation behavior.

The behavior with respect to a disturbance is given by the output sensitivity function

$$S_{yp}(q^{-1}) = \frac{A(q^{-1}) S(q^{-1})}{A(q^{-1}) S(q^{-1}) + q^{-d} B(q^{-1}) R(q^{-1})} = \frac{A(q^{-1}) S(q^{-1})}{P(q^{-1})} \qquad (3.3.6)$$

It can thus be seen that $P(q^{-1})$ corresponds to the denominator of the output sensitivity function and, thereby, it defines for a large extent the regulation behavior.

3.3.2 Choice of the Closed Loop Poles (P(q⁻¹))

We have seen for the case of a digital PID controller that one can specify a polynomial $P(q^{-1})$ defining the closed loop poles on the basis of a second-order continuous-time system with the desired natural frequency and damping (see Section 3.2.2). One can also directly specify the polynomial $P(q^{-1})$ from the desired performances. To illustrate the last statement, consider the following example.

Let

$$P(q^{-1}) = 1 + p_1 \, q^{-1} \quad \text{with} \quad p_1 = -0.5$$

When there is no reference, the free output response is defined by

$$y(t+1) = -p_1 \, y(t) = 0.5 \, y(t)$$

One thus obtains a relative decrease of 50% for the output amplitude at each sampling instant (see Figure 3.12).

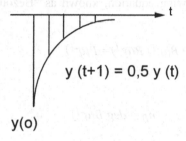

y(o)

Figure 3.12. Responses for $P(q^{-1}) = 1 - 0.5 \, q^{-1}$

Choosing p_1 between -0.2 and -0.8, it is clear that the disturbance rejection speed can be controlled.

Nevertheless, generally speaking, $P(q^{-1})$ is chosen in the form of a second-order polynomial by discretization of a second-order continuous time system, specifying ω_0, ζ and assuring that the condition

$$0.25 \leq \omega_0 \, T_s \leq 1.5 \quad ; \quad 0.7 \leq \zeta \leq 1$$

is satisfied (see Section 3.2.2)[5].

The polynomial chosen from the desired closed loop performances will define the dominant closed loop poles and it will be named $P_D(q^{-1})$.

---

[5] The functions *fd2pol.sci*, *omega_dmp.sci* (Scilab) and *fd2pol.m*, *omega_dmp.m* (MATLAB®) can be used.

If it is desired to introduce a filtering action in certain frequency regions (or to reduce the effect of the noise on the measure, or to smooth the variations of the control signal, or to improve the robustness), the poles of the corresponding filter, defined by a polynomial $P_F(q^{-1})$, should also be the poles of the closed loop. As a consequence, the polynomial $P(q^{-1})$ defining the desired closed loop poles will be the product of the polynomials $P_D(q^{-1})$ and $P_F(q^{-1})$ specifying dominant and auxiliary closed loop poles, respectively.

$$P(q^{-1}) = P_D(q^{-1}) \cdot P_F(q^{-1}) \qquad (3.3.7)$$

As a general rule, the poles named "auxiliary poles" are faster than the "dominant poles". That is expressed, for the case of discrete-time models, by the property that the roots of $P_F(q^{-1})$ should have a real part smaller than the real part of the roots of $P_D(q^{-1})$.

3.3.3 Regulation (Computation of R(q^{-1}) and S(q^{-1}))

Once $P(q^{-1})$ is specified, in order to compute $R(q^{-1})$ and $S(q^{-1})$ according to Equation 3.3.4 , the following equation, known as "Bezout identity" (equation), must be solved:

$$A(q^{-1}) S(q^{-1}) + q^{-d} B(q^{-1}) R(q^{-1}) = P(q^{-1}) \qquad (3.3.8)$$

Defining

$$n_A = deg\ A(q^{-1}) \quad ; \quad n_B = deg\ B(q^{-1}) \qquad (3.3.9)$$

this polynomial equation has a unique solution with minimal degree (when ($A(q^{-1})$ and $B(q^{-1})$ do not have common factors) for

$$n_P = deg\ P(q^{-1}) \leq n_A + n_B + d - 1\ ;$$
$$n_S = deg\ S(q^{-1}) = n_B + d - 1 \quad ; \quad n_R = deg\ R(q^{-1}) = n_A - 1 \qquad (3.3.10)$$

in which

$$S(q^{-1}) = 1 + s_1 q^{-1} + ... + s_{n_S} q^{-n_S} = 1 + q^{-1} S^*(q^{-1}) \qquad (3.3.11)$$
$$R(q^{-1}) = r_0 + r_1 q^{-1} + ... + r_{n_R} q^{-n_R} \qquad (3.3.12)$$

In order to solve effectively Equation 3.3.8 , this latter is often put in the matrix form

$$M x = p \qquad (3.3.13)$$

in which

$$x^T = [1, s_1, ..., s_{n_S}, r_0, ..., r_{n_R}]$$ (3.3.14)

$$p^T = [1, p_1, ..., p_i, ..., p_{n_p}, 0, ..., 0]$$ (3.3.15)

and the matrix M has the following form

$$
\underbrace{\quad\quad\quad\quad n_B + d \quad\quad\quad\quad}
\underbrace{\quad\quad\quad\quad n_A \quad\quad\quad\quad}
$$

$$
\left.
\begin{bmatrix}
1 & 0 & \cdots & 0 & 0 & \cdots & \cdots & 0 \\
a_1 & 1 & & . & b'_1 & & & \\
a_2 & & 0 & & b'_2 & & & b'_1 \\
& & 1 & & & & & b'_2 \\
& & a_1 & & . & & & \\
a_{n_A} & & a_2 & & b'_{n_B} & & & . \\
0 & & . & & 0 & . & . & . \\
0 & \cdots & 0 & a_{n_A} & 0 & 0 & 0 & b'_{n_B}
\end{bmatrix}
\right\} n_A + n_B + d
$$

$$n_A + n_B + d$$

where:

$$b'_i = 0 \quad for \quad i = 0, 1 ... d \quad ; \quad b'_i = b_{i-d} \quad for \quad i \geq d+1$$

The vector x, which contains the coefficients of the polynomials $R(q^{-1})$ and $S(q^{-1})$, is obtained, after the inversion of the matrix M, by the formula

$$x = M^{-1} p$$ (3.3.16)

where M^{-1} is the matrix inverse of M. This inverse exists if the determinant of the matrix M is different from zero. One can prove that this is verified if, and only if, $A(q^{-1})$ and $B(q^{-1})$ are coprime polynomials (no simplifications between zeros and poles).

Several methods are used to solve Equation 3.3.8. These methods give better numerical performances with respect to a simple matrix inversion.

For different reasons the polynomials $R(q^{-1})$ and $S(q^{-1})$ may contain, in general, fixed parts specified before the resolution of Equation 3.3.8 . For example, if zero steady state error is required for a step on the reference, or for a step disturbance, the presence of an integrator in the open loop transfer function is required. This corresponds to the introduction of a term $(1-q^{-1})$ in the polynomial $S(q^{-1})$ (see Section 2.4).

In order to take into account these pre-specified fixed parts, the polynomials $R(q^{-1})$ and $S(q^{-1})$ are factorized in the form

$$R(q^{-1}) = R'(q^{-1}) H_R (q^{-1})$$ (3.3.17)

$$S(q^{-1}) = S'(q^{-1}) H_S (q^{-1})$$ (3.3.18)

where $H_R (q^{-1})$ and $H_S (q^{-1})$ are pre-specified polynomials and

$$R'(q^{-1}) = r'_0 + r'_1 q^{-1} + ... + r'_{n_{R'}} q^{-n_{R'}}$$ (3.3.19)

$$S'(q^{-1}) = 1 + s'_1 q^{-1} + ... + s'_{n_{S'}} q^{-n_{S'}}$$ (3.3.20)

For this parameterization of polynomials $R(q^{-1})$ and $S(q^{-1})$, the closed loop transfer function will be

$$H_{CL}(q^{-1}) = \frac{q^{-d}T(q^{-1})B(q^{-1})}{A(q^{-1})S'(q^{-1})H_S(q^{-1}) + q^{-d}B(q^{-1})R'(q^{-1})H_R(q^{-1})}$$

$$= \frac{q^{-d}T(q^{-1})B(q^{-1})}{P(q^{-1})}$$ (3.3.21)

and, instead of Equation 3.3.8 , one needs to solve the equation

$$A(q^{-1}) H_S(q^{-1}) S'(q^{-1}) + q^{-d} B(q^{-1}) H_R(q^{-1}) R'(q^{-1}) = P(q^{-1})$$ (3.3.22)

In order to solve Equation 3.3.22, one needs to solve Equation 3.3.8 after replacing: $A(q^{-1})$ by $A'(q^{-1}) = A(q^{-1})H_S(q^{-1})$ and $B(q^{-1})$ and $B'(q^{-1}) = q^{-d}B(q^{-1})$ $H_R(q^{-1})$ with the restriction that polynomials $[A(q^{-1}) H_S(q^{-1})]$ and $[B(q^{-1}) H_R(q^{-1})]$ are coprime.

The condition of Equation 3.3.10 on the orders of the polynomials that allow one to get a unique solution of minimal order, become in this case

$$n_P = deg\ P(q^{-1}) \le n_A + n_{H_S} + n_B + n_{H_R} + d - 1\ ;$$

$$n_{S'} = deg\ S'(q^{-1}) = n_B + n_{H_R} + d - 1\ ;\ n_{R'} = deg\ R'(q^{-1}) = n_A + n_{H_S} - 1$$ (3.3.23)

For the controller implementation, $S(q^{-1})$ will be given by $S'(q^{-1})H_S(q^{-1})$ and $R(q^{-1})$ by $R'(q^{-1}) H_R(q^{-1})$.

Equation 3.3.22 or 3.3.8 can be solved by means of the functions *bezoutd.sci* (Scilab) and *bezoutd.m* (MATLAB®) available on the book website.

Use of Fixed Parts of the Controller (H_R and H_S): Examples

Steady-State Error

As illustrated in the previous sections, $S(q^{-1})$ must contain a term $(1-q^{-1})$ in order to have a zero steady state error for a step input or disturbance (S_{yp} (q^{-1}) must be zero in steady state, *i.e.* for $q = 1$). Thus

$$S_{yp}(q^{-1}) = \frac{A(q^{-1})H_S(q^{-1})S'(q^{-1})}{P(q^{-1})} \qquad (3.3.24)$$

and then one needs to choose:

$$H_S(q^{-1}) = 1 - q^{-1} \qquad (3.3.25)$$

Rejection of a Sinusoidal Disturbance

If a perfect rejection of a sinusoidal disturbance is required at a specified frequency, it isessential that S_{yp} (q^{-1}) is zero at this frequency, which is equivalent to the requirement that $H_S(q^{-1})$ has a pair of *undamped* complex zeros at this frequency.

In this case:

$$H_S(q^{-1}) = 1 + \alpha\, q^{-1} + q^{-2} \qquad (3.3.26)$$

with $\alpha = - 2\ cos\ \omega\ T_s = - 2\ cos\ (2\ \pi\ f/f_s\ )$.

If one requires only a given attenuation, $H_S(q^{-1})$ must introduce a pair of *damped* complex zeros with a damping factor depending on the desired attenuation.

Signal Blocking

In some applications the measured signal contains signal components at particular frequencies for which the controller should not react (these signals may serve for process technological operation). This implies that at these frequencies one should have $S_{yp}(q^{-1})=1$. It results from Equation 3.3.24   and 3.3.6   that at these frequencies it is essential that $H_R(q^{-1})$ be null. As a consequence, the input sensitivity function $S_{up}(q^{-1})$ must be null (no effect of the disturbance upon the control signal).

The expression of $S_{up}(q^{-1})$ is given by

$$S_{up}(q^{-1}) = -\frac{A(q^{-1})H_R(q^{-1})R'(q^{-1})}{P(q^{-1})} \tag{3.3.27}$$

and one must choose the fixed part $R(q^{-1})$ named $H_R(q^{-1})$ such that it has a null gain at this frequency.

The fixed part of $R(q^{-1})$ will be of the form

$$H_R(q^{-1}) = 1 + \beta q^{-1} + q^{-2} \tag{3.3.28}$$

where $\beta = -2 \cos \omega_0 T_s = -2 \cos (2 \pi f / f_s)$.

This introduces a pair of undamped complex zeros at the frequency f or, more generally, in the form of a second order polynomial corresponding to a damped complex zeros pair, if a desired attenuation is accepted.

In many applications it is required that the controller does not react to signals close to $0.5 f_s$ (where the gain of the cascade actuator-plant is generally low). In this case one imposes:

$$H_R(q^{-1}) = (1 + \beta q^{-1})^n \qquad n=1,2 \tag{3.3.29}$$

with $0 < \beta \le 1$.

Note that $(1 + \beta q^{-1})^2$ corresponds to a damped second-order system with a resonant frequency equal to πf_s (see Section 2.3.2):

$$\omega_0 \sqrt{1 - \zeta^2} = \pi f_s$$

and the corresponding damping factor depends upon β through the relation

$$\beta = e^{-\frac{\zeta}{\sqrt{1-\zeta^2}}\pi}$$

For $\beta = 1$ it follows that $\zeta = 0$ and the closed loop system works in open loop at $0.5 f_s$ (even for $n=1$).

Robustness
In order to guarantee both robustness margins and specified closed loop performances, fixed parts $H_R(q^{-1})$ and $H_S(q^{-1})$ (as well as auxiliary poles) should be introduced for shaping the frequency characteristics of the sensitivity functions $S_{yp}(q^{-1})$ and $S_{up}(q^{-1})$ in specified frequency regions. This will be discussed in Section 3.6.

3.3.4 Tracking (Computation of $T(q^{-1})$)

Ideally, when the reference changes, it is desired that the system output $y(t)$ follows a desired trajectory $y^*(t)$. This trajectory may be stored or generated each time the reference is changed by using a reference model (as indicated in Figure 3.13).

The transfer function of the reference model is

$$H_m(q^{-1}) = \frac{q^{-1}B_m(q^{-1})}{A_m(q^{-1})} \qquad (3.3.30)$$

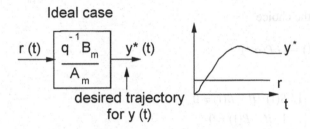

Figure 3.13. Desired trajectory y*(t) generation

It is generally determined from desired performances (rising time, overshoot, settling time). For example, a second-order normalized continuous-time model (parameters ω_0, ζ) can be defined by means of the curves given in Figure 1.10 starting from the desired performances. Once the continuous-time transfer function and the sampling period T_s, are known, the pulse transfer function of the reference model is obtained by discretization.

The pulse transfer function of the reference model will be of the form

$$H_m(q^{-1}) = \frac{q^{-1}B_m(q^{-1})}{A_m(q^{-1})} = \frac{q^{-1}(b_{m0} + b_{m1}q^{-1})}{1 + a_{m1} + a_{m2}q^{-2}} \qquad (3.3.31)$$

This is the transfer function that the controller must achieve between the reference r and the output y, eventually multiplied by q^{-d} in the case of a time delay d in the plant model (the delay cannot be compensated). In the case of the *pole placement* this cannot be obtained because the plant zeros are maintained (polynomial $B(q^{-1})$).

The objective is then to approach the delayed model reference trajectory

$$y^*(t) = \frac{q^{-(d+1)}B_m(q^{-1})}{A_m(q^{-1})}r(t) \qquad (3.3.32)$$

For this, firstly $y^*(t+d+1)$ is generated from $r(t)$:

$$y^*(t+d+1) = \frac{B_m(q^{-1})}{A_m(q^{-1})} r(t) \qquad (3.3.33)$$

and one chooses $T(q^{-1})$ to impose:

- Unit static gain between y^* and y
- Compensation of the regulation dynamics defined by $P(q^{-1})$ (because the regulation dynamics is in general different from the tracking dynamics $A_m(q^{-1})$)

This leads to the choice

$$T(q^{-1}) = G\,P(q^{-1}) \qquad (3.3.34)$$

where

$$G = \begin{cases} 1/\,B(1) & \text{if } B(1) \neq 0 \\ 1 & \text{if } B(1) = 0 \end{cases} \qquad (3.3.35)$$

The control law equation becomes

$$S(q^{-1})\,u(t) + R(q^{-1})\,y(t) = T(q^{-1})\,y^*(t+d+1) \qquad (3.3.36)$$

The full diagram for the pole placement is given in Figure 3.14.

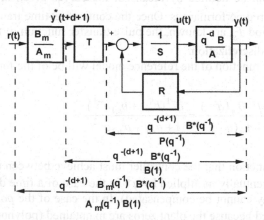

Figure 3.14. Pole placement - tracking and regulation

The transfer function (operator) between the reference and the output will be

$$H_{CL}(q^{-1}) = \frac{q^{-d+1} B_m(q^{-1})}{A_m(q^{-1})} \cdot \frac{B^*(q^{-1})}{B(1)} \qquad (3.3.37)$$

In some cases one can consider a simplification of the polynomial T by taking into account just the dominant poles (as auxiliary poles are often at high frequencies with a small influence on the time response). In this case

$$H_{CL}(q^{-1}) = \frac{q^{-d+1}B_m(q^{-1})}{A_m(q^{-1})} \cdot \frac{B^*(q^{-1})}{B(1)} \cdot \frac{P_F(1)}{P_F(q^{-1})} \qquad (3.3.38)$$

and

$$T(q^{-1}) = GP_D(q^{-1}) \qquad (3.3.39)$$

where

$$G = \begin{cases} \dfrac{P_F(1)}{B(1)} & \text{if} \quad B(1) \neq 0 \\ 1 & \text{if} \quad B(1) = 0 \end{cases}$$

If the regulation dynamics is the same as the tracking dynamics, there is no need for a reference model of the form of Equation 3.3.33 and the polynomial $T(q^{-1})$ is replaced by a gain[6]:

$$T(q^{-1}) = G = \begin{cases} \dfrac{P(1)}{B(1)} & \text{if} \quad B(1) \neq 0 \\ 1 & \text{if} \quad B(1) = 0 \end{cases}$$

which guarantees a unit static gain between the reference trajectory and the output (if $B(1)$ is not null).

The controller equations for the pole placement under different forms are summed up in Table 3.5 (the recursive equations needed for the implementation are boxed).

---

[6] If $S(q^{-1})$ contains an integrator ($S(1)=0$) it follows from Equation 3.3.21 that $B(1)R(1) = P(1)$ and $P(1)/B(1) = R(1)$, respectively.

3.3.5 Pole Placement: Examples

Table 3.6 gives the results for a pole placement design. The considered example is the control of a plant model characterized by a second-order discrete-time model with two real poles (at 0.6 and 0.7) and an unstable zero.

The tracking dynamics (polynomials $A_m(q^{-1})$ and $B_m(q^{-1})$) has been obtained by discretization of a second-order normalized continuous-time model, with $\omega_0 = 0.5$ rad/s and $\zeta = 0.9$, ($T_s = 1s$). The regulation dynamics (polynomials $P(q^{-1})$) has been obtained by discretization of a second-order normalized continuous-time model with $\omega_0 = 0.4$ rad/s and $\zeta = 0.9$. The controller includes an integrator. Satisfactory robustness margins are obtained. The performances are shown in Figure 3.15.

Table 3.5. Pole placement - control law equations

$$u(t) = \frac{T(q^{-1})y^*(t+d+1) - R(q^{-1})y(t)}{S(q^{-1})}$$

$$S(q^{-1})u(t) + R(q^{-1})y(t) = GP(q^{-1})y^*(t+d+1) = T(q^{-1})y^*(t+d+1)$$

$$S(q^{-1}) = 1 + q^{-1}S^*(q^{-1})$$

$$\boxed{u(t) = P(q^{-1})Gy^*(t+d+1) - S^*(q^{-1})u(t-1) - R(q^{-1})y(t)}$$

$$y^*(t+d+1) = \frac{B_m(q^{-1})}{A_m(q^{-1})}r(t)$$

$$A_m(q^{-1}) = 1 + q^{-1}A_m^*(q^{-1})$$

$$\boxed{y^*(t+d+1) = -A_m^*(q^{-1})y(t+d) + B_m(q^{-1})r(t)}$$

$$B_m(q^{-1}) = b_{m0} + b_{m1}q^{-1} + \ldots$$

$$A_m(q^{-1}) = 1 + a_{m1}q^{-1} + a_{m2}q^{-2} + \ldots$$

Exercise:
What are the specifications for the "pole placement" technique that will lead to a PID 2 controller ?

Table 3.6. Pole placement results

Plant:

- $d=0$
- $B(q^{-1}) = 0.1\,q^{-1} + 0.2\,q^{-2}$
- $A(q^{-1}) = 1 - 1.3\,q^{-1} + 0.42\,q^{-2}$

Tracking dynamics:

- $B_m(q^{-1}) = 0.0927 + q^{-1} + 0.0687\,q^{-2}$
- $A_m(q^{-1}) = 1 - 1.2451q^{-1} + 0.4066\,q^{-2}$
- $T_S = 1s$, $\omega_0 = 0.5\,rad/s$, $\zeta = 0.9$

Regulation dynamics → $P(q^{-1}) = 1 - 1.3741\,q^{-1} + 0.4867\,q^{-2}$
$T_S = 1s$, $\omega_0 = 0.4\,rad/s$, $\zeta = 0.9$

Pre-specifications: Integrator

*** CONTROL LAW ***

$$S(q^{-1})\,u(t) + R(q^{-1})\,y(t) = T(q^{-1})\,y^*(t+d+1)$$

$$y^*(t+d+1) = [B_m(q^{-1})/A_m(q^{-1})]\,r(t)$$

Controller:

- $R(q^{-1}) = 3 - 3.94\,q^{-1} + 1.3141\,q^{-2}$
- $S(q^{-1}) = 1 - 0.3742\,q^{-1} - 0.6258\,q^{-2}$
- $T(q^{-1}) = 3.333 - 4.5806\,q^{-1} + 1.6225\,q^{-2}$

Gain margin: 2.703	Phase margin: 65.4 deg
Modulus margin: 0.618 (- 4.19 dB)	Delay margin: 2.1 s

3.4 Tracking and Regulation with Independent Objectives

This controller design method makes it possible to obtain the desired tracking behavior (changing of reference) independently of the desired regulation behavior (rejection of a disturbance). For example, the performance specifications illustrated in Figure 3.16 correspond to a situation for which the desired regulation response time (for a step disturbance) is significantly smaller than the desired tracking response time (for a step reference). Note that the reverse situation may be encountered. This method is a generalization of the so-called "model reference control".

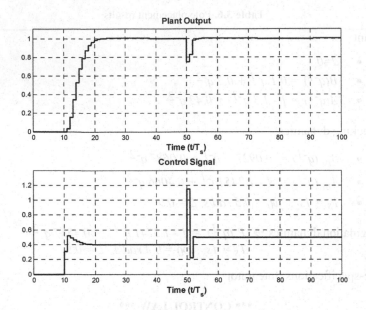

Figure 3.15. Pole placement performances

Unlike the "pole placement" method (Section 3.3), *this method leads to the simplification of the zeros of the discrete-time plant model.* This enables the tracking and regulation performances to be achieved without approximation.

This control strategy permits a RST digital controller to be designed for both stable and unstable systems:

- Without restriction on the degrees of the polynomials $A(q^{-1})$ and $B(q^{-1})$ characterizing the pulse transfer function of the plant
- Without restriction on the time delay d of the system

As a result of the simplification of the zeros, however, this strategy can only be applied to discrete time models with stable zeros.

This method cannot be applied to systems with fractional delay greater than $0.5T_s$. In the case of a fractional delay greater than $0.5T_s$, however, one can consider approximating (in the identification phase) the model with a fractional delay by a model with an augmented integer delay.

Note that unstable zeros can be the consequence of a too fast sampling of continuous time systems characterized by a difference greater than 2 between the numerator and denominator degree of the continuous time transfer function (even if the continuous time zeros are stable) (Åström and Wittenmark 1997).

This method can be considered as a particular case of the "pole placement" method. This equivalence can be obtained by imposing that the closed loop poles contain the zeros of the discrete time plant model (defined by the polynomial

$B^*(q^{-1})$). This is the reason for which the zeros of the plant model must be stable (see Section 3.4.2)[7].

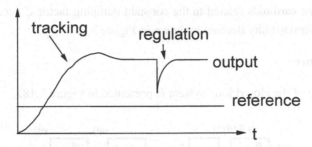

Figure 3.16. Tracking and regulation performances

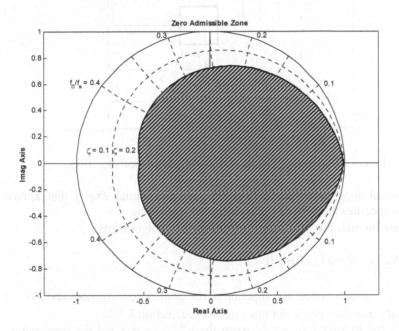

Figure 3.17. Admissible domain (hatching zone) for the zeros of the discrete-time plant model (tracking and regulation with independent objectives)

---

[7] An approximation of the tracking and regulation with independent objectives, for the case of unstable zeros, can be obtained by pole placement, where the unstable zeros, which should be specified as closed loop poles, are approximated by stable zeros. This technique is presented in the stochastic context in Section 4.3.

It is convenient then to verify that, before the application of this method, the zeros of $B^*(q^{-1})$ are stable, and, moreover, that complex zeros have a sufficiently high damping factor ($\zeta > 0.2$). In other words, the zeros should lie inside a region defined by the cardioids related to the constant damping factor $\zeta = 0.2$ (see Figure 2.21). This admissibility domain is shown in Figure 3.17.

3.4.1 Structure

The structure of the closed loop system is presented in Figure 3.18.

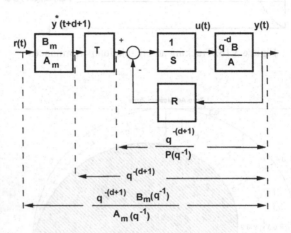

Figure 3.18. Tracking and regulation with independent objectives

The closed loop poles are defined by the polynomial $P(q^{-1})$ that almost completely specifies the desired behavior for the regulation.

As a general rule, $P(q^{-1})$ is the product of the two polynomials

$$P(q^{-1}) = P_D(q^{-1}) \cdot P_F(q^{-1})$$

where $P_D(q^{-1})$ is determined as a function of the desired performances and $P_F(q^{-1})$ represents the auxiliary poles (for more details see Section 3.3.2).

The desired transfer function between the reference $r(t)$ and the plant output $y(t)$, which defines the tracking dynamics, is

$$q^{-(d+1)} B_m(q^{-1}) / A_m(q^{-1})$$

The output of the tracking model $B_m(q^{-1}) / A_m(q^{-1})$ specifies the desired trajectory y^* with $d+1$ steps ahead.

The plant to be controlled is characterized by the pulse transfer function given in Equation 3.3.1, where the polynomials $A(q^{-1})$ and $B(q^{-1})$ are specified by

Equations 3.3.2 and 3.3.3. Note that in this case polynomials $A(q^{-1})$ and $B(q^{-1})$ can have common factors.

The computation of $R(q^{-1})$, $S(q^{-1})$, $T(q^{-1})$ is done in two stages. First, by means of polynomials $R(q^{-1})$ and $S(q^{-1})$, the closed loop poles will be placed at the desired values specified by the polynomial $P(q^{-1})$, and the zeros of the discrete-time plant model will be simplified. Second, the pre-filter $T(q^{-1})$ is computed in order to obtain the tracking of the reference trajectory y^*, delayed by d steps. Note that for $P(q^{-1}) = 1$ this method corresponds to the "model reference control".

3.4.2 Regulation (Computation of R(q⁻¹) and S(q⁻¹))

Without the pre-filter $T(q^{-1})$, the closed loop transfer function is:

$$H_{CL}(q^{-1}) = \frac{q^{-d+1}B^*(q^{-1})}{A(q^{-1})S(q^{-1}) + q^{-d+1}B^*(q^{-1})R(q^{-1})} = \frac{q^{-d+1}}{P(q^{-1})}$$

$$= \frac{q^{-d+1}B^*(q^{-1})}{B^*(q^{-1})P(q^{-1})}$$

$$(3.4.1)$$

The closed loop poles should be those defined by $P(q^{-1})$ and the system zeros should be cancelled (in order to obtain a perfect tracking at a later stage).

From Equation 3.4.1, it results that the closed loop poles must contain in addition the zeros of the plant model.

Once $P(q^{-1})$ is specified, it results from Equation 3.4.1 that, in order to compute $R(q^{-1})$ and $S(q^{-1})$, the following equation must be solved:

$$A(q^{-1}) S(q^{-1}) + q^{-(d+1)} B^*(q^{-1}) R(q^{-1}) = B^*(q^{-1}) P(q^{-1}) \qquad (3.4.2)$$

Equation 3.4.2 corresponds to the pole placement with a particular choice of the desired closed loop poles (that contain $n_B - 1$ additional poles corresponding to the plant zeros).

However, in order to solve Equation 3.4.2, one observes that $S(q^{-1})$ should have $B^*(q^{-1})$ as a common factor:

$$S(q^{-1}) = s_0 + s_1 q^{-1} + ... + s_{n_S} q^{-n_S} = B^*(q^{-1})S'(q^{-1}) \qquad (3.4.3)$$

Introducing the expression of $S(q^{-1})$ given by Equation 3.4.3 in Equation 3.4.2, and after simplification by $B^*(q^{-1})$, one obtains

$$A(q^{-1}) S'(q^{-1}) + q^{-(d+1)} R(q^{-1}) = P(q^{-1}) \qquad (3.4.4)$$

This polynomial equation has a unique solution for

$$n_P = deg\ P(q^{-1}) = n_A + d\ ;\quad deg\ S'(q^{-1}) = d\ ;\quad deg\ R(q^{-1}) = n_A - 1 \quad (3.4.5)$$

in which

$$R(q^{-1}) = r_0 + r_1 q^{-1} + \dots + r_{n_A-1} q^{-n_A+1} \quad (3.4.6)$$

and

$$S'(q^{-1}) = 1 + s'_1 q^{-1} + \dots + s'_d q^{-d} \quad (3.4.7)$$

Equation 3.4.4 corresponds, under the conditions of Equation 3.4.5, to the matrix equation

$$Mx = p \quad (3.4.8)$$

where M is a lower triangular matrix of dimension $(n_A+d+1) \times (n_A+d+1)$

$$n_A + d + 1$$

$$
\begin{bmatrix}
1 & 0 & & & & & & & & 0 \\
a_1 & 1 & & & & & & & & \\
a_2 & a_1 & & 0 & & & & & & \vdots \\
\vdots & \vdots & & 1 & & & & & & \\
a_d & a_{d-1} & \dots & a_1 & 1 & & & & & \vdots \\
a_{d+1} & a_d & & & a_1 & 1 & & & & \\
a_{d+2} & a_{d+1} & & & a_2 & & 0 & & & \vdots \\
& & & & & & & & & \\
& & & & & & & & 0 & \\
0 & 0 & \dots & 0 & a_{n_A} & & 0 & 0 & 1 &
\end{bmatrix}
\left. \phantom{\begin{matrix}1\\1\\1\\1\\1\\1\\1\\1\\1\\1\end{matrix}}\right\} n_A + d + 1
$$

$$d + 1 \qquad\qquad\qquad n_A$$

$$x^T = [1, s'_1, \dots, s'_d, r_0, r_1, \dots, r_{n-1}] \quad (3.4.9)$$

$$p^T = [1, p_1, p_2, \dots, p_{n_A}, p_{n_A+1}, \dots, p_{n_A+d}] \quad (3.4.10)$$

Some coefficients p_i can be zero. Since M is a lower triangular matrix, Equation 3.4.8 (and respectively Equation 3.4.4) always has a solution.

In order to solve Equation 3.4.4 one can use *predisol.sci* (Scilab) and *predisol.m* (MATLAB®) functions available on the book website.

One observes that, because of the nature of the design methodology, $S(q^{-1})$ already contains a fixed part (Equation 3.4.3) specified before solving Equation 3.4.4.

Thus, it is convenient to consider, as in the pole placement case, a parameterization of $S(q^{-1})$ and $R(q^{-1})$ in the form

$$R(q^{-1}) = H_R (q^{-1}) R'(q^{-1}) \tag{3.4.11}$$
$$S(q^{-1}) = H_S (q^{-1}) S'(q^{-1}) \tag{3.4.12}$$

In which $H_R(q^{-1})$ and $H_S(q^{-1})$ represent the pre-specified parts of $R(q^{-1})$ and $S(q^{-1})$.

In the case of tracking and regulation with independent objectives, $H_S(q^{-1})$ will have the form

$$H_S(q^{-1}) = B^*(q^{-1}) . H'_S(q^{-1}) \tag{3.4.13}$$

and Equation 3.4.4, in the general case, becomes

$$A(q^{-1}) H'_S(q^{-1}) S'(q^{-1}) + q^{-(d+1)} H_R(q^{-1}) R'(q^{-1}) = P(q^{-1}) \tag{3.4.14}$$

For solving 3.4.14 one can also use *predisol.sci* (Scilab) and *predisol.m* (MATLAB®) functions. Note also that pole placement equations can be used for computing $S'(q^{-1})$ and $R'(q^{-1})$ by introducing $B^*(q^{-1})$ in the expression of $P(q^{-1})$.

Steady-State Error
In order to have a zero steady state error for a step input or a step disturbance, the open loop cascade must contain a digital integrator, *i.e.* the polynomial $S(q^{-1})$ must contain a term $(1-q^{-1})$:

$$S(q^{-1}) = B^*(q^{-1}) (1 - q^{-1}) S'(q^{-1}) = B^*(q^{-1}) H'_S(q^{-1}) S'(q^{-1}) \tag{3.4.15}$$

Introducing this expression in Equation 3.4.2 and after simplification by $B^*(q^{-1})$ one gets

$$A(q^{-1}) (1-q^{-1}) S'(q^{-1}) + q^{-(d+1)} R(q^{-1}) = P(q^{-1}) \tag{3.4.16}$$

which must be solved in order to obtain the corresponding coefficients of $S'(q^{-1})$ and $R(q^{-1})$ when an integrator is used.

3.4.3 Tracking (Computation of T(q^{-1}))

The pre-filter $T(q^{-1})$ is computed in order to achieve (in accordance with Figure 3.17), between reference $r(t)$ and $y(t)$, a transfer function:

$$H_{CL}(q^{-1}) = \frac{q^{-(d+1)}B_m(q^{-1})}{A_m(q^{-1})} = \frac{B_m(q^{-1})T(q^{-1})q^{-(d+1)}}{A_m(q^{-1})P(q^{-1})} \qquad (3.4.17)$$

From Equation 3.4.17

$$T(q^{-1}) = P(q^{-1}) \qquad (3.4.18)$$

The input of $T(q^{-1})$ is the $(d+1)$ steps ahead prediction of the desired trajectory $y^*(t+d+1)$, obtained by filtering $r(t)$ with the tracking model $B_m(q^{-1})/A_m(q^{-1})$.

$$y^*(t+d+1) = \frac{B_m(q^{-1})}{A_m(q^{-1})}r(t) \qquad (3.4.19)$$

and the controller equation will be given by

$$S(q^{-1})u(t) + R(q^{-1})y(t) = P(q^{-1})y^*(t+d+1) \qquad (3.4.20)$$

Equation 3.4.20 may also take the form

$$u(t) = \frac{P(q^{-1})y^*(t+d+1) - R(q^{-1})y(t)}{S(q^{-1})} \qquad (3.4.21)$$

Since $S(q^{-1})$ is of the form

$$S(q^{-1}) = s_0 + s_1 q^{-1} + ... + s_{n_S} q^{-n_S} = s_0 + q^{-1}S^*(q^{-1})$$
$$= B^*(q^{-1})S'(q^{-1}) \qquad (3.4.22)$$

and considering the expressions of $B^*(q^{-1})$ and $S'(q^{-1})$ (given by Equation 3.4.7):

$$s_0 = b_1 \qquad (3.4.23)$$

Therefore Equation 3.4.20 may also take the form[8]

---

[8] One can also normalize the parameters of the controller by dividing all parameters by $s_0 = b_1$. This avoids a multiplication operation in real time implementation.

$$u(t) = \frac{1}{b_1}\left[P(q^{-1})y^*(t+d+1) - S^*(q^{-1})u(t-1) - R(q^{-1})y(t)\right] \qquad (3.4.24)$$

As for the pole placement case, if the desired dynamics is the same both for tracking and regulation, the reference model is no longer necessary and the polynomial $T(q^{-1})$ is replaced by a simple gain to guarantee unit static gain between the reference trajectory and the output:

$$T(q^{-1}) = G = P(1)$$

If $S(q^{-1})$ contains an integrator, then the polynomial $T(q^{-1})$ becomes

$$T(q^{-1}) = G = R(1)$$

3.4.4 Tracking and Regulation with Independent Objectives: Examples

Table 3.7 gives the results of the design for tracking and regulation with independent objectives.

The example considered here concerns the control of a plant characterized by a second-order discrete-time model with two poles at $z = 0.6$ and at $z = 0.7$, and with a stable zero. The desired tracking dynamics (polynomials A_m (q^{-1}) and B_m (q^{-1})) has been obtained by discretization of a second-order normalized continuous-time model with $\omega_0 = 0.5$ rad/s and $\zeta = 0.9$ $(T_s = 1s)$. The desired regulation dynamics (polynomial $P(q^{-1})$) has been obtained by discretization of a second-order normalized continuous-time model with $\omega_0 = 0.4$ rad/s and $\zeta = 0.9$. The controller contains an integrator. The computed values of $R(q^{-1})$ and $S(q^{-1})$ are given in the lower part of Table 3.3, and the simulation results, for a reference tracking and a load disturbance (step) on the output, are presented in Figure 3.19. Note that the position of the zero is close to the boundaries of the admissibility domain ($z_0 = -0.5$ corresponds to $\zeta = 0.2$), and this explains the damped oscillations observed on the control signal when a step disturbance occurs.

Table 3.8 gives the computed values of the controller parameters for a plant having the same dynamics of the previous one but with a time delay $d = 3$ (instead of $d = 0$). All the desired performances have been maintained. Polynomial $S(q^{-1})$ has a higher order. Closed loop system responses are presented in Figure 3.20. One observes that the output response is the same as in the case of $d = 0$ (Figure 3.20) except that they are shifted by d steps. Moreover, by comparing Figures 3.19 and 3.20, one observes that the control signal is the same. In other words, even if the same output value is recovered with a delay of three sampling periods, the controller computes a signal that is the same as in the case of $d = 0$. In fact, *for the case with time delay, the controller contains a three step ahead predictor* (for details see Appendix B).

Table 3.7. Tracking and regulation with independent objectives (d=0)

Plant:

- $d = 0$
- $B(q^{-1}) = 0.2\ q^{-1} + 0.1\ q^{-2}$
- $A(q^{-1}) = 1 - 1.3\ q^{-1} + 0.42\ q^{-2}$

Tracking dynamics:

- $B_m\ (q^{-1}) = 0.0927q^{-1} + 0.0687\ q^{-2}$
- $A_m\ (q^{-1}) = 1 - 1.2451q^{-1} + 0.4066\ q^{-2}$
- $T_s = 1s$, $\omega_0 = 0.5\ rad/s$, $\zeta = 0.9$

Regulation dynamics → $P\ (q^{-1}) = 1 - 1.3741\ q^{-1} + 0.4867\ q^{-2}$
$T_s = 1s$, $\omega_0 = 0.4\ rad/s$, $\zeta = 0.9$

Pre-specifications: Integrator

*** CONTROL LAW ***

$$S\ (q^{-1})\ u(t) + R\ (q^{-1})\ y(t) = T\ (q^{-1})\ y^*(t+d+1)$$

$$y^*(t+d+1) = [B_m\ (q^{-1})/A_m\ (q^{-1})]\ .\ r(t)$$

Controller:

- $R(q^{-1}) = 0.9258 - 1.2332\ q^{-1} + 0.42\ q^{-2}$
- $S(q^{-1}) = 0.2 - 0.1\ q^{-1} - 0.1\ q^{-2}$
- $T(q^{-1}) = P(q^{-1})$

Gain margin: 2.109   Phase margin: 65.3 deg
Modulus margin: 0.526 (- 5.58 dB)   Delay margin: 1.2 s

One observes that the delay margin is smaller than one sampling period. In this case, one should specify high frequency auxiliary poles in $P(q^{-1})$ that lightly affect the low frequency behavior but improve the robustness at high frequencies (in particular the delay margin).

The maximum number of closed loop poles (the maximum degree of polynomial $P(q^{-1})$) is 5. Having already specified a pair of dominant poles, one can impose, for example, $P_F(q^{-1}) = (1 - 0.1\ q^{-1})^3$. In this case, one obtains a gain margin of 2.157, a modulus margin of 0.534 (-5.45dB), a phase margin of 58.5 degrees and a delay margin of 1.19s (greater than one sampling period).

The closed loop system responses are presented in Figure 3.21. Tracking performances are the same as for the case without auxiliary pole, and regulation performances are almost unchanged. One also remarks that the introduction of these poles reduces the stress on the actuator in the transient for the disturbance rejection.

Table 3.8. Tracking and regulation with independent objectives (d=3)

Plant:

- $d = 3$
- $B(q^{-1}) = 0.2\,q^{-1} + 0.1\,q^{-2}$
- $A(q^{-1}) = 1 - 1.3\,q^{-1} + 0.42\,q^{-2}$

Tracking dynamics:

- $B_m\,(q^{-1}) = 0.0927 + 0.0687\,q^{-1}$
- $A_m\,(q^{-1}) = 1 - 1.2451q^{-1} + 0.4066\,q^{-2}$
- $T_s = 1s$, $\omega_0 = 0.5\,rad/s$, $\zeta = 0.9$

Regulation dynamics $\rightarrow$ $P\,(q^{-1}) = 1 - 1.3741\,q^{-1} + 0.4867\,q^{-2}$
$$T_s = 1s\ ,\ \omega_0 = 0.4\,rad/s,\ \zeta = 0.9$$

Pre-specifications: Integrator

*** CONTROL LAW ***

$$S\,(q^{-1})\,u(t) + R\,(q^{-1})\,y(t) = T\,(q^{-1})\,y^*(t+d+1)$$

$$y^*(t+d+1) = [B_m\,(q^{-1})/A_m\,(q^{-1})]\,.ref(t)$$

Controller:

- $R(q^{-1}) = 0.8914 - 1.1521\,q^{-1} + 0.3732\,q^{-2}$
- $S(q^{-1}) = 0.2 + 0.0852\,q^{-1} - 0.0134\,q^{-2} - 0.0045\,q^{-3} - 0.1785\,q^{-4}$
  $- 0.0888\,q^{-5}$
- $T(q^{-1}) = P(q^{-1})$

Gain margin : 2.078	Phase margin: 58 deg
Modulus margin: 0.518 (- 5.71 dB)	Delay margin: 0.7 s

As a general rule, when the plant has a time delay, one needs to specify not only the dominant poles of $P(q^{-1})$, but also auxiliary poles at values other than 0 in order to improve the robustness with respect to the possible variation of the delay (this remark is also valid for the other methods presented in this chapter).

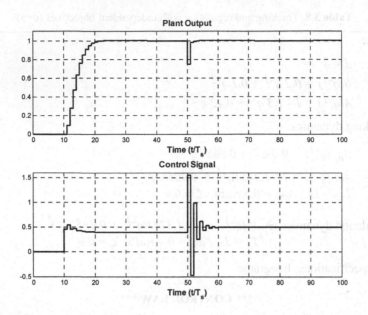

Figure 3.19. Performances of tracking and regulation with independent objectives (d = 0)

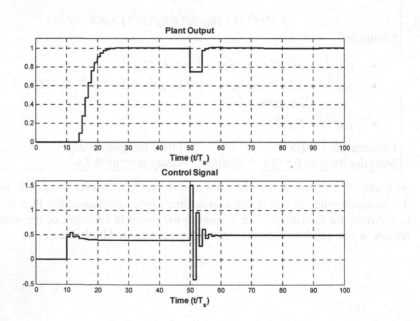

Figure 3.20. Performances of tracking and regulation with independent objectives in presence of a time delay (d = 3)

3.5 Internal Model Control (Tracking and Regulation)

The internal model control (IMC) strategy (not to be confused with the *internal model principle*) is a particular case of the pole placement technique. In the internal mode control the poles of the plant model are chosen as the desired dominant poles of the closed loop. As a consequence, the closed loop system will not be faster than the open loop system. Since there will not be an acceleration of the closed loop time response with respect to the open loop time response, one can expect a better closed loop robustness with respect to plant model uncertainties. As the closed loop poles include the poles of the plant model, *this technique can be applied only to stable and well damped discrete time plant models.* This technique is often used for the control of stable plants with a large time delay with respect to the open loop rise time (the rise time does not include the time delay!).

The control structure is the same as for the pole placement (Figure 3.14).

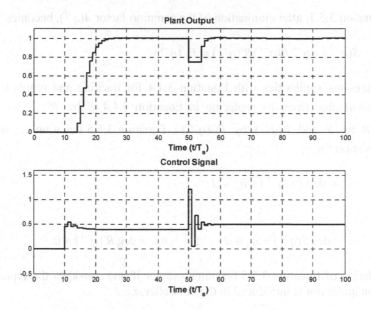

Figure 3.21. Performances of tracking and regulation with independent objectives with auxiliary poles in presence of a pure time delay (d = 3)

3.5.1 Regulation

In the case of the internal model control Equation 3.3.22 of the pole placement becomes

$$A(q^{-1})S(q^{-1}) + q^{-d}B(q^{-1})R(q^{-1}) = A(q^{-1})P_F(q^{-1}) = P(q^{-1}) \qquad (3.5.1)$$

Often in practice one selects

$$P_F(q^{-1}) = \left(1 + \alpha q^{-1}\right)^{n_{P_F}} \qquad (3.5.2)$$

or

$$P_F(q^{-1}) = \left(1 + \alpha_1 q^{-1}\right)\left(1 + \alpha q^{-1}\right)^{n_{P_F} - 1} \qquad (3.5.3)$$

By examining the structure of Equation 3.5.1 one observes that $R(q^{-1})$ should have as a factor $A(q^{-1})$:

$$R(q^{-1}) = A(q^{-1})R'(q^{-1}) \qquad (3.5.4)$$

and Equation 3.5.1, after elimination of the common factor $A(q^{-1})$, becomes

$$S(q^{-1}) + q^{-d}B(q^{-1})R'(q^{-1}) = P_F(q^{-1}) \qquad (3.5.5)$$

which presents similarities with Equation 3.4.4 for tracking and regulation with independent objectives (by replacing in Equation 3.4.4 S' by R', A by $q^{-d}B$, $q^{-(d+1)}R$ by S and $P(q^{-1})$ by $P_F(q^{-1})$). Equation 3.5.5 has a unique minimal order solution for

$$n_{P_F} = \deg P_F(q^{-1}) \le n_B + d$$

$$n_S = \deg S(q^{-1}) = n_B + d \quad ; \quad n_{R'} = \deg R'(q^{-1}) = 0$$

The solution of Equation 3.5.5 becomes simpler if one considers the typical case where an integrator is introduced in the controller, *i.e.*:

$$S(q^{-1}) = \left(1 - q^{-1}\right)S'(q^{-1}) \qquad (3.5.6)$$

In this case, for $q=1$ Equation 3.5.5 becomes

$$B(1)R'(1) = P_F(1) \qquad (3.5.7)$$

since $S(1)=0$ and one obtains

$$R'(q^{-1}) = R'(1) = \frac{P_F(1)}{B(1)} \qquad (3.5.8)$$

and

$$R(q^{-1}) = A(q^{-1})\frac{P_F(1)}{B(1)} \qquad (3.5.9)$$

$$S(q^{-1}) = (1-q^{-1})S'(q^{-1}) = P_F(q^{-1}) - q^{-d}B(q^{-1})\frac{P_F(1)}{B(1)} \qquad (3.5.10)$$

From Equation 3.5.9 it follows that in this design the poles of the plant model (defined by $A(q^{-1})$) will be simplified by the zeros introduced by the controller. This implies that *the poles of the plant model should be stable* (see Chapter 2, Section 2.5.2).

One interesting aspect of this design is the possibility to characterize the set of all the controllers as a function of $P_F(q^{-1})$ without having to solve a polynomial equation.

If, in addition to the integrator, a fixed part $H_R(q^{-1})$ is imposed in $R(q^{-1})$, $S(q^{-1})$ has the structure given in Equation 3.5.6 and $R(q^{-1})$ will have the structure

$$R(q^{-1}) = A(q^{-1})H_R(q^{-1})R'(q^{-1}) \qquad (3.5.11)$$

Equation 3.5.5 becomes

$$S(q^{-1}) + q^{-d}B(q^{-1})H_R(q^{-1})R'(q^{-1}) = P_F(q^{-1}) \qquad (3.5.12)$$

with the condition

$$P_F(1) = B(1)H_R(1)R'(1) \qquad (3.5.13)$$

from which one obtains

$$R'(q^{-1}) = R'(1) = \frac{P_F(1)}{B(1)H_R(1)} \qquad (3.5.14)$$

$$R(q^{-1}) = A(q^{-1})H_R(q^{-1})\frac{P_F(1)}{B(1)H_R(1)} \qquad (3.5.15)$$

$$S(q^{-1}) = P_F(q^{-1}) - q^{-d}B(q^{-1})H_R(q^{-1})R'(q^{-1}) \qquad (3.5.16)$$

3.5.2 Tracking

In the case of internal model control the polynomial $T(q^{-1})$ used for tracking is given by (like in the pole placement)

$$T(q^{-1}) = A(q^{-1})P_F(q^{-1})/B(1) \tag{3.5.17}$$

But if one chooses the same dynamics for tracking and regulation, the tracking reference model can be eliminated and $T(q^{-1})$ will be given by

$$T(q^{-1}) = T(1) = \frac{A(1)P_F(1)}{B(1)} \tag{3.5.18}$$

which guarantees a unit steady state gain between the desired tracking trajectory and the controlled output.

3.5.3 An Interpretation of the Internal Model Control

Let us consider first the case $H_R(q^{-1})=1$. The control law equation (Table 3.5) is written using the expressions of R, S, T given by Equations 3.5.9, 3.5.10 and 3.5.17:

$$S(q^{-1})u(t) = \left[P_F(q^{-1}) - \frac{P_F(1)}{B(1)} q^{-d} B(q^{-1}) \right] u(t) =$$

$$\left[\frac{1}{B(1)} A(q^{-1})P_F(q^{-1})y^*(t+d+1) - \frac{P_F(1)}{B(1)} A(q^{-1})y(t) \right] \tag{3.5.19}$$

which can be re-written as

$$P_F(q^{-1})u(t) = \frac{1}{B(1)} A(q^{-1})P_F(q^{-1})y^*(t+d+1) -$$

$$\frac{P_F(1)}{B(1)} \left[A(q^{-1})y(t) - q^{-d} B(q^{-1})u(t) \right] \tag{3.5.20}$$

Taking into account that by hypothesis $A(q^{-1})$ is asymptotically stable one obtains

$$S_0(q^{-1})u(t) = P_F(q^{-1})u(t) = \frac{1}{B(1)} A(q^{-1})P_F(q^{-1})y^*(t+d+1) -$$

$$\frac{P_F(1)}{B(1)} A(q^{-1}) \left[y(t) - \frac{q^{-d}B(q^{-1})}{A(q^{-1})} u(t) \right] =$$

$$T_0(q^{-1})y^*(t+d+1) - R_0(q^{-1}) \left[y(t) - \frac{q^{-d}B(q^{-1})}{A(q^{-1})} u(t) \right] \tag{3.5.21}$$

where

$$R_0(q^{-1}) = \frac{P_F(1)}{B(1)} A(q^{-1})$$

$$S_0(q^{-1}) = P_F(q^{-1})$$

$$T_0(q^{-1}) = \frac{1}{B(1)} P(q^{-1}) = \frac{1}{B(1)} A(q^{-1}) P_F(q^{-1})$$

This leads to the equivalent representation of the closed loop behavior shown in Figure 3.22.

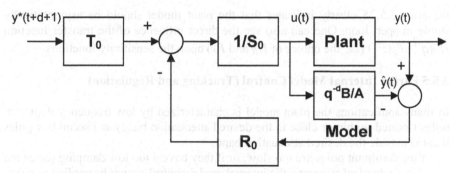

Figure 3.22. Equivalent diagram of the internal model control

One can see that an equivalent implementation of the RST controller can be considered. This implementation explicitly features the model of the plant (prediction model) as a component of the control scheme. It is the error signal between the plant output $y(t)$ and the predictor output $\hat{y}(t)$, which is fed-back.

Note also that the computation of R_0 and S_0 does not require the solution of a polynomial equation.

For the case of $H_R(q^{-1}) \neq 1$, using Equations 3.5.14, 3.5.15, 3.5.16 and 3.5.17 one obtains a similar result. Only the expression of $R_0(q^{-1})$ changes (it will contain in addition the factor $H_R(q^{-1})/H_R(1)$).

3.5.4 The Sensitivity Functions

In the case of internal model control the sensitivity functions have a particular structure, as a consequence of the choice for the dominant poles of the closed loop (they are equal to the plant model poles):

$$S_{yp}(z^{-1}) = \frac{S(z^{-1})}{P_F(z^{-1})} = 1 - \frac{z^{-d} B(z^{-1}) H_R(z^{-1}) P_F(1)}{B(1) H_R(1) P_F(z^{-1})} \qquad (3.5.22)$$

$$S_{yb}(z^{-1}) = -\frac{z^{-d}B(z^{-1})R(z^{-1})}{A(z^{-1})P_F(z^{-1})} = -\frac{z^{-d}B(z^{-1})H_R(z^{-1})P_F(1)}{B(1)H_R(1)P_F(z^{-1})} \tag{3.5.23}$$

$$S_{up}(z^{-1}) = -\frac{R(z^{-1})}{P_F(z^{-1})} = -\frac{A(z^{-1})H_R(z^{-1})P_F(1)}{B(1)H_R(1)P_F(z^{-1})} \tag{3.5.24}$$

$$S_{yv}(z^{-1}) = -\frac{z^{-d}B(z^{-1})S(z^{-1})}{A(z^{-1})P_F(z^{-1})} = -\frac{z^{-d}B(z^{-1})}{A(z^{-1})}S_{yp} \tag{3.5.25}$$

Equation 3.5.25 clearly indicates that the plant model should be asymptotically stable in open loop. One can also see the direct influence of the transfer function $H_R(q^{-1})/P_F(z^{-1})$ (*i.e.* the choice of H_R and P_F) upon the sensitivity functions.

3.5.5 Partial Internal Model Control (Tracking and Regulation)

In many applications the plant model is characterized by low frequency dominant poles (located within or close to the desired attenuation band) and secondary poles located outside the desired attenuation band.

If the dominant poles are too slow, or if they have a too low damping (often the case for mechanical systems), the internal model control cannot be applied as such.

In this case one uses a mixture of pole placement and internal model control. The pole placement will be used to assign the desired dominant poles of the closed loop but it will not move the secondary poles of the plant model (which will become poles of the closed loop).

Suppose $A(q^{-1})$ has the form

$$A(q^{-1}) = A_1(q^{-1})A_2(q^{-1}) \tag{3.5.26}$$

where $A_1(q^{-1})$ defines the dominant poles of the plant model. In the case of partial internal model control, the equation defining the closed loop poles will be

$$\begin{aligned} & A_1(q^{-1})A_2(q^{-1})S(q^{-1}) + q^{-d}B(q^{-1})R(q^{-1}) \\ & = P_D(q^{-1})A_2(q^{-1})P_F(q^{-1}) \end{aligned} \tag{3.5.27}$$

Examining the structure of Equation 3.5.27 it is seen that $R(q^{-1})$ will be of the form

$$R(q^{-1}) = A_2(q^{-1})R'(q^{-1}) \tag{3.5.28}$$

and Equation 3.5.27 (after elimination of the common factor $A_2(q^{-1})$) becomes

$$A_1(q^{-1})S(q^{-1}) + q^{-d}B(q^{-1})R'(q^{-1}) = P_D(q^{-1})P_F(q^{-1}) \qquad (3.5.29)$$

This technique can also be interpreted as a simplification of the plant model secondary poles (defined by $A_2(q^{-1})$) by the zeros of the controller.

3.5.6 Internal Model Control for Plant Models with Stable Zeros

If $B(q^{-1})$ has all its zeros inside the unit circle (and these zeros are sufficiently damped) one can use the tracking and regulation with independent objectives design but using as desired poles the dominant poles of the plant model.

One considers that $S(q^{-1})$ has the structure given by Equation 3.4.15 (presence of an integrator) and one supposes $H_R(q^{-1})=1$ (in order to simplify the presentation).

Equation 3.4.16 becomes

$$A(q^{-1})\,(1-q^{-1})\,S'(q^{-1}) + q^{-(d+1)}\,R(q^{-1}) = A(q^{-1})\,P_F(q^{-1}) \qquad (3.5.30)$$

This implies that $R(q^{-1})$ will have the form

$$R(q^{-1})=A(q^{-1})R'(q^{-1}) \qquad (3.5.31)$$

and Equation 3.5.30 becomes

$$(1-q^{-1})\,S'(q^{-1}) + q^{-(d+1)}\,R'(q^{-1}) = P_F(q^{-1}) \qquad (3.5.32)$$

For $q=1$ one has

$$R'(1) = R'(q^{-1}) = P_F(1) \qquad (3.5.33)$$

It results that

$$(1-q^{-1})\,S'(q^{-1}) = P_F(q^{-1}) - q^{-(d+1)}\,P_F(1) \qquad (3.5.34)$$

from which one gets

$$S(q^{-1}) = B^*(q^{-1})\,[P_F(q^{-1}) - q^{-(d+1)}\,P_F(1)] \qquad (3.5.35)$$

$$R(q^{-1}) = A(q^{-1})\,P_F(1) \qquad (3.5.36)$$

3.5.7 Example: Control of Systems with Time Delay

Internal model control is often used for the control of systems with large time delay compared to the rise time of the system without time delay. If $d > t_R$ (where t_R is expressed in sampling periods), the reduction of the rise time in closed loop does not produce a significant improvement of the total response time, since this is mainly determined by d. Therefore, one is able to use internal model control. However, even when one takes $P_D(q^{-1}) = A(q^{-1})$, the introduction of the auxiliary poles and/or of the fixed part $H_R(q^{-1})$ is necessary in order to assure the robustness of the closed loop with respect to the time delay variations.

Consider the case $P_F(q^{-1}) = 1$, $H_R(q^{-1}) = 1$ and $B(q^{-1}) = b_1 q^{-1}$ $(b_1 > 0)$. From Equations 3.5.22 and 3.5.23 one gets

$$S_{yp}(z^{-1}) = 1 - z^{-d-1} = (1 - z^{-1})(1 + z^{-1} + z^{-2} + \ldots + z^{-d}) \tag{3.5.37}$$

$$S_{yb}(z^{-1}) = z^{-d-1} \tag{3.5.38}$$

$S_{yp}(z^{-1})$ has a zero at $z = 1$ $(\omega = 0)$ and d zeros on the unit circle if d is even, or $d-1$ zeros on the unit circle plus a zero at $z = -1$ $(\omega = \pi)$ if d is odd. From Equation 3.5.37, since $|z^{-d-1}|$ is always equal to 1, it results that

$$\left| S_{yp}(e^{-j\omega}) \right|_{max} \leq 2 \tag{3.5.39}$$

and then the modulus margin $\Delta M = 0.5$ is always achieved.

However

$$\left| S_{yb}(e^{-j\omega}) \right| \equiv 1 \qquad\qquad \text{for } 0 \leq \omega \leq \pi \tag{3.5.40}$$

and therefore the frequency characteristic of $| S_{yb} |$ will cross the template for $\Delta \tau = T_s$ (see Figure 2.39: $| S_{yb} |$ should be smaller than 1 starting from $f = 0.17 f_s$).

The condition for satisfying the delay margin in Figure 2.39 is in this case

$$\left| z^{-d-1} \right| \leq \frac{1}{\left| z^{-1} - 1 \right|} \tag{3.5.41}$$

and clearly this will not be satisfied at all frequencies.

We will examine next the beneficial effect of the introduction of the auxiliary poles $P_F(q^{-1})$ and of a fixed part $H_R(q^{-1})$ in the controller.

Auxiliary Poles $(P_F(q^{-1}) \neq 1, H_R(q^{-1}) = 1)$
Let us select:

$$P_F(q^{-1}) = (1 + \alpha q^{-1}) \qquad\qquad -1 < \alpha < 0 \qquad\qquad (3.5.42)$$

and search for the value of α assuring the delay margin $\Delta \tau = T_s$. In this case S_{yb} is given by Equation 3.5.23 and the condition for the delay margin $\Delta \tau = T_s$ takes the form

$$\left| S_{yb}(z^{-1}) \right| = \left| \frac{z^{-d-1} P_F(1)}{P_F(z^{-1})} \right| < \frac{1}{\left| 1 - z^{-1} \right|} \qquad z = e^{j\omega} \quad 0 \le \omega \le \pi \qquad (3.5.43)$$

which for $P_F(q^{-1})$ given by Equation 3.5.42 becomes

$$\left| \frac{1+\alpha}{1+\alpha z^{-1}} \right| < \frac{1}{\left| 1 - z^{-1} \right|} \qquad z = e^{j\omega} \quad 0 \le \omega \le \pi \qquad (3.5.44)$$

Replacing z by $z = e^{j\omega}$ one observes that the worst situation occurs at $\omega = \pi$ (i.e. $f=0.5f_s$) where 3.5.44 becomes:

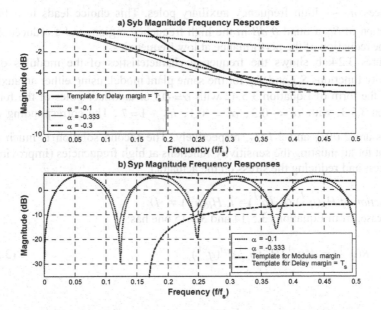

Figure 3.23a,b. Auxiliary poles effect on the delay margin: **a** noise- input sensitivity function (S_{yb}); **b** output sensitivity function (S_{yp})

$$\left|\frac{1+\alpha}{1-\alpha}\right| < 0.5 \quad \Rightarrow \quad \alpha \le -0.333 \tag{3.5.45}$$

Figure 3.23a,b shows the frequency characteristics of $|S_{yb}|$ and $|S_{yp}|$ (controller with integrator) for a plant model characterized by $d = 7$, $B(q^{-1}) = q^{-1}$ ($b_1 = 1$), $A(q^{-1}) = 1 + a_1 q^{-1}$ with $a_1 = -0.2$ and for three values of α: $\alpha = -0.1$; $\alpha = -0.3$; $\alpha = -0.333$. For $\alpha = -0.1$ and $\alpha = -0.3$ one gets the delay margin $\Delta\tau = 0.52T_s$ and $\Delta\tau = 0.91T_s$, respectively, and the corresponding curves of S_{yb} and S_{yp} intersect the robustness template for $\Delta\tau = T_s$. For $\alpha = -0.333$ the effective delay margin is $\Delta\tau = T_s$ and it can be seen that the corresponding sensitivity functions do not cross the robustness template for the delay margin $\Delta\tau = T_s$.

If one uses auxiliary poles $P_F(z^{-1})$ of the form Equation 3.5.3:

$$P_F(q^{-1}) = (1 + \alpha q^{-1})(1 + \alpha' q^{-1})^{n_{P_F} - 1}$$

one selects $-1 < \alpha \le 0$ and $-0.25 < \alpha' \le -0.05$ with $n_{P_F} \le n_B + d$. In this case one introduces $n_F - 1$ high frequency auxiliary poles. This choice leads to a further contraction of S_{yp} around 0 dB in the high frequency region, and a reduced effect upon the reduction of the performance at low frequencies.

Figures 3.24a,b shows the frequency characteristics of the modulus of the sensitivity functions S_{yb} and S_{yp} for the same plant model, using either an auxiliary pole of the form of Equation 3.5.42 with $\alpha = -0.5$ or auxiliary poles of the form of Equation 3.5.3, with $\alpha = -0.3$, $\alpha' = -0.1$, $n_{P_F} - 1 = 7$. The corresponding delay margins are $2.09T_s$ and $2.14T_s$, respectively. The second solution is much more efficient for attenuating the sensitivity functions at high frequencies (improving the robustness and reducing the stress on the actuator).

Introduction of $H_R(q^{-1})$ $(P_F(q^{-1}) = 1, H_R(q^{-1}) \ne 1)$
In this case, from Equations 3.5.14 and 3.5.15 one has:

$$R(q^{-1}) = A(q^{-1})H_R(q^{-1})R'(q^{-1}) \tag{3.5.46}$$

with

$$H_R(q^{-1}) = 1 + \beta q^{-1} \tag{3.5.47}$$

which leads to

$$R'(q^{-1}) = \frac{1}{B(1)(1+\beta)}$$

(3.5.48)

and (using Equation 3.5.16):

$$S(q^{-1}) = 1 - \frac{q^{-d} B(q^{-1})(1+\beta q^{-1})}{B(1)(1+\beta)}$$

(3.5.49)

For $B(q^{-1}) = b_1 q^{-1}$ one gets

$$S_{yb}(z^{-1}) = -\frac{z^{-d-1}(1+\beta z^{-1})}{1+\beta}$$

(3.5.50)

and the condition for getting a delay margin $\Delta\tau = T_s$ becomes

$$\left| \frac{1+\beta z^{-1}}{1+\beta} \right| < \frac{1}{\left| 1 - z^{-1} \right|} \qquad z = e^{j\omega} \quad 0 \le \omega \le \pi$$

(3.5.51)

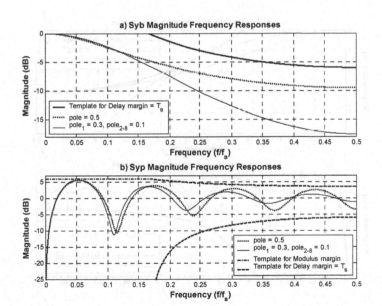

Figure 3.24a,b. Effect of high frequency auxiliary poles on the sensitivity functions of plant models with time delay: **a** noise- input sensitivity function (S_{yb}); **b** output sensitivity function (S_{yp})

The worst situation occurs at $\omega = \pi/2$ ($f=0.25f_s$) where the condition of Equation 3.5.51 becomes (replacing $<$ by $\leq$)

$$\frac{1+\beta^2}{(1+\beta)^2} \leq 0.5 \quad \Rightarrow \quad \beta = 1 \qquad (3.5.52)$$

Therefore one should choose in this case

$$H_R(q^{-1}) = 1 + q^{-1} \qquad (3.5.53)$$

in order to assure a delay margin of (almost) one sampling period.

It is interesting to note that Equation 3.5.53 corresponds to the opening of the closed loop at $0.5f_s$ (see Section 3.3). This solution leads to $S_{up} = 0$ at $f = 0.5f_s$ and therefore a significant increase of the robustness at high frequencies. Figure 3.25a,b presents comparatively the frequency characteristics of the modulus of S_{yb} and S_{yp} for the cases: 1) $H_R(q^{-1}) = 1 + q^{-1}$, $P_F(q^{-1}) = 1$ and 2) $H_R(q^{-1}) = 1$, $P_F(q^{-1}) = 1 - 0.333z^{-1}$.

Of course one can use simultaneously $P_F(q^{-1}) \neq 1$ and $H_R(q^{-1}) \neq 1$.

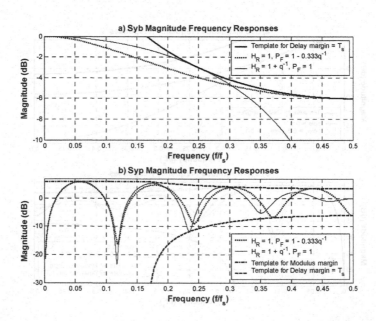

Figure 3.25a,b. Effect of the fixed part $H_R(q^{-1}) = 1 + q^{-1}$ on the sensitivity function of a plant model with time delay: **a** noise- input sensitivity function (S_{yb}); **b** output sensitivity function (S_{yp})

3.6 Pole Placement with Sensitivity Function Shaping

In many applications, in order to satisfy simultaneously the imposed performance in regulation and the robustness margins, one is obliged to shape the modulus of the output and input sensitivity functions in the frequency domain. The shaping of the sensitivity functions is done by the appropriate selection of the desired closed loop poles and the introduction of pre-specified filters in the controllers.

The modulus of the output sensitivity function $|S_{yp} (z^{-1})|$[9] is a significant indicator of both disturbance rejection properties and robustness properties of the closed loop (see Chapter 2, Section 2.6).

The modulus of the input sensitivity functions $|S_{up} (z^{-1})|$ is an indicator of the actuator stress in various frequency regions as well as an indicator of the tolerance to the additive uncertainties upon the plant model (see Chapter 2, Section 2.6).

Using a digital RST controller, the output sensitivity function has the expression

$$S_{yp}(z^{-1}) = \frac{A(z^{-1})S(z^{-1})}{A(z^{-1})S(z^{-1}) + z^{-d}B(z^{-1})R(z^{-1})} \tag{3.6.1}$$

and the input sensitivity function has the expression

$$S_{up}(z^{-1}) = -\frac{A(z^{-1})R(z^{-1})}{A(z^{-1})S(z^{-1}) + z^{-d}B(z^{-1})R(z^{-1})} \tag{3.6.2}$$

where

$$R(z^{-1}) = H_R(z^{-1}) R'(z^{-1}) \tag{3.6.3}$$

$$S(z^{-1}) = H_S(z^{-1}) S'(z^{-1}) \tag{3.6.4}$$

and

$$A(z^{-1}) S(z^{-1}) + z^{-d} B(z^{-1}) R(z^{-1}) = P_D(z^{-1}) \cdot P_F(z^{-1}) = P(z^{-1}) \tag{3.6.5}$$

In Equations 3.6.3 and 3.6.4, $H_R(z^{-1})$ and $H_S(z^{-1})$ correspond to pre-specified fixed filters incorporated in $R(z^{-1})$ and $S(z^{-1})$ respectively. $S(z^{-1})$ and $R(z^{-1})$ (more precisely $R'(z^{-1})$ and $S'(z^{-1})$) are the solutions of Equation 3.6.5 where the polynomial $P(z^{-1})$ defines the desired closed loop poles. The polynomial $P(z^{-1})$ is factorized in order to emphasize the dominant poles defined by $P_D(z^{-1})$ and the auxiliary poles defined by $P_F(z^{-1})$.

---

[9] In this section the notation "z" (complex variable) will be used instead of "q" when we will examine the properties of the various transfer functions in the frequency domain.

In what follows we will examine the properties of the sensitivity functions in the frequency domain $(q = z = e^{j\omega}, \omega = 2\pi \, f / f_s)$.

The various properties of the sensitivity functions will be illustrated by means of the following example $(T_s = 1s)$.

Plant model:

$$A(q^{-1}) = 1 - 0.7 \, q^{-1}, \, B(q^{-1}) = 0.3 \, q^{-1}, \, d = 2.$$

Specified performance (polynomial $P(q^{-1})$):

Defined by the discretization of a continuous time second-order system with

- $\omega_0 = 0.4 \text{ or } 0.6 \text{ or } 1 \text{ rad/s}$
- $\zeta = 0.9$ (constant)

3.6.1 Properties of the Output Sensitivity Function

Property 1
The modulus of the output sensitivity function at a certain frequency gives the amplification or the attenuation of the disturbance.

At the frequencies where $|Syp(\omega)| = 1$ (0 dB), there is neither amplification nor attenuation of the disturbance (operation like in open loop).

At the frequencies where $|Syp(\omega)| < 1$ (0 dB), the disturbance is attenuated.

At the frequencies where $|Syp(\omega)| > 1$ (0 dB), the disturbance is amplified.

Property 2
For asymptotically stable closed loop system, and stable open loop, the integral of the logarithm of the modulus of the output sensitivity function from 0 to 0.5 f_s satisfies[10]

$$\int_0^{0.5 f_s} \log \left| S_{yp}(e^{-j2\pi f/f_s}) \right| df = 0$$

In other terms, the sum of the areas between the modulus of the sensitivity function and the 0 dB axis, taken with their sign, is null. As a consequence, the attenuation of disturbances in a certain frequency region implies necessarily the amplification of disturbances in other frequency regions. Figure 3.26 illustrates this phenomenon.

---

[10] For a proof of this property see Sung and Hara (1988). In the case of unstable open loop systems but stable in closed loop, the value of this integral is positive.

The output sensitivity functions shown in Figure 3.26 correspond to the example mentioned earlier for various values of ω_0 *(0.4; 0.6; 1 rad/s)* but ζ = *constant (0.9)*.

It follows that increasing the value of attenuation in a frequency region, or widening the attenuation band, will lead to a higher amplification of disturbances outside the attenuation band. Figure 3.26 clearly emphasizes this phenomenon.

Property 3
The inverse of the maximum of the modulus of the output sensitivity function corresponds to the modulus margin ΔM:

$$\Delta M = (|S_{yp}(e^{-j\omega})|_{max})^{-1} \qquad (3.6.6)$$

The modulus margin is defined as the minimal distance between the Nyquist plot of the open loop transfer function and the critical point [-1, j0]. The typical values for the modulus margin (see Section 2.6) are

$$\Delta M \geq 0.5 \; (-6 \; dB) \; [min: 0.4 \; (-8 \; dB)]$$

Recall that a modulus margin $\Delta M \geq 0.5$ implies a gain margin $\Delta G \geq 2$ and a phase margin $\Delta\phi > 29^o$. To assure a safe modulus margin, it is necessary that:

$$|S_{yp}(e^{-j\omega})|_{max} \leq 6 \; dB \; (or \; exceptionally \; 8 \; dB)$$

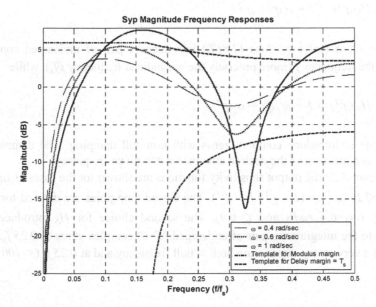

Figure 3.26. Modulus of the output sensitivity function for different attenuation bands of the disturbance

From property 2, it is seen that the increase of the attenuation in a certain frequency region, or the increase of the attenuation band, will in general lead to the increase of $|S_{yp}(e^{-j\omega})|_{max}$ and therefore a reduction of the modulus margin (and of the system robustness).

Property 4
Cancellation of the disturbance effect on the output ($|S_{yp}|= 0$) is obtained at the frequencies where

$$A(e^{-j\omega})S(e^{-j\omega}) = A(e^{-j\omega})H_S(e^{-j\omega})S'(e^{-j\omega}) = 0 \; ; \; \omega = 2\pi f / f_s \; (3.6.7)$$

This results immediately from Equation 3.6.1. Equation 3.6.7 with $q= z= e^{j\omega}$ defines the zeros of the output sensitivity function in the frequency domain.

The fixed pre-specified part of $S(q^{-1})$, denoted $H_S(q^{-1})$, allows one to introduce the zeros at the desired frequencies.

For example

$$H_S(q^{-1}) = 1 - q^{-1}$$

introduces a zero at the zero frequency and allows a perfect rejection of constant disturbances

$$H_S(q^{-1}) = 1 + \alpha q^{-1} + q^{-2}$$

with $\alpha = -2cos(\omega T_s) = -2cos(2\pi f/f_s)$ introduces a pair of undamped complex zeros at the frequency f (more precisely the normalized frequency f/f_s), while

$$H_S(q^{-1}) = 1 + \alpha_1 q^{-1} + \alpha_2 q^{-2}$$

allows one to introduce complex zeros with non-null damping. The damping is selected as a function of the desired attenuation at a given frequency.

In Figure 3.27 the output sensitivity functions are shown for the cases $H_S(q^{-1}) = 1 - q^{-1}$ and $H_S(q^{-1}) = (1 - q^{-1})(1 + q^{-1})$. The closed loop poles are defined for both cases by ($\omega_0=0.6$ rad/s and $\zeta=0.9$). The second choice for H_S introduces, in addition to the integrator, a pair of undamped ($\zeta=0$) complex zeros at $0.25 f_s$. One observes a very strong attenuation both at null frequency and at $0.25 f_s$ (<-100 dB).

Property 5
The modulus of the output sensitivity functions is equal to 1, that is

$$|S_{yp}(e^{j\omega})| = 1 \ (0 \ dB)$$

at the frequencies where

$$B^*(e^{-j\omega})R(e^{-j\omega}) = B^*(e^{-j\omega})H_R(e^{-j\omega})R'(e^{-j\omega}) = 0 \ ; \ \omega = 2\pi f / f_s \qquad (3.6.8)$$

This results immediately from Equation 3.6.1 since under the condition of Equation 3.6.8 one gets $S_{yp}(j\omega) = 1$.

The specified fixed part of $R(q^{-1})$, denoted $H_R(q^{-1})$, allows one to obtain a null gain for $R(q^{-1})$ at certain frequencies, assuring at these frequencies $|S_{yp}(e^{j\omega})| = 1$ (open loop type operation).

For example,

$$H_R(q^{-1}) = 1 + q^{-1}$$

introduces a zero at $0.5 f_s$ implying $|S_{yp}(e^{-j\pi})| = 1$.

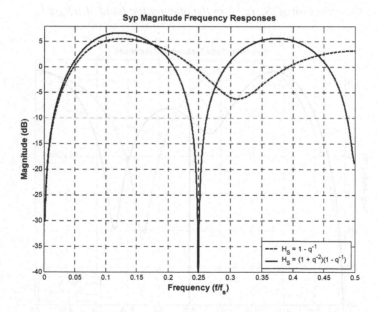

Figure 3.27. Output sensitivity function for the case $H_S(q^{-1}) = (1 - q^{-1})$ and $H_S(q^{-1}) = (1 - q^{-1})(1 + q^{-2})$

$$H_R(q^{-1}) = 1 + \beta q^{-1} + q^{-2}$$

with $\beta = -2cos\,(\omega\,T_s) = -2cos\,2\pi f/f_s$ introduces a pair of undamped complex zeros at the normalized frequency f/f_s leading to $|S_{yp}(e^{-j\pi f/f_s})| = 1$.

$$H_R(q^{-1}) = 1 + \beta_1 q^{-1} + \beta_2 q^{-2}$$

introduces a pair of complex zeros with a non null damping, allowing to influence the attenuation of the disturbance at a certain frequency.

Figure 3.28 illustrates the effect of the $H_R(q^{-1}) = 1 + q^{-2}$, which introduces a pair of complex zeros with null damping at $f = 0.25\,f_s$. One can see that, in the presence of $H_R(q^{-1})$, one has $|S_{yp}(e^{j\omega})| = 1$ (0 dB) at this frequency, while, without introducing $H_R(q^{-1})$ one has at $f = 0.25\,f_s$ a gain $|S_{yp}(e^{j\omega})| = 3$ dB.

Note also that $R(z^{-1})$ defines some of the zeros of the input sensitivity function $S_{up}(z^{-1})$ (given in Equation 3.6.2). Therefore at the frequencies where $R(z^{-1}) = 0$, this sensitivity function will be null.

Property 6
   The introduction of asymptotically stable auxiliary poles $P_F(z^{-1})$ leads in general to the reduction of $|S_{yp}(z^{-1})|$ in the attenuation band of $1/P_F(z^{-1})$.

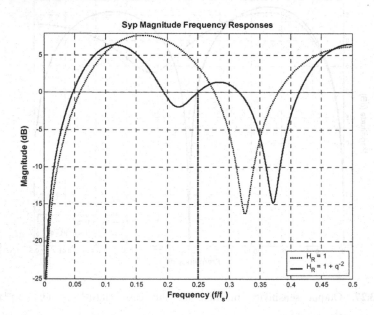

Figure 3.28. Output sensitivity function for the case $H_R(q^{-1}) = 1$ and $H(q^{-1}) = 1 + q^{-2}$

From the expressions of the sensitivity function in Equation 3.6.1 and of the closed loop poles, one can see that $1/(P_D(z^{-1}) \, P_F(z^{-1}))$ will introduce a stronger attenuation in the frequency domain than $1/P_D(z^{-1})$, provided that the auxiliary poles defined by $P_F(z^{-1})$ are asymptotically stable aperiodic poles. However, since $S'(z^{-1})$ will depend upon these poles through Equation 3.6.4, this property cannot be guaranteed for all possible values of $P_F(z^{-1})$.

The auxiliary poles are in general selected as real poles located at high frequencies, and they take the form

$$P_F(q^{-1}) = (1 + p'q^{-1})^{n_{P_F}} \qquad -0.5 \le p' \le -0.05$$

where

$$n_{P_F} \le n_P - n_{P_D} \quad ; \quad n_P = (\deg P)_{max} \quad ; \quad n_{P_D} = \deg P_D$$

The effect of auxiliary poles is illustrated in Figure 3.29.

Remark: in many applications, the introduction of high frequency auxiliary poles is enough in order to assure the imposed robustness margins.

Property 7
Simultaneous introduction of a fixed part H_{S_i} and of a pair of auxiliary poles P_{F_i} in the form

$$\frac{H_{S_i}(z^{-1})}{P_{F_i}(z^{-1})} = \frac{1 + \beta_1 z^{-1} + \beta_2 z^{-2}}{1 + \alpha_1 z^{-1} + \alpha_2 z^{-2}} \tag{3.6.9}$$

resulting from the discretization of the continuous-time filter

$$F(s) = \frac{s^2 + 2\zeta_{num}\omega_0 s + \omega_0^2}{s^2 + 2\zeta_{den}\omega_0 s + \omega_0^2} \tag{3.6.10}$$

using the bilinear transformation[11]

$$s = \frac{2}{T_s} \frac{1 - z^{-1}}{1 + z^{-1}} \tag{3.6.11}$$

---

[11] The bilinear transformation assures a better approximation of a continuous-time model by a discrete-time model in the frequency domain than the replacement of differentiation by a difference, *i.e.* $s = (1 - z^{-1}) \, T_S$ (see Equations 2.3.6 and 2.5.2).

introduces an attenuation (a "hole") at the normalized discretized frequency

$$\omega_{disc} = 2\arctan\left(\frac{\omega_0 T_s}{2}\right) \tag{3.6.12}$$

as a function of the ratio $\zeta_{num} / \zeta_{den} < 1$. The attenuation at ω_{disc} is given by

$$M_t = 20\log\left(\frac{\zeta_{num}}{\zeta_{den}}\right) \quad ; \quad (\zeta_{num} < \zeta_{den}) \tag{3.6.13}$$

The effect upon the frequency characteristics of S_{yp} at frequencies $f << f_{disc}$ and $f >> f_{disc}$ is negligible.

Figure 3.30 illustrates the effect of the simultaneous introduction of a fixed part H_S and a pair of poles in P, corresponding to the discretization of a resonant filter of the form of Equation 3.6.10. One observes its weak effect on the frequency characteristics of S_{yp}, far from the resonance frequency of the filter.

This pole-zero filter is essential for an accurate shaping of the modulus of the sensitivity functions in the various frequency regions in order to satisfy the constraints. It allows one to reduce the interaction between the tuning in different regions.

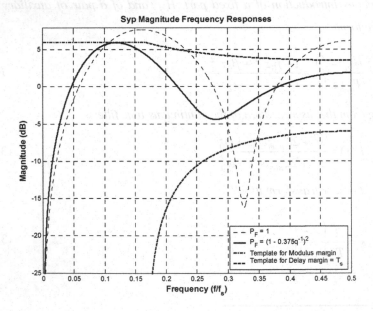

Figure 3.29. Effects of auxiliary poles on the output sensitivity function

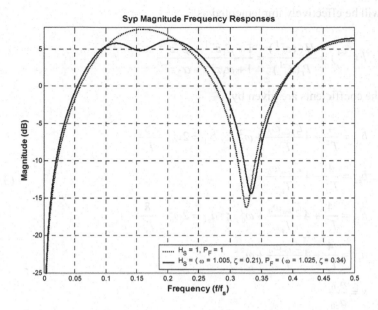

Figure 3.30. Effects of a resonant filter H_{S_i} / P_{F_i} on the output sensitivity functions.

Design of the Resonant Pole-Zero Filter H_{S_i} / P_{F_i}

The computation of the coefficients of H_{S_i} and P_{F_i} is done in the following way:

Specifications:

- Central normalized frequency f_{disc} ($\omega_{disc} = 2\pi f_{disc}$)
- Desired attenuation at frequency f_{disc}: M_t dB
- Minimum accepted damping for auxiliary poles

$$P_{F_i} : (\zeta_{den})_{min} \ (\geq 0.3)$$

Step I: Design of the continuous-time filter

$$\omega_0 = \frac{2}{T_s} \tan\left(\frac{\omega_{disc}}{2}\right) \quad 0 \leq \omega_{disc} \leq \pi \quad \zeta_{num} = 10^{M_t/20} \zeta_{den}$$

Step II: Design of the discrete-time filter using the bilinear transformation of Equation 3.6.11.

Using Equation 3.6.11 one gets:

$$F(z^{-1}) = \frac{b_{z0} + b_{z1}z^{-1} + b_{z2}z^{-2}}{a_{z0} + a_{z1}z^{-1} + a_{z2}z^{-2}} = \gamma \frac{1 + \beta_1 z^{-1} + \beta_2 z^{-2}}{1 + \alpha_1 z^{-1} + \alpha_2 z^{-2}} \tag{3.6.14}$$

which will be effectively implemented as[12]

$$F(z^{-1}) = \frac{H_S(z^{-1})}{P_i(z^{-1})} = \frac{1 + \beta_1 z^{-1} + \beta_2 z^{-2}}{1 + \alpha_1 z^{-1} + \alpha_2 z^{-2}}$$

where the coefficients are given by

$$b_{z0} = \frac{4}{T_s^2} + 4\frac{\zeta_{num}\omega_0}{T_s} + \omega_0^2 \quad ; \quad b_{z1} = 2\omega_0^2 - \frac{8}{T_s^2}$$

$$b_{z2} = \frac{4}{T_s^2} - 4\frac{\zeta_{num}\omega_0}{T_s} + \omega_0^2$$

$$a_{z0} = \frac{4}{T_s^2} + 4\frac{\zeta_{den}\omega_0}{T_s} + \omega_0^2 \quad ; \quad a_{z1} = 2\omega_0^2 - \frac{8}{T_s^2}$$ (3.6.15)

$$a_{z2} = \frac{4}{T_s^2} - 4\frac{\zeta_{den}\omega_0}{T_s} + \omega_0^2$$

$$\gamma = \frac{b_{z0}}{a_{z0}}$$

$$\beta_1 = \frac{b_{z1}}{b_{z0}} \quad ; \quad \beta_2 = \frac{b_{z2}}{b_{z0}}$$ (3.6.16)

$$\alpha_1 = \frac{a_{z1}}{a_{z0}} \quad ; \quad \alpha_2 = \frac{a_{z2}}{a_{z0}}$$

The resulting filters H_{S_i} and P_{F_i} can be characterized by the undamped resonance frequency ω_0 and the damping ζ. Therefore, first we will compute the roots of numerator and denominator of $F(z^{-1})$. One gets

$$z_{n1,2} = \frac{-\beta_1 \pm j\sqrt{4\beta_2 - \beta_1^2}}{2} = A_n e^{j\varphi_n}$$

$$z_{d1,2} = \frac{-\alpha_1 \pm j\sqrt{4\alpha_2 - \alpha_1^2}}{2} = A_d e^{j\varphi_d}$$ (3.6.17)

From Table 2.4 and expressions given in Section 2.3.8, one can establish the relation between the filter and the undamped resonance frequency and damping of an equivalent continuous-time filter (discretized with a ZOH). The roots of the second-order monic polynomial in z^{-1} have the expression

$$z_{1,2} = e^{-\zeta_{disc}\omega_{0disc}T_s} e^{\pm j\omega_{0disc}T_s\sqrt{1-\zeta_{disc}^2}}$$ (3.6.18)

---

[12] The factor γ has no effect on the final result (coefficients of R and S). It is possible, however, to implement the filter without normalizing the numerator coefficients.

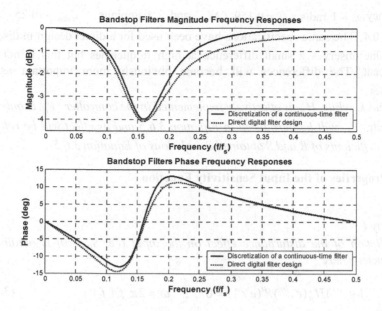

Figure 3.31. Frequency characteristics of the resonant filter H_S / P_F used in the example presented in Figure 3.30

One gets therefore for the numerator and denominator of $F(z^{-1})$

$$\omega_{0num} = \sqrt{\frac{\varphi_n^2 + \ln^2 A_n}{T_s^2}} \quad ; \quad \zeta_{numd} = -\frac{\ln A_n}{\omega_{0num} T_s}$$

$$\omega_{0den} = \sqrt{\frac{\varphi_d^2 + \ln^2 A_d}{T_s^2}} \quad ; \quad \zeta_{dend} = -\frac{\ln A_d}{\omega_{0den} T_s} \tag{3.6.19}$$

where the indexes "num" and "den" correspond to H_S and P_F, respectively. These filters can be computed using the functions *filter22.sci* (Scilab) *filter22.m* (MATLAB®) and also with *ppmaster* (MATLAB®)[13].

Remark: for frequencies below 0.17 f_s the design can be done with a very good precision directly in discrete-time. In this case, $\omega_0 = \omega_{0,den} = \omega_{0,num}$ and the damping of the discrete time filters H_{S_i} and P_{F_i} is computed as a function of the attenuation directly using Equation 3.6.13.

Figure 3.31 gives the frequency characteristics of a filter H_S / P_F obtained by the discretization of a continuous-time filter and used in Figure 3.30 (continuous line) as well as the characteristics of the discrete-time filter directly designed in discrete time (dashed line). The continuous-time filter is characterized by a natural

---

[13] To be download from the book website (*http://landau-bookic.lag.ensieg.inpg.fr*).

frequency $\omega_0 = 1$ rad/s ($f_0 = 0.159 f_s$),   and   dampings $\zeta_{num} = 0.25$ and $\zeta_{den} = 0.4$. The same specifications have been used for a direct design in discrete-time. One observes a small difference at high frequencies but this is not very significant. The differences will become obviously more important as ω_0 increases.

Remark: while H_S is effectively implemented in the controller, P_F is only used indirectly. P_F will be introduced in Equation 3.6.5 and its effect will be reflected in the coefficients of R and S obtained as solutions of Equation 3.6.5.

3.6.2 Properties of the Input Sensitivity Function

Property 1
Cancellation of the disturbance effect on the input (i.e. $S_{up}= 0$) is obtained at frequencies where

$$A(e^{-j\omega})H_R(e^{-j\omega})R'(e^{-j\omega}) = 0 \quad ; \quad \omega = 2\pi f / f_s \qquad (3.6.20)$$

At these frequencies $S_{yp}= 1$ (see Property 5 of the output sensitivity function) and the system operates in open loop.

Figure 3.32 illustrates the effect of $H_R(q^{-1})$ on $| S_{up} |$ for

$$H_R(q^{-1}) = 1 + \beta q^{-1} \qquad 0 < \beta \le 1 \qquad (3.6.21)$$

for $\beta=1$, one has $| S_{up} | = 0$ at $0.5 f_s$. Using $0< \beta <1$ one could reduce more or less the modulus of S_{up} around $0.5 f_s$. Note that this structure for $H_R(q^{-1})$ is systematically used for reducing the modulus of the input sensitivity function at high frequencies.
One can also use

$$H_R(q^{-1}) = (1 + \beta q^{-1})^n \qquad 0 < \beta \le 1$$

(usually $n=1$ or 2).

Property 2
At frequencies where

$$A(e^{-j\omega})H_S(e^{-j\omega})S'(e^{-j\omega}) = 0 \quad ; \quad \omega = 2\pi f / f_s \qquad (3.6.22)$$

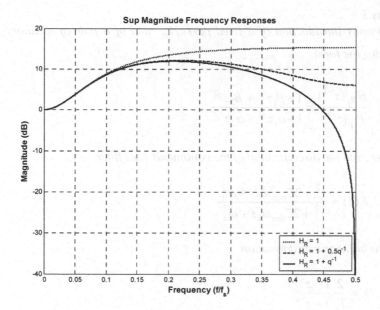

Figure 3.32. Effect of the filter $H_R(q^{-1}) = 1 + \beta q^{-1}$ $0 < \beta \leq 1$ on the modulus of the input sensitivity function: a) $\beta = 0$; b) $\beta = 0.5$; c) $\beta = 1$

which correspond to frequencies where a perfect rejection of disturbances is achieved ($S_{yp} = 0$), one has

$$\left| S_{up}(e^{-j\omega}) \right| = \frac{\left| A(e^{-j\omega}) \right|}{\left| B(e^{-j\omega}) \right|} \tag{3.6.23}$$

Equation 3.6.23 corresponds to the inverse of the gain of the system to be controlled. The implication of Equation 3.6.23 is that *cancellation (or in general an important attenuation) of disturbances on the output should be done only in frequency regions where the system gain is large enough*. If the gain of the controlled system is to low, $|S_{up}|$ will be large at these frequencies. Therefore, the robustness *vs* additive term uncertainties will be reduced and the stress on the actuator will become important. Equation 3.6.23 also implies that serious problems will occur if B(z^{-1}) has complex zeros close to the unit circle (stable or unstable zeros) at frequencies where an important attenuation of disturbances is required. It is mandatory to avoid attenuation of disturbances at these frequencies

Property 3
Simultaneous introduction of a fixed part H_{R_i} *and of a pair of auxiliary poles* P_{F_i} *having the form*

$$\frac{H_{R_i}(z^{-1})}{P_{F_i}(z^{-1})} = \frac{1 + \beta_1 z^{-1} + \beta_2 z^{-2}}{1 + \alpha_1 z^{-1} + \alpha_2 z^{-2}}$$

resulting from the discretization of the continuous time filter

$$F(s) = \frac{s^2 + 2\zeta_{num}\omega_0 s + \omega_0^2}{s^2 + 2\zeta_{den}\omega_0 s + \omega_0^2}$$

using the bilinear transformation

$$s = \frac{2}{T_s}\frac{1 - z^{-1}}{1 + z^{-1}}$$

introduces an attenuation (a " hole") at the normalized discretized frequency

$$\omega_{disc} = 2\arctan\left(\frac{\omega_0 T_s}{2}\right)$$

as a function of the ratio $\zeta_{num} / \zeta_{den} < 1$.
The attenuation at ω_{disc} *is given by*

$$M_t = 20\log\left(\frac{\zeta_{num}}{\zeta_{den}}\right) \quad ; \quad (\zeta_{num} < \zeta_{den})$$

The effect on the frequency characteristics of S_{yp} *at frequencies* $f << f_{disc}$ *and* $f >> f_{disc}$ *is negligible.*
   The design of these filters is done with the method described in Section 3.6.1.

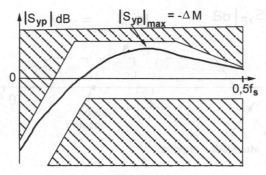

Figure 3.33. Desired template for the output sensitivity function (case of disturbances rejection at low frequencies)

3.6.3 Definition of the "Templates" for the Sensitivity Functions

The performance and robustness specifications lead to the definition of desired "templates" for the frequency characteristics of the sensitivity functions. Most often we are interested in attenuating the disturbances at low frequencies, while assuring certain values for the modulus and delay margins, in order to achieve both good robustness and low amplification of disturbances outside the attenuation band. Figure 3.33 shows a typical template for the modulus of the output sensitivity function.

This template is defined by accepted maximum values for the modulus of the output sensitivity function ("upper template"). One can also consider introducing a "lower template" since on one hand a too large attenuation in a certain frequency region will induce a too important increase of the modulus of the sensitivity function at other frequency regions (and $|S_{yp}|_{max}$ will increase) and, on the other hand, the delay margin requires a "lower template" to be considered.

However, more complex templates will result if, for example, the attenuation should occur in several frequency regions and if, in addition, it is required that the system operate in open loop at a specified frequency (neither amplification, nor attenuation of the disturbance). Such a type of template is shown in Figure 3.34. The constraints on robustness and on the actuator also lead to the definition of an upper template for the modulus of the input sensitivity function S_{up}.

Recall first that the inverse of the modulus of S_{up} gives at each frequency the tolerated additive uncertainty for preserving the stability of the loop (see Chapter 2, Section 2.6). Therefore, *in the frequency regions where there are uncertainties upon the plant model used for the design, one should impose a very low acceptable value for the modulus of the input sensitivity function S_{up}*.

We recall that opening the loop ($S_{up} = 0$) may be required at specified frequencies either because the system should not react to a particular disturbance (see Section 3.6.6), or because one would not like to excite the plant at certain frequencies (for example the plant is characterized by high frequency vibration modes).

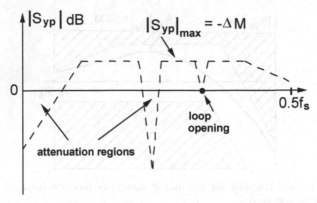

Figure 3.34. Desired templates for the modulus of the output sensitivity function (case of disturbances rejection at certain frequencies and open loop operation required at a specified frequency)

Moreover, for robustness reasons and insensitivity to noise at high frequencies, one systematically reduces the modulus of S_{up} close to $0.5\,f_s$ (often by opening the loop at $0.5\,f_s$).

The limitation of the actuator stress in certain frequency regions (often at high frequencies) induces a limitation of acceptable values of the modulus of S_{up} in these frequency regions.

An example of a template for $|S_{up}|$ is given in Figure 3.35.

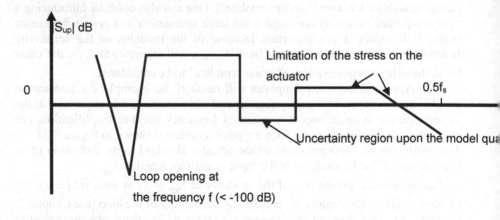

Figure 3.35. Example of a desired template for the modulus of the input sensitivity function

3.6.4 Shaping of the Sensitivity Functions

In order to achieve the shaping of the modulus of the sensitivity functions such that the constraints defined by the template are satisfied, one has the following possibilities:

1. Selection of the desired dominant and auxiliary poles of the closed loop
2. Selection of the fixed parts of the controller ($H_R(q^{-1})$ and $H_S(q^{-1})$)
3. Simultaneous selection of the auxiliary poles and of the fixed parts of the controller

Despite the coupling between the shaping of the output sensitivity function and the input sensitivity function, it is possible, in practice, to deal with these two problems almost independently.

In particular, it is useful to remember that at frequencies where $|S_{yp}|$ is close to 1 (0dB), $|S_{up}|$ is close to 0 (<-60dB).

Automatic methods using convex optimization techniques (Langer and Landau 1999; Langer and Constantinescu 1999; Adaptech 1998a) are available in order to solve the shaping problem[14]. However, in most of practical situations, it is relatively easy to calibrate the sensitivity functions using the tools previously indicated, and taking into account the properties of the sensitivity functions.

The iterative procedure that will be presented next is very efficient and helps to understand the operation of the controller. However, the use of a CACSD tool like *ppmaster*[15] (MATLAB®) (Prochazka and Landau 2001) will considerably accelerate the procedure.

Shaping the Output Sensitivity Function.
The shaping objective can be summarized as follows: given a desired attenuation band, select the closed loop poles and the fixed parts $H_R(q^{-1})$ and $H_S(q^{-1})$ in order to "flatten" $|S_{yp}(e^{-j\omega})|$ outside the attenuation band, and to reduce $|S_{yp}(e^{-j\omega})|_{max}$ while satisfying Property 2.

Often the selection of auxiliary poles and of $H_R(q^{-1})$ and $H_S(q^{-1})$ is sufficient for matching the performance and robustness specifications. However, for a fine tuning, it may be also necessary to select simultaneously the fixed part $H_S(q^{-1})$ and the auxiliary poles. (see Property 7: second-order pole-zero filter).

The iterative procedure for the shaping of the output sensitivity function can be summarized as follows[16]:

---

[14] To apply convex optimization techniques, the Youla-Kucera parametrization of the controller is used. See Appendix E.

[15] Available on the book website (*http://landau-bookic.lag.ensieg.inpg.fr*).

[16] The principles of the methodology for shaping the output sensitivity function can be extended to the shaping of other sensitivity functions.

Step I

- Selection of $P(q^{-1})$ and fixed parts $H_R(q^{-1})$ and $H_S(q^{-1})$ from the performance specifications. (Example: zero steady-state error will require the introduction of $H_S(q^{-1}) = 1 - q^{-1}$, opening the loop at a certain frequency will require $H_R(q^{-1}) = 1 + \alpha q^{-1} + q^{-2}$)
- Controller computation
- Analysis of the resulting output sensitivity function.

If the upper template corresponding to the desired modulus margin $\Delta M = 0.5$ and the delay margin $\Delta \tau = T_s$ are violated, one can distinguish three different solutions:

1. The maximum of the modulus of the sensitivity function is at high frequencies (the resulting controller is often unstable as a consequence of Property 2, since the area between the modulus of the sensitivity function and the $0\ dB$ axis is positive).
2. A local maximum of the modulus of the sensitivity function is located in a frequency region close to the attenuation band.
3. The modulus of the sensitivity function presents a maximum both at low and high frequencies.

Case 1 and 3

Step II
One introduces auxiliary poles at high frequencies:

$$P_F(q^{-1}) = (1 + p_1 q^{-1})^{n_F} \quad ; \quad -0.05 \geq p_1 \geq -0.5$$

with increasing values of $|p_1|$ starting from 0.05.

The number of auxiliary poles that can be introduced without increasing the size of the controller is given by

$$n_{P_F} \leq n_P - n_{P_D} \quad ; \quad n_P = (deg\ P)_{max} \quad ; \quad n_{P_D} = deg\ P_D$$

The value of $|p_1|$ is increased in order to avoid a violation of the template of the sensitivity function at high frequencies (*i.e.* over $0.25\ f_s$ or $0.30\ f_s$).

Usually, the maximum of the modulus of the sensitivity function will move towards low frequencies. If this maximum is above the template we are in case 2 which will be discussed next.

Case 2

Step II – Use of second-order pole-zero resonant filters H_{S_i}/P_i.

One identifies first the frequency region where $|S_{yp}|$ is above the "upper template". One finds the value of the frequency at which the maximum of $|S_{yp}|$ occurs and the attenuation which should be introduced in order to be below the upper template. Then one computes the corresponding filter H_{S_i} / P_i on the basis of these specifications. The selection of the damping of the denominator depends to some extent upon the width of the frequency region where $|S_{yp}|$ is above the template.

Some Hints

- If the modulus of the sensitivity function presents a too strong minimum at high frequencies, or in the central frequency region, this can be raised by a few dB by adding a pair of complex auxiliary poles placed at the frequency corresponding to the minimum
- If the attenuation band increases, then one should impose slower dominant poles
- If the attenuation band is reduced as a consequence of the introduction of the auxiliary poles, then one can either raise the natural frequency of the dominant poles, or introduce a further zero-pole filter around the frequency corresponding to the desired attenuation band.

In order to match the constraints in the high frequency region corresponding to the imposed delay margin, one can either use auxiliary poles or impose the opening of the loop at $0.5 f_s$ ($H_R(q^{-1}) = 1 + q^{-1}$). In practice, a simultaneous application of both solutions can be considered.

Step III
One iterates around the values of the dominant and auxiliary poles selected in step I and II in order to get the best result in terms of robustness and performance constraints matching.

Shaping the Input Sensitivity Function
The shaping of the input sensitivity function is usually performed after the shaping of the output sensitivity function. If the value of the modulus of the input sensitivity function is not enough low at high frequencies (frequencies close to 0.5 f_s), one introduces a fixed part $H_R(q^{-1})$ of the form

$$H_R(q^{-1}) = (1 + \beta q^{-1})^n \qquad 0.5 < \beta < 1 \qquad n = 1,2$$

This will have, in general, a weak interaction with the shaping of the output sensitivity function.

In the frequency regions where there are uncertainties upon the plant model, $|S_{up}|$ should be low (upper bound defined by the template). If the resulting sensitivity function does not fulfill the specifications, one has to introduce a pole-zero resonant filter H_{R_i} / P_{F_i} . In order to match the template imposed on $|S_{up}|$ the resonance frequency of the filter is chosen close to that corresponding to the local maximum of $|S_{up}|$, and an appropriate damping is selected in order to provide the required attenuation.

Once the shaping of $|S_{up}|$ is achieved, one should go back to check if $|S_{yp}|$ still match the template (an appropriate CACSD like *ppmaster* allows to shape simultaneously both sensitivity functions).

3.6.5 Shaping of the Sensitivity Functions: Example 1

The considered plant model is characterized by

$$A(q^{-1}) = 1 - 0.7 \, q^{-1} \quad ; \quad B(q^{-1}) = 0.3 \, q^{-1} \quad ; \quad d = 2 \; ; \; T_s = 1s$$

An integrator is imposed in the controller. One considers the roots of the polynomial obtained from the discretization of a second-order continuous time system with $\omega_0 = 1 \, rad/s$, and $\zeta = 0.9$ as the imposed dominant poles for the closed loop. The controller is designed by means of *pole placement*. The output sensitivity function corresponding to this design is shown in Figure 3.37 (curve A). It crosses the standard robustness template defined by $\Delta M = - \, 6dB$ and $\Delta\tau = T_s$. Both obtained margins (modulus and delay) are lower than specifications (see Table 3.5). The upper frequency of the attenuation band is at 0.058 Hz.

The objective is to obtain the same attenuation band but with

$$\Delta M \geq - \, 6dB \; (|S_{yp}|_{max} \leq 6dB) \; and \; \Delta\tau \geq T_s$$

First we will add auxiliary real poles. From Equations 3.3.22 and 3.3.23 it results that one can assign a number of closed loop poles equal to:

$$n_P = deg \, P(q^{-1}) \leq n_A + n_{HS} + n_B + n_{HR} + d - 1$$

without increasing the size of the controller. Since in this example $n_A = 1$, $n_B = 1$, $n_{HS} = 1$, $d = 2$, one can assign four poles. Two poles have been already assigned as dominant poles and therefore we will add two auxiliary poles of the form

$$P_F(q^{-1}) = (1 - 0.4 \, q^{-1})^2$$

The resulting sensitivity function is shown in Figure 3.36 (curve B). The constraints on the modulus margin and the delay margin are satisfied but the attenuation band has been reduced (0.045 Hz instead of 0.058 Hz).

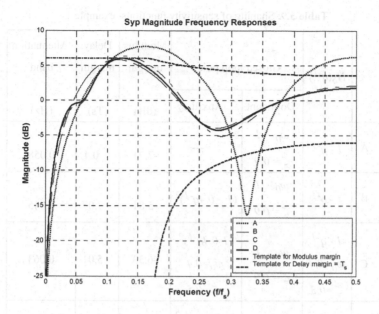

Figure 3.36. Output sensitivity functions corresponding to different RST controllers (see details in Table 3.5)

In order to increase the attenuation in this frequency region without influencing high frequency regions, one chooses a pole-zero filter H_S/P'_F centered at $\omega_0=0.4$ rad/s ($f=0.064$ Hz) with H_S resulting from the discretization of a second-order continuous-time system with $\omega_0 = 0.4$ rad/s and $\zeta = 0.3$, and P'_F resulting from the discretization of a second-order continuous-time system with $\omega_0 = 0.4$ rad/s and $\zeta = 0.5$ (we are below 0.17 f_s and the direct design of the discrete filter is possible). Since $\zeta_{num} = 0.3$ and $\zeta_{den} = 0.5$, an attenuation will result at 0.064 Hz. The corresponding sensitivity function is shown in Figure 3.36 (curve C). One observes that the attenuation band matches the specifications but the maximum value of $|S_{yp}|$ is slightly higher than 6 dB. In order to match this constraint the value of the real auxiliary pole is increased from 0.4 to 0.44. The final sensitivity function is shown in Figure 3.36 (curve D).

3.6.6 Shaping of the Sensitivity Functions: Example 2[17]

The considered plant model is characterized by

$$A(q^{-1}) = 1 - q^{-1} \;\; ; \;\;\; B(q^{-1}) = 0.5\, q^{-1} \;\; ; \;\;\; d = 2 \;\; ; \;\; T_s = 1s$$

---

[17] This example and the corresponding specifications are related to the continuous casting of steel.

Table 3.9. Shaping of sensitivity functions- example 1

$H_S(q^{-1})$	Closed loop poles		Modulus margin (dB)	Delay margin (Ts)	Attenuation band (Hz)
	Dominant	Auxiliary			
A $\quad 1 - q^{-1}$	$\omega_0 = 1$ $\zeta = 0.9$	---	-7.71	0.4	0.058
B $\quad 1 - q^{-1}$	$\omega_0 = 1$ $\zeta = 0.9$	$(1 - 0.4q^{-1})^2$	-5.81	3.07	0.045
C $\quad \begin{array}{l}1 - q^{-1}\\ \omega_0 = 0.4\\ \zeta = 0.3\end{array}$	$\omega_0 = 1$ $\zeta = 0.9$	$\begin{array}{l}(1-0.4q^{-1})^2\\ \omega_0 = 0.4\\ \zeta = 0.5\end{array}$	-6.33	5.01	0.063
D $\quad$ idem	$\omega_0 = 1$ $\zeta = 0.9$	$\begin{array}{l}(1-0.44q^{-1})^2\\ \omega_0 = 0.4\\ \zeta = 0.5\end{array}$	-5.99	5.34	0.060

The plant has an integrator behavior with delay. The plant is subject to two types of disturbances:

- Low frequency disturbances that should be attenuated
- A sinusoidal disturbance at 0.25 Hz that should not be compensated by the action of the controller

The block diagram related to the plant model is shown in Figure 3.37.

The robustness and performance specifications are as follows:

1. No attenuation of the sinusoidal disturbance at $f = 0.25$ Hz ($|S_{yp}| = 0$ dB at *0.25 Hz*)
2. Attenuation band at low frequencies (frequency region corresponding to $|S_{yp}| < 0dB$): 0 to 0.03 Hz
3. Disturbances amplification at 0.07 Hz less than 3 dB ($|S_{yp}| < 3$ dB at *0.07 Hz*)
4. Modulus margin $\geq$ -6 dB ($|S_{yp}|_{max} \leq 6$ dB)
5. Delay margin $\geq$ 1s (T_s)
6. No integrator in the controller

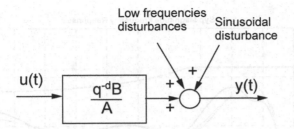

Figure 3.37. Block diagram for plant model and disturbances

The presence of an integrator has not been imposed because the plant model has already an integrator behavior.

One starts by designing the fixed parts of the controller, $H_R(q^{-1})$ and $H_S(q^{-1})$. In order to obtain $|S_{yp}| = 0dB$ at $0.25\ Hz = 0.25\ f_s$, one should introduce $H_R(q^{-1})$ with a pair of undamped complex zeros at this frequency. This is obtained by

$$H_R(q^{-1}) = 1 + \beta q^{-1} + q^{-2}$$

with

$$\beta = -2cos\ (\omega T_s) = -2\ cos\ (2\pi f/f_s)$$

From the condition

$$H_R(e^{-j\omega})\Big|_{\omega=\omega_s/4} = 0$$

it results that

$$\beta = -2cos\ (2\pi.\ 0.25) = 0$$

and then

$$H_R(q^{-1}) = 1 + q^{-2}$$

Since there is no integrator imposed in the controller, as a first choice one chooses $H_S(q^{-1}) = 1$. One selects for the dominant poles a polynomial $P(q^{-1})$ resulting from the discretization of a second-order continuous-time system with $\omega_0 = 0.628\ rad/s$ ($f_0 = 0.1\ Hz$) and $\zeta = 0.9$.

The corresponding sensitivity function is shown in Figure 3.38 (curve A).

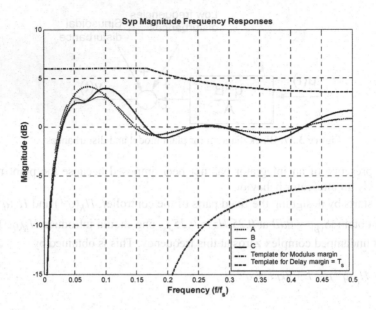

Figure 3.38. Output sensitivity functions corresponding to different RST controllers (see details in Table 3.6)

One observes that the specifications at 0.07 Hz are not satisfied. On the other hand one can remark that $|S_{yp}|$ is equal to 1 (0dB) for $f = 0.25 f_s = 0.25$ Hz, as required.

In order to reduce the modulus of the output sensitivity function at $f = 0.07$ Hz, one considers the introduction of a narrow-band pole-zero discrete filter H_S / P_F centered at $\omega_0 = 0.44$ rad/s ($f_0 = 0.07$ Hz). Since the frequency f_0 is below $0.17 f_s$, one can directly design the discrete-time filter. One chooses for H_S a damping $\zeta = 0.3$ and for P_F a damping $\zeta = 0.4$, corresponding to a filter attenuation of about 2.4 dB around ω_0. Using this filter (that is H_S and P_F) one obtains the sensitivity function shown in Figure 3.38 (curve B). The attenuation introduced by this filter at 0.07 Hz is sufficient ($|S_{yp}|_{0.07Hz} = 2.6$ dB) but the width of the attenuation band is smaller than the specified one. In order to correct that, it is sufficient to increase the natural frequency of the dominant poles. With $\omega_0 = 0.9$ rad/s (instead of 0.628) one obtains the sensitivity function shown in Figure 3.38 (curve C). All the specifications are fulfilled. The results are summarized in Table 3.10.

Note that the choice of H_S / P_F is not critical. The specifications can also be satisfied with a discrete-time filter H_S / P_F centered at $\omega_0 = 0.5$ rad/s with $\zeta_{num} = 0.35$ and $\zeta_{den} = 0.45$.

The same design steps will be followed for the case when an integrator is introduced in the controller.

Table 3.10. Shaping of the sensitivity functions - Example 2

		Closed loop poles		Attenuation band (Hz)	Modulus margin (dB)	Delay margin (Ts)	$\|S_{yp}\|$ 0.07 Hz (dB)
H_R	H_S	Dominants	Auxiliaries				
A	$1+q^{-2}$	---	$\omega_0 = 0.628$ $\zeta = 0.9$		0.03	-4.12	6.52	4.11
B	$1+q^{-2}$	$\omega_0 = 0.44$ $\zeta = 0.3$	$\omega_0 = 0.628$ $\zeta = 0.9$	$\omega_0 = 0.44$ $\zeta = 0.4$	0.026	-3.06	7.61	2.6
C	$1+q^{-2}$	idem	$\omega_0 = 0.9$ $\zeta = 0.9$		0.03	-3.94	6.62	2.6

3.7 Concluding Remarks

In this chapter several digital control designs in a deterministic environment have been presented. All the digital controllers have a three-branched structure (RST), corresponding to a control law of the form

$$S(q^{-1})\, u(t) + R(q^{-1})\, y(t) = T(q^{-1})\, y^*(t+d+1)$$

where u is the control input, y is the output of the plant, y^* is the desired tracking trajectory and d is the time delay.

The design of the controller involves essentially two stages:

1. Computation of the polynomials $S(q^{-1})$ and $R(q^{-1})$ in order to match the desired regulation performances
2. Computation of the polynomial $T(q^{-1})$ in order to approach (or to match) the desired tracking performances

The digital controllers are perfectly suited for the control of plants characterized by high order models with time delay and/or resonant modes. The controller complexity (*i.e.* the degrees of the polynomials $R(q^{-1})$, $S(q^{-1})$, $T(q^{-1})$) depends upon the complexity of the polynomials of the plant model transfer function. The discrete-time plant model used for design can be either obtained directly by an identification technique, or by computation from the continuous-time model.

The *pole placement* control strategy is applied to plants having discrete-time models both with stable and unstable zeros. It makes possible to match the desired regulation performance and the desired tracking performance filtered by the plant zeros.

The *tracking and regulation with independent objectives* allows to match perfectly the desired tracking and regulation performances, but, in order to be applied, the pulse transfer function of the discrete-time plant model must have stable zeros.

The *internal model control* strategy is a particular case of the pole placement in which the plant model poles are chosen as closed loop system poles. This strategy can only be applied to enough damped stable systems. The closed loop dynamics is not faster than the open loop dynamics.

The digital PID controllers are directly designed by *pole placement* on the basis of discrete-time models of the plants to be controlled. This technique can be applied to plants characterized by discrete-time models of lower order ($n_{max} \leq 2$). Two structures of digital PID controllers have been examined. The difference is only in the choice of the polynomial $T(q^{-1})$. It is advisable to use the choice $T(q^{-1})$ = $R(1)$ (corresponding to the digital PID 2), which offers better tracking performances. The digital controllers designed by the methods presented in this chapter implement *a predictive control* in the time domain. They implicitly contain a *predictor* of the plant to be controlled.

Independently of the design method used, it is required to verify the *robustness margins* of the closed loop (mainly the modulus margin and the delay margin). Taking into account simultaneously performance and robustness margins may require the shaping of the sensitivity functions (output and input sensitivity functions), by imposing pre-specified fixed parts in the controller and choosing appropriate closed loop system poles.

An iterative methodology for the shaping of the sensitivity functions has been presented in Section 3.6.4. Moreover, it is possible to obtain in an automatic way the controller, fulfilling the various constraints on the sensitivity functions, by means of a convex optimization procedure (see Langer and Landau 1999; Adaptech 1998b).

3.8 Notes and References

For different types of digital PID controllers and their computation by means of the pole placement see:

Åström K.J., Wittenmark B. (1997) Computer Controlled Systems Theory and Design, 3rd edition, Prentice Hall, N.J., U.S.A.

Åström K.J., Hägglund I. (1995) PID Controllers Theory, Design and Tuning, 2nd edition ISA, Research Triangle Park, N.C., U.S.A.

The pole placement and the solution of the Bezout identity is discussed in:

Goodwin G.C., Sin K.S. (1984) Adaptive Filtering Prediction and Control, Prentice-Hall, Englewood Cliffs, N.J.

Kailath T. (1980) Linear systems, Prentice Hall, Englewood Cliffs, N.J.

For the solution of the Bezout equation by means of recursive least squares see:

Lozano R., Landau I.D. (1982) Quasi-direct adaptive control for nonminimum phase systems, Transactions A.S.M.E., Journal of D.S.M.C., vol. 104, n°4, pp. 311-316, December.

For the numerical solutions of the Bezout equation see:

Press W.H., Vetterling W.T., Teukolsky S., Flanery B. (1992) Numerical recipes in C (The art of scientific computing),2nd edition, Cambridge University Press, Cambridge, Mass.

For the links between the pole placement and the optimal control with quadratic criterion see (Åström and Wittemark 1997).

Tracking and regulation with independent objectives is discussed in:

Landau I.D., Lozano R. (1981) Unification of Discrete-Time Explicit Model Reference Adaptive Control Designs, Automatica, vol. 12, pp. 593-611.

The interpretation and the design of RST controllers in the time domain are discussed in appendix B.

Robust control of systems with delay and links with the Smith predictor and the internal model control are discussed in:
Landau I.D. (1995) Robust digital control of systems with time delay (the Smith predictor revisited), Int. J. of Control, vol. 62, pp. 325-347.

For properties of the integral of the sensitivity functions in discrete-time see:

Sung H.K., Hara S. (1988) Properties of sensitivity and complementary sensitivity functions in single-input, single-output digital systems, Int. J. of Cont., vol. 48, n°6, pp. 2429-2439.

For examples about the iterative design methodology using pole placement with shaping of the sensitivity functions see:

Landau I.D., Langer J., Rey D., Barnier J. (1996) Robust control of a 360° flexible arm using the combined pole placement / sensitivity function shaping method, IEEE Trans. on Control Systems Tech., vol. 4, no. 4, pp. 369-383.
Landau I.D., Karimi A. (1998) Robust digital control using pole placement with sensitivity function shaping method, Int. J. of Robust and Nonlinear Control, vol. 8, pp. 191-210.
Prochazka H., LandauI.D. (2003) Pole placement with sensitivity function shaping using 2nd order digital notch filters, Automatica, Vol. 39, 6, pp. 1103-1107.

For an "automatic" solution based on convex optimization for designing robust digital controllers by means of the pole placement with shaping of the sensitivity functions see:

Langer J., Landau I.D. (1999) Combined pole placement / sensitivity function shaping method using convex optimization criteria, Automatica, vol. 35, no. 6, pp. 1111-1120.
Langer J., Constantinescu A. (1999) Pole placement design using convex optimization criteria for the flexible transmission benchmark, European Journal of Control, vol. 5, no. 2-4, pp. 193-207.
Adaptech (1998) Optreg – Software for automated design of robust digital controllers using convex optimization (for MATLAB®), Adaptech, St. Martin d'Hères, France.

4

Design of Digital Controllers in the Presence of Random Disturbances

A large number of systems are subject to disturbances of a random nature. In the first part of this chapter models suitable for the representation of these stochastic disturbances (the ARMAX models) will be presented and their properties analyzed. Three design methods will be presented: 1. minimum variance tracking and regulation (which minimizes the mean-square difference between the reference and the controlled variable); 2. the approximation of the minimum variance tracking and regulation by means of the pole placement (for the case of systems with unstable zeros); 3. generalized minimum variance tracking and regulation.

4.1 Models for Random Disturbances

4.1.1 Description of the Disturbances

First consider the basic deterministic disturbances, namely the Dirac pulse, the step and the ramp, represented in Figure 4.1.

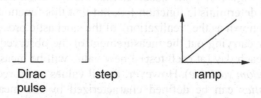

Dirac pulse   step   ramp

Figure 4.1. Main deterministic disturbances

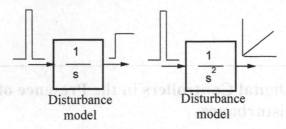

Figure 4.2. Deterministic disturbance models

Note that the step and the ramp may be described as resulting from the passing of the Dirac pulse through a filter, as indicated in Figure 4.2. The corresponding filters are known as "disturbance models". Any deterministic disturbance may be obtained by passing a Dirac pulse through a "disturbance model" (filter) having an appropriate structure. *Knowledge of the "disturbance model" is equivalent to knowledge of the disturbance.*

What follows is an attempt to extend this concept of the "disturbance model" to the description of random disturbances.

By *random or stochastic disturbances*, is meant those disturbances, which cannot be described in a deterministic way, given as they are not reproducible.

To provide an example of a random (stochastic) process, one can consider the evolution of the controlled output of a plant in regulation on a significant horizon (one day) and during several tests (several days). This is illustrated in Figure 4.3.

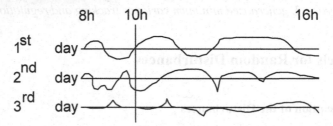

Figure 4.3. Recording of a controlled variable in regulation

By examining Figure 4.3, one observes that the evolution during one day may be described by a deterministic function f(t), but that this function will be different every day (*f(t)* is known as the "realization" of the stochastic process).

If the time for carrying out the measurement of the observed variable is fixed (e.g. at 10 a.m.), each day (at each test) a new value will be measured (this it what is known as a *random variable*). However, for all values measured every day at the same time, *statistics* can be defined characterized by the mean value and the variance of the measurements. The probabilities of the occurrence of different values may be defined as well.

The stochastic process (partially) represented in Figure 4.3, is dependent on the time (during a day) and on the experiment (first, second...fourth day).

More formally, a *stochastic process* may be described as a function $f(t, \xi)$ where t represents the time and belongs to the set T of real variables, and ξ

represents the stochastic variable (the outcome of an experiment), which belongs to a *probability space* S^1. For a given $\xi = \xi_0$, the function $f(t, \xi_0)$ is a regular time function called a *realization*. For fixed $t = t_0$ the function $f(t_0, \xi)$ is a *random variable*. The argument ξ is often omitted.

If the stochastic (random) process is *ergodic*, the statistics related to an experiment (in our example over one day) are significant, *i.e.* the result obtained is identical to that obtained from measurements taken on several experiments when the time is maintained constant (at the same time of the day). If, in addition, the stochastic process is *gaussian*, the knowledge of the mean value and of the variance allows the probability of occurrence of a given value to be specified (Gauss's bell – see Appendix A).

In practice, the majority of random disturbances occurring in automatic control systems may be accurately described as a *discrete-time white noise* passed through a filter. This *discrete-time white noise* is a random signal having an energy uniformly distributed at all frequencies between 0 and $0.5 f_s$. Note that the discrete-time white noise has a physical realization, since it is a finite energy signal (the frequency band is finite), whereas the continuous-time white noise does not correspond to a physical reality since the energy is constant over an infinite frequency range (infinite energy signal).

The filters that will constitute the *random disturbance models* will modify the frequency spectrum of the energy distribution of the white noise in order to obtain a distribution corresponding to the frequency distribution of energy of the various random disturbances encountered.

The *white noise* has, in the random case, the same role as the Dirac pulse in the deterministic case. It constitutes the *fundamental generator signal*.

The *gaussian discrete-time white noise* will henceforward be considered as the generator signal. This is a sequence of independent equally distributed gaussian random variables of zero mean value and variance σ^2. This sequence will be noted $\{e(t)\}$ and will be characterized by the parameters $(0, \sigma)$, in which the first term indicates the mean value and σ is the standard deviation (square root of the variance). A part of such a sequence is represented in Figure 4.4.

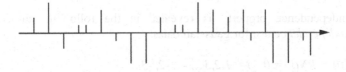

Figure 4.4. Discrete-time white noise sequence

The mean value (or expectation) is given by

$$M.V. = E\{e(t)\} = \lim_{N \to \infty} \frac{1}{N} \sum_{t=1}^{N} e(t) = 0 \qquad (4.1.1)$$

and the variance will be given by

$$var = E\{e^2(t)\} = \lim_{N \to \infty} \frac{1}{N} \sum_{t=1}^{N} e^2(t) = \sigma^2 \qquad (4.1.2)$$

(note that sequence *{e(t)}* is an ergodic process, see above).

The *autocorrelation or covariance function R(i)* is defined by the following expression:

$$R(i) = E\{e(t)e(t-i)\} = \lim_{N \to \infty} \frac{1}{N} \sum_{t=1}^{N} e(t)e(t-i) \qquad (4.1.3)$$

or, in other words, by the product between the sequence *{e(t)}* and the same sequence shifted by *i*-steps.

Note that

$$R(0) = var = \sigma^2 \qquad (4.1.4)$$

The *normalized covariance (or autocorrelation)* is defined as

$$RN(i) = \frac{R(i)}{R(0)} \qquad (RN(0) = 1) \qquad (4.1.5)$$

In the case of the gaussian white noise, which is an ergodic process, since this is a sequence of independent random variables and taking into consideration the ergodic nature of the process, it results that the knowledge of *e(t)* does not allow an approximation of *e(t +1), e(t + 2)*... to be predicted, the best prediction being *0*.

This independence property is revealed in the following autocorrelation property (*independence test* for gaussian data):

$$R(i) = RN(i) = 0 \quad i = 1,2,3..., -1, -2, -3.... \qquad (4.1.6)$$

If the evolution of the normalized covariance for the white noise is plotted, the curve shown in Figure 4.5 is obtained[2].

---

[2] This whiteness test is used even when a finite length sequence is available. In this case, $|RN(i)| \neq 0$, $i = 1,2,3,...$ but they must be smaller than a specified value. The whiteness test for a sequence of length N is discussed in Chapter 6, Section 6.2.

Since $E\{e(t)\} = const.$ and $E\{e(t)\ e(t\text{-}i)\}$ is only a function of i, the *gaussian discrete-time white noise* sequence has the properties of a weakly stationary stochastic process (in fact it is a real stationary stochastic process since the properties of the sequence $\{e(t)\}$ are the same as those of the sequence $\{e(t+\tau)\}$). The knowledge of the covariances $R(i)$ for a weakly stationary stochastic process makes it possible to compute the energy distribution in the frequency domain, known as the *spectral density function*.

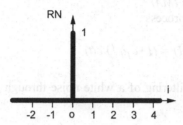

Figure 4.5. Normalized autocorrelations of the white noise

This is given by (discrete Fourier transform of the covariance function)

$$\phi(\omega) = \frac{1}{2\pi} \sum_{i=-\infty}^{\infty} R(i) e^{-ji\omega}$$

Since in the case of the discrete-time white noise all the $R(i) = 0$ for $i \neq 0$, it results that the spectral density of the discrete-time white noise is constant and equal to

$$\phi_e(\omega) = \frac{R(0)}{2\pi} = \frac{\sigma^2}{2\pi}$$

The spectral density of the white noise is represented in Figure 4.6. A uniform energy distribution between 0 and $0.5\,f_s$ is observed.

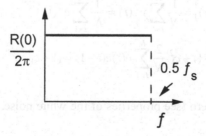

Figure 4.6. Spectral density of the discrete-time white noise

4.1.2 Models of Random Disturbances

As it has already been mentioned in Section 4.1, different types of random disturbances, whose spectral density can be approximated by a rational function of the frequency, can be considered as resulting from the filtering of a white noise through a shaping filter. Several types of processes thus obtained will be examined.

"Moving Average" Process (MA)
Consider, for example, the process

$$y(t) = e(t) + c_1 e(t-1) = (1+c_1 q^{-1})\, e(t) \qquad (4.1.7)$$

which corresponds to the filtering of a white noise through a filter $(1+c_1 q^{-1})$, as shown in Figure 4.7a.

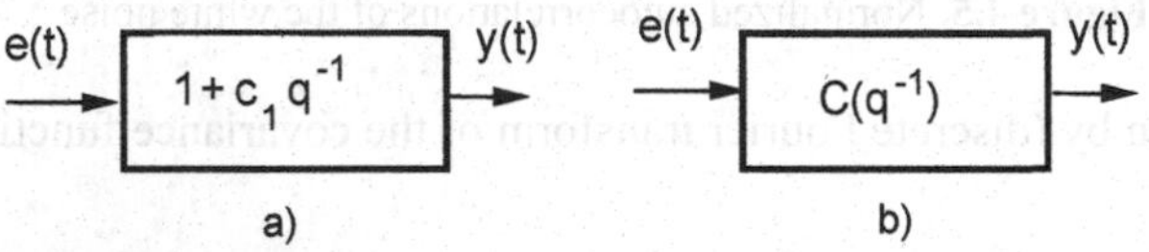

a)   b)

Figure 4.7a,b. Generation of a "Moving Average" random process: **a** first order; **b** general case

The mean value of $y(t)$ is [3]

$$V.M. = E\{y(t)\} = \frac{1}{N}\sum_{t=1}^{N} y(t) = \frac{1}{N}\sum_{t=1}^{N} e(t) + c_1 \frac{1}{N}\sum_{t=1}^{N} e(t-1) = 0 \qquad (4.1.8)$$

The variance of the process $y(t)$ is

$$R_y(0) = E\{y^2(t)\} = \frac{1}{N}\sum_{t=1}^{N} y^2(t) = \frac{1}{N}\sum_{t=1}^{N} e^2(t) +$$

$$c_1^2 \frac{1}{N}\sum_{t=1}^{N} e^2(t-1) + c_1^2 \frac{2}{N}\sum_{t=1}^{N} e(t)e(t-1) = (1+c_1^2)\sigma^2 \qquad (4.1.9)$$

since the third term is zero (see properties of the white noise, Equations 4.1.3 and 4.1.6).

---

[3] In the following we omit $\lim_{N\to\infty}$ but one should recall that all formulas are rigorously valid only for a large value of N ($\to \infty$).

From Equation 4.1.7, one obtains by shifting

$$y(t-1) = e(t-1) + c_2\, e(t-2) \tag{4.1.10}$$

which enables $R_y(1)$ to be computed:

$$R_y(1) = E\{y(t)y(t-1)\} = \frac{1}{N}\sum_{t=1}^{N} y(t)y(t-1) = \frac{1}{N}c_1\sum_{t=1}^{N} e^2(t) = c_1^2\sigma^2 \tag{4.1.11}$$

all the other terms being zero (expectations of products of shifted white noise sequences).

In the same way one verifies that

$$R_y(2) = 0\; ;\; R_y(3) = 0 \ldots R_y(n) = 0 \tag{4.1.12}$$

The plot of the normalized autocorrelations for this first-order "moving average" process is represented in Figure 4.8.

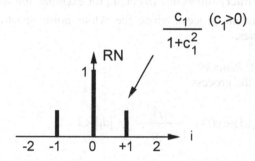

Figure 4.8. The normalized autocorrelation functions for a first-order "moving average" process

The general form of a "moving average" process is

$$y(t) = e(t) + \sum_{i=1}^{n_c} c_i e(t-i) = C(q^{-1})e(t) \tag{4.1.13}$$

where

$$C(q^{-1}) = 1 + \sum_{i=1}^{n_c} c_i q^{-i} = 1 + q^{-1}C^*(q^{-1}) \tag{4.1.14}$$

The corresponding filter representation is given in Figure 4.7b.

The "moving average" processes are characterized by the property

$$R(i) = 0 \quad i \geq n_C+1 \quad ; \quad i \leq -(n_C+1) \tag{4.1.15}$$

The spectral density is given by the formula

$$\phi_y(\omega) = C(e^{j\omega})C(e^{-j\omega})\frac{\sigma^2}{2\pi} = \left|C(e^{j\omega})\right|^2 \frac{\sigma^2}{2\pi} \tag{4.1.16}$$

$\sigma^2/2\pi$ being the spectral density of the white noise (noted $\phi_e(e^{j\omega})$).

This allows one to obtain the general relation between the spectrum of the input signal (white noise) of the generator filter and the spectrum of the moving average process:

$$\phi_y(z) = C(z)\, C(z^{-1})\, \phi_e(z) \tag{4.1.17}$$

the frequency distribution being obtained for $z = e^{j\omega}$.

From Equations 4.1.16 and 4.1.17 it can be deduced that the knowledge of $C(q^{-1})$ (the generator filter) allows one to obtain, for example, the spectral energy distribution of the random process, since the white noise spectral density is constant at all frequencies.

"Auto-regressive" (AR) Process
Consider, for example, the process

$$y(t) = -a_1 y(t-1) + e(t) = \frac{e(t)}{1 + a_1 q^{-1}} \quad |a_1| < 1 \tag{4.1.18}$$

which corresponds to a white noise passed through a stable filter $1/(1+a_1\, q^{-1})$ as is shown in Figure 4.9a. The general form of an "auto-regressive" process is given by

$$y(t) = -\sum_{i=1}^{n_A} a_i y(t-1) + e(t) \tag{4.1.19}$$

which is also written as

$$A(q^{-1})\, y(t) = e(t) \tag{4.1.20}$$

in which

$$A(q^{-1}) = 1 + \sum_{i=1}^{n_A} a_i q^{-i} = 1 + q^{-1} A^*(q^{-1}) \qquad (4.1.21)$$

is a polynomial with all its roots inside the unit circle ($A(z^{-1}) = 0 \Rightarrow |z| < 1$).

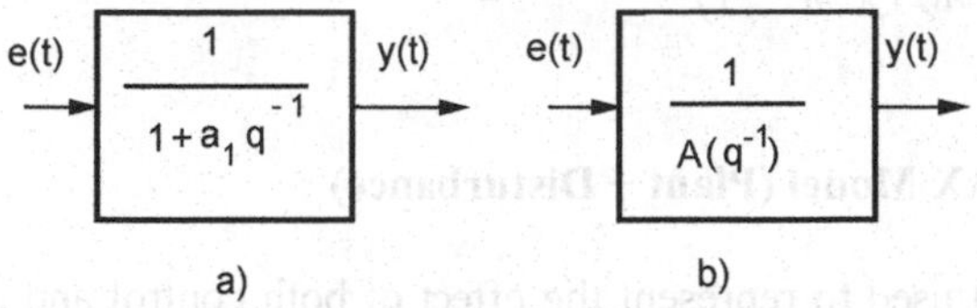

a)   b)

Figure 4.9a,b. Generation of an "Auto-Regressive" process: **a)** 1st order ; **b)** general case

The spectral density of the auto-regressive process is given by the expression

$$\phi_y(z) = \frac{1}{A(z)} \frac{1}{A(z^{-1})} \phi_e(z) \qquad (4.1.22)$$

in which $\phi_e(z) = \sigma^2/2\pi$ and the frequency distribution is given by

$$\phi_y(\omega) = \phi_y(z)\big|_{z=e^{j\omega}} \qquad (4.1.23)$$

"ARMA" Process (Auto-Regressive Moving Average)
This process is obtained by passing a white noise through a stable filter with poles and zeros as is illustrated in Figure 4.10.

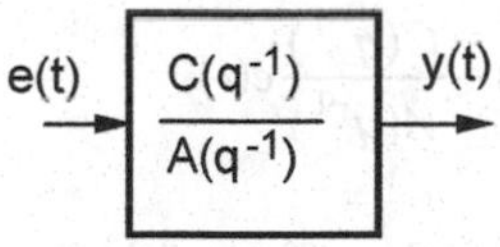

Figure 4.10. Generation of an ARMA random process

This process is described, in the general case, by

$$y(t) = -\sum_{i=1}^{n_A} a_i y(t-i) + \sum_{i=1}^{n_C} c_i e(t-i) + e(t) \qquad (4.1.24)$$

or also in the form

$$A(q^{-1}) y(t) = C(q^{-1}) e(t) \qquad (4.1.25)$$

where $A(q^{-1})$ is a polynomial (having all its roots inside of the unit circle) given by Equation 4.1.21 and $C(q^{-1})$ is given by Equation 4.1.14.

The spectral density of the ARMA process is given by

$$\phi_y(z) = \left(\frac{C(z)}{A(z)}\right)\left(\frac{C(z^{-1})}{A(z^{-1})}\right)\phi_e(z) \tag{4.1.26}$$

4.1.3 The ARMAX Model (Plant + Disturbance)

This is the model used to represent the effect of both control and disturbances on the plant output. ARMAX means ARMA process with *exogenous* (external) input, which in our case is $u(t)$. The generation of the ARMAX process is illustrated in Figure 4.11.

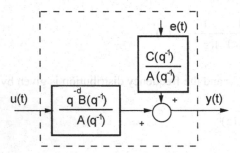

Figure 4.11. Generation of the ARMAX random process

The disturbed output of the plant $y(t)$ is written as

$$y(t) = \frac{q^{-d}B(q^{-1})}{A(q^{-1})}u(t) + \frac{C(q^{-1})}{A(q^{-1})}e(t) \tag{4.1.27}$$

in which the first term represents the effect of the control and the second term the effect of the disturbance. By multiplying both sides of Equation 4.1.27 by $A(q^{-1})$, one obtains

$$A(q^{-1})\,y(t) = q^{-d}B(q^{-1}) + C(q^{-1})\,e(t)\; ; (B(q^{-1}) = q^{-1}B^*(q^{-1})) \tag{4.1.28}$$

which, taking into account the expressions of $A(q^{-1})$, $B(q^{-1})$ and $C(q^{-1})$, is also written as

$$y(t+1) = -\sum_{i=1}^{n_A} a_i y(t+1-i) + \sum_{i=1}^{n_B} b_i u(t+1-d-i) + \sum_{i=1}^{n_C} c_i e(t+1-i) \tag{4.1.29}$$

$$+ e(t+1) = -A^*(q^{-1})y(t) + B^*(q^{-1})u(t-d) + C(q^{-1})e(t+1)$$

The question that may arise concerns the presence of the denominator $A(q^{-1})$, both in the transfer function of the plant and in the transfer function of the generator filter of the disturbance (see Equations 4.1.27). Is this a loss of generality as compared to the case where the two denominators are different, as shown in Figure 4.12 ?

In this case, output $y(t)$ is given by

$$y(t) = \frac{q^{-d} B_1(q^{-1})}{A_1(q^{-1})} u(t) + \frac{C_2(q^{-1})}{A_2(q^{-1})} e(t)$$ (4.1.30)

Reducing to the same denominator, one obtains

$$y(t) = \frac{q^{-d} B_1 A_2}{A_1 A_2} u(t) + \frac{C_2 A_1}{A_1 A_2} e(t) = \frac{q^{-d} B}{A} u(t) + \frac{C}{A} e(t)$$ (4.1.31)

where

$$A = A_1 A_2 \ ; \ B = B_1 A_2 \ ; \ C = C_2 A_1$$ (4.1.32)

which, generally speaking, enables the model given by Equation 4.1.27 to be used.

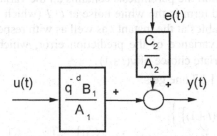

Figure 4.12. ARMAX model with different poles

4.1.4 Optimal Prediction

Taking into account the random nature of the disturbances, it is not possible for ARMAX models to compute exactly the future value of $y(t)$ at instant $t+1$ knowing all the values of y and u up to and including the instant t.

For this reason one considers the concept of *optimal prediction* of the future value of y at the instant $t+1$, computed from the knowledge of y and u up to and including the instant t, which will be denoted $\hat{y}(t+1/t)$ or simply $\hat{y}(t+1)$.

We define a *prediction error* as

$$\varepsilon(t+1) = y(t+1) - \hat{y}(t+1) \tag{4.1.33}$$

The objective will be to construct an optimal predictor as a function of the available information at instant t:

$$\hat{y}(t+1/t) = \hat{y}(t+1) = f(y(t), y(t-1),..., u(t), u(t-1),...) \tag{4.1.34}$$

such that the variance of the prediction error be minimized *i.e.*

$$E\left\{[y(t+1) - \hat{y}(t+1)]^2\right\} = \min \tag{4.1.35}$$

For example, let consider the plant and the disturbance represented by the following ARMAX model:

$$y(t+1) = -a_1 y(t) + b_1 u(t) + c_1 e(t) + e(t+1) \tag{4.1.36}$$

The prediction error is given by

$$\begin{aligned}\varepsilon(t+1) &= y(t+1) - \hat{y}(t+1) \\ &= [-a_1 y(t) + b_1 u(t) + c_1 e(t) - \hat{y}(t+1)] + e(t+1)\end{aligned} \tag{4.1.37}$$

where the first term within the parenthesis contains all the variables available at the instant t and the second term is the white noise at $t+1$ (which is independent with respect to all other variables at the instant t as well as with respect to $e(t)$).

One computes the variance of the prediction error, which we would like to minimize by an appropriate choice of $\hat{y}(t+1)$.

Using the expression 4.1.37 one gets:

$$\begin{aligned}&E\left\{[y(t+1) - \hat{y}(t+1)]^2\right\} \\ &= E\left\{[-a_1 y(t) + b_1 u(t) + c_1 e(t) - \hat{y}(t+1)]^2\right\} + E\left\{e^2(t+1)\right\} \\ &+ 2E\left\{e(t+1)[-a_1 y(t) + b_1 u(t) + c_1 e(t) - \hat{y}(t+1)]\right\}\end{aligned} \tag{4.1.38}$$

The third term in the right hand side of the Equation 4.1.38 is null since $e(t+1)$, the white noise at $t+1$, is independent with respect to all signals at the instants t, $t-1$,... The term $E\left\{e^2(t+1)\right\}$ does not depend upon the choice of the predictor. It results that the choice of $\hat{y}(t+1)$ will affect only the first term, which can be only positive or null. The optimal value of the prediction will be the one that makes this term null at each instant i.e.:

$$E\left\{[-a_1 y(t) + b_1 u(t) + c_1 e(t) - \hat{y}(t+1)]^2\right\} = 0 \tag{4.1.39}$$

From this condition one gets the expression of the optimal predictor:

$$\hat{y}(t+1)\big|_{opt} = -a_1 y(t) + b_1 u(t) + c_1 e(t) \tag{4.1.40}$$

With this choice of the predictor, the prediction error becomes:

$$\varepsilon(t+1)\big|_{opt} = y(t+1) - \hat{y}(t+1)\big|_{opt} = e(t+1) \tag{4.1.41}$$

which is a white noise.
Observe that:

$$\varepsilon(t) = e(t) \tag{4.1.42}$$

which allows to rewrite the optimal predictor as:

$$\hat{y}(t+1)\big|_{opt} = -a_1 y(t) + b_1 u(t) + c_1 \varepsilon(t) \tag{4.1.43}$$

This result can be generalized for ARMAX models 4.1.29 of any order. For the general case the optimal predictor has the expression:

$$\hat{y}(t+1) = -A^*(q^{-1})y(t) + B^*(q^{-1})u(t-d) + C^*(q^{-1})e(t) \tag{4.1.44}$$

and one immediately concludes that the prediction error is a white noise:

$$\varepsilon(t+1)\big|_{opt} = y(t+1) - \hat{y}(t+1) = e(t+1) \tag{4.1.45}$$

This allows to rewrite the optimal predictor 4.1.44 under the form:

$$\hat{y}(t+1) = -A^*(q^{-1})y(t) + B^*(q^{-1})u(t-d) + C^*(q^{-1})\varepsilon(t) \tag{4.1.46}$$

For the case of the multi-step optimal prediction see Equation 4.2.32 and Appendix B.

4.2 Minimum Variance Tracking and Regulation

This strategy concerns optimal controller design, ensuring a minimum variance of the controlled variable around the reference, in the case of systems subject to random disturbances. *It can be applied only to discrete time plant models with stable zeros.*

The objective of the *minimum variance tracking and regulation* is to reduce the variance (standard deviation, mean square error) of the controlled output around the reference value either for a constant value (minimum variance control) or for a variable value (minimum variance tracking). The effect of a minimum variance control is illustrated in Figure 4.13.

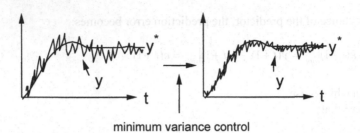

minimum variance control

Figure 4.13. Effect of a minimum variance control in the presence of random disturbances.

The interest of the variance minimization of the controlled output clearly results from the output measures histogram.

If the variance of the controlled output is large, one obtains a distribution of the measures having the form shown in Figure 4.14. In this case, an important percentage of the measured values of the controlled output is far from the reference value. Since in several applications a minimum value should be assured for the controlled output (i.e.: coat thickness, water content of paper, etc ...) one is obliged to set the reference to a value significantly greater than the necessary minimum.

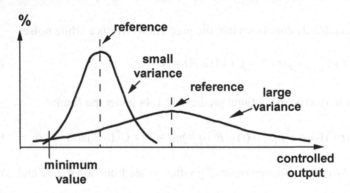

Figure 4.14. Histograms of the controlled output

On the other hand, if the controller reduces significantly the variance of the controlled output, one obtains a distribution of the measures narrowed around the reference. In this case, one can not only improve the quality of the product (better uniformity), but also reduce the reference value to approach the desired minimum value (see Figure 4.14). This implies, in general, a very important reduction of costs (see Chapter 8).

Taking into account the definition of the variance of a random process (see Equation 4.1.9), it results that the objective is to compute $u(t)$ which minimizes the following criterion:

$$J(u(t)) = E\left\{[y(t) - y^*(t)]^2\right\} \approx \frac{1}{N}\sum_{t=1}^{N}[y(t) - y^*(t)]^2 = \min \qquad (4.2.1)$$

where $y(t) - y^*(t)$ represents the difference between the output and the desired value y^* at the instant t.

For solving this problem, disturbance models must be considered in addition to plant models. The structure considered is the ARMAX model which incorporates both the plant and disturbance models (see Section 4.1.3). *Consequently, when identifying a system, both plant model and disturbance model should be identified, in order to apply this control strategy.*

4.2.1 An Example

Let the plant and disturbance be represented by the following ARMAX model:

$$y(t+1) = -a_1 y(t) + b_1 u(t) + b_2 u(t-1) + c_1 e(t) + e(t+1) \qquad (4.2.2)$$

The reference trajectory will be $y^*(t+1)$ (it is either stored or generated by a dynamic model from the reference).

The variance of the difference $y(t+1) - y^*(t+1)$ is computed, and it represents the performance criterion to be minimized:

$$E\{[y(t+1) - y^*(t+1)]^2\} = E\{[-a_1y(t) + b_1u(t) + b_2u(t-1) + c_1e(t) - y^*(t+1)]^2\}$$
$$+E\{e^2(t+1)\} + 2E\{e(t+1)[-a_1y(t) + b_1u(t) + b_2u(t-1) + c_1e(t) - y^*(t+1)]\} \qquad (4.2.3)$$

The third term of the right hand side of Equation 4.2.3 is zero since $e(t + 1)$, the white noise at instant $t + 1$ is independent of all signals appearing at instants t, $t-1$,... (note that $y^* (t + 1)$ depends upon the reference $r(t)$, $r(t-1)$,... only, see for example Section 3.3. Equation 3.3.33). Of the two terms which then remain in the criterion given in Equation 4.2.3, $E \{(e^2 (t + 1)\}$ does not depend upon $u(t)$. It results that the choice of $u(t)$ will only affect the first term which can only be positive or zero. It follows that the minimization of the criterion 4.2.3 corresponds to finding $u(t)$ such that:

$$E\{[-a_1 y(t) + b_1 u(t) + b_2 u(t-1) + c_1 e(t) - y^*(t+1)]^2\} = 0 \qquad (4.2.4)$$

which may be obtained by making the bracketed expression zero. The (theoretical) control law that results is:

$$u(t) = \frac{y^*(t+1) - c_1 e(t) + a_1 y(t)}{b_1 + b_2 q^{-1}} \qquad (4.2.5)$$

Introducing this expression into the plant output equation given by Equation 4.2.2, one obtains that:

$$y(t+1) - y^*(t+1) = e(t+1)$$ (4.2.6)

and respectively

$$y(t) - y^*(t) = e(t)$$ (4.2.7)

This leads to the following remarks:

a) the application of the control law given by Equation 4.2.5 leads to a minimum variance for the difference $y(t) - y^*(t)$ which becomes a white noise ;

b) The controller cancels the zeros of the discrete time plant model (the zeros must be stable) ;

c) $e(t)$ can be replaced in Equation 4.2.5 by the measurable expression given in Equation 4.2.7. This results in the control law:

$$u(t) = \frac{(1 + c_1 q^{-1})y^*(t+1) - (c_1 - a_1)y(t)}{b_1 + b_2 q^{-1}}$$

$$= \frac{T(q^{-1})y^*(t+1) - R(q^{-1})y(t)}{S(q^{-1})}$$ (4.2.8)

The structure of the minimum variance control law is the same as that for *tracking and regulation with independent objectives* in the deterministic case (Chapter 3, Section 3.4) if $P(q^{-1}) = C(q^{-1})$ (desired closed loop poles).

In fact the transfer operator between $y^*(t+1)$ and $y(t+1)$ is:

$$H_{CL}(q^{-1}) = \frac{T(q^{-1})q^{-1}B^*(q^{-1})}{A(q^{-1})S(q^{-1}) + q^{-1}B^*(q^{-1})R(q^{-1})}$$ (4.2.9)

In the example considered above:

$$A(q^{-1}) = 1 + a_1 q^{-1}$$

$$B(q^{-1}) = q^{-1} B^*(q^{-1}) \; ; \; B^*(q^{-1}) = b_1 + b_2 q^{-1} \; ; \quad d = 0$$

$$T(q^{-1}) = C(q^{-1}) = 1 + c_1 q^{-1}$$

$$S(q^{-1}) = B^*(q^{-1}) = b_1 + b_2\, q^{-1} \; ; \; R(q^{-1}) = r_0 = c_1 - a_1$$

and one obtains:

$$H_{CL}(q^{-1}) = \frac{T(q^{-1})q^{-1}}{A(q^{-1}) + q^{-1}R(q^{-1})} = \frac{T(q^{-1})q^{-1}}{C(q^{-1})} = q^{-1} \qquad (4.2.10)$$

The closed loop poles are effectively defined by $C(q^{-1})$ which characterizes the disturbance.

In other words, *an optimum choice exists for closed loop poles (regulation behavior), and this choice directly depends upon the zeros of the disturbance model.*

Finally, an optimal performance test can be carried out for controller tuning by means of a whiteness test applied to the sequence {y(t) - y(t)} for cases without time delay. In cases with a time delay of d samples at the optimum, {y(t) - y*(t)} is a MA process of order d and thus the autocorrelation functions R(i) will be zero for indexes $i \geq d + 1$ (see Equation 4.2.30).*

An Interpretation of the Minimum Variance Control
Let consider the optimal predictor for the ARMAX process given by 4.2.2:

$$\hat{y}(t+1) = -a_1 y(t) + b_1 u(t) + b_2 u(t-1) + c_1 e(t)$$

Let find the value $u(t)$ that imposes

$$\hat{y}(t+1) = y^*(t+1)$$

Then one exactly obtains the expression of $u(t)$ given by 4.2.5 and 4.2.8 respectively.

Thus, one can consider the minimum variance control law as obtained in two stages[4]:

1. Computation of an optimal output predictor;
2. Computation of a control law such that the output prediction be equal to the reference (or in general in order to satisfy a specified deterministic criterion).

Note that the computation of the control law for the predictor is a deterministic problem as all the variables are known at the computation stage.

This computation strategy in two stages for the control in a stochastic environment is very general (separation theorem). It can also be summarized as follows: first compute the best output prediction and afterwards consider the

---

[4] As in the deterministic case, it is a *predictive control* (see Appendix B).

problem as a deterministic control problem, by replacing the real measured output with its prediction.

4.2.2 General Case

Taking into account the similarity with the deterministic *tracking and regulation with independent objectives,* the controller computation will be exactly the same by choosing:

$$P(q^{-1}) = C(q^{-1}) \tag{4.2.11}$$

and this is summarized in Figure 4.15.

The process plus the disturbance is described in the general case by:

$$A(q^{-1})y(t) = q^{-d}B(q^{-1})u(t) + C(q^{-1})e(t) \tag{4.2.12}$$

in which

$$A(q^{-1}) = 1 + a_1 q^{-1} + \ldots + a_{n_A} q^{-n_A} \tag{4.2.13}$$

$$B(q^{-1}) = b_1 q^{-1} + \ldots + b_{n_B} q^{-n_B} = q^{-1} B^*(q^{-1}) \tag{4.2.14}$$

$$C(q^{-1}) = 1 + c_1 q^{-1} + \ldots + c_{n_C} q^{-n_C} \tag{4.2.15}$$

Note that $B(q^{-1})$ must have stable zeros as well as $C(q^{-1})$ that specify the closed loop poles. Moreover $C(q^{-1})$ is always stable if the disturbance is stationary.
The controller is given by:

$$u(t) = \frac{T(q^{-1})y^*(t+d+1) - R(q^{-1})y(t)}{S(q^{-1})} \tag{4.2.16}$$

in which the reference trajectory is defined by:

$$y^*(t+d+1) = \frac{B_m(q^{-1})}{A_m(q^{-1})}r(t) \tag{4.2.17}$$

$$B_m(q^{-1}) = b_{m_0} + b_{m_1}q^{-1} + \ldots \tag{4.2.18}$$

$$A_m(q^{-1}) = 1 + a_{m_1}q^{-1} + a_{m_2}q^{-2} + \ldots \tag{4.2.19}$$

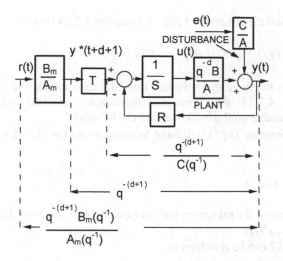

Figure 4.15. Minimum variance tracking and regulation

The closed loop transfer function without $T(q^{-1})$) is (see Figure 4.15):

$$H_{BF_1}(q^{-1}) = \frac{q^{-(d+1)}B^*(q^{-1})}{A(q^{-1})S(q^{-1}) + q^{-(d+1)}B^*(q^{-1})R(q^{-1})} = \frac{q^{-(d+1)}}{C(q^{-1})}$$
$$= \frac{q^{-(d+1)}B^*(q^{-1})}{B^*(q^{-1})C(q^{-1})} \tag{4.2.20}$$

As for *tracking and regulation with independent objectives*:

$$S(q^{-1}) = B^*(q^{-1})\, S'(q^{-1}) \tag{4.2.21}$$

with

$$S'(q^{-1}) = 1 + s'_1 q^{-1} + \ldots + s'_d q^{-d} \tag{4.2.22}$$

and:

$$R(q^{-1}) = r_0 + r_1 q^{-1} + \ldots + r_{n_A-1} q^{-n_A+1} \tag{4.2.23}$$

In order to compute $R(q^{-1})$ and $S(q^{-1})$ the following equation must be solved:

$$A(q^{-1})\, S(q^{-1}) + q^{-(d+1)}\, B^*(q^{-1})\, R(q^{-1}) = B^*(q^{-1})\, C(q^{-1}) \tag{4.2.24}$$

Taking into account the structure of $S(q^{-1})$, Equation 4.2.24 becomes:

$$A(q^{-1}) S'(q^{-1}) + q^{-(d+1)} R(q^{-1}) = C(q^{-1}) \qquad (4.2.25)$$

and the solution is the same as for the deterministic case, taking $P(q^{-1}) = C(q^{-1})$.

For solving 4.2.25 the functions *predisol.sci* (Scilab) and *predisol.m* (MATLAB®) available on the book website can be used.

The precompensator $T(q^{-1})$ will have to compensate the closed loop poles and thus:

$$T(q^{-1}) = C(q^{-1}) \qquad (4.2.26)$$

Let study the effect of the minimum variance control law, given by Equation 4.2.16 on the error $[y(t) - y^*(t)]$.

Equation 4.2.12 can be rewritten as:

$$A(q^{-1}) y(t+d+1) = B^*(q^{-1}) u(t) + C(q^{-1}) e(t+d+1) \qquad (4.2.27)$$

Introducing the control law $u(t)$ given by Equation 4.2.16 and multiplying by $S(q^{-1})$ both sides, one obtains:

$$A(q^{-1}) S(q^{-1}) y(t+d+1) = B^*(q^{-1}) C(q^{-1}) y^*(t+d+1) +$$
$$- q^{-(d+1)} B^*(q^{-1}) R(q^{-1}) y(t+d+1) + S(q^{-1}) C(q^{-1}) e(t+d+1) \qquad (4.2.28)$$

Regrouping the terms in $y(t+d+1)$ and considering Equations 4.2.24 and 4.2.21, Equation 4.2.28 becomes:

$$B^*(q^{-1}) C(q^{-1}) y(t+d+1) = B^*(q^{-1}) C(q^{-1}) y^*(t+d+1) +$$
$$+ S'(q^{-1}) B^*(q^{-1}) C(q^{-1}) e(t+d+1) \qquad (4.2.29)$$

and dividing by $B^*(q^{-1}) C(q^{-1})$ one obtains:

$$y(t+d+1) - y^*(t+d+1) = S'(q^{-1}) e(t+d+1) \qquad (4.2.30)$$

in other words, *the tracking (or regulation) error is a MA process of order d* (for $d=0$, $[y(t+d+1) - y^*(t+d+1)]$ is a white noise).

This corresponds to the minimization of the variance of the error $[y(t) - y^*(t)]$. In fact, using Equation 4.2.25, it is possible to write:

$$C(q^{-1}) y(t+d+1) = [A(q^{-1}) S'(q^{-1}) + q^{-(d+1)} R(q^{-1})] y(t+d+1) \qquad (4.2.31)$$

Taking into account also Equation 4.2.27 and dividing by $C(q^{-1})$ both sides, one obtains[5]:

$$y(t+d+1) = \frac{R(q^{-1})}{C(q^{-1})} y(t) + \frac{S'(q^{-1})B^*(q^{-1})}{C(q^{-1})} u(t) + S'(q^{-1})e(t+d+1) \quad (4.2.32)$$

from which it results:

$$E\left\{\left[y(t+d+1) - y^*(t+d+1)\right]^2\right\} =$$

$$E\left\{\left[\frac{R(q^{-1})}{C(q^{-1})} y(t) + \frac{S(q^{-1})}{C(q^{-1})} u(t) - y^*(t+d+1)\right]^2\right\}$$

$$+ E\left\{\left[S'(q^{-1})e(t+d+1)\right]^2\right\} \quad (4.2.33)$$

$$+ 2E\left\{\left[\frac{R(q^{-1})}{C(q^{-1})} y(t) + \frac{S(q^{-1})}{C(q^{-1})} u(t) - y^*(t+d+1)\right] \cdot \left[S'(q^{-1})e(t+d+1)\right]\right\}$$

The third term of the right hand side member will be zero since $S'(q^{-1})\,e(t+d+1)$ contains $e(t+1)$, $e(t+2)$.....$e(t+d+1)$, which are all independent of $y(t)$, $y(t-1)$,..., $u(t)$, $u(t-1)$ ... $y^*(t+d+1)$, $y^*(t+d)$.... The second term does not depend on $u(t)$ and, finally, by using the control law given by Equation 4.2.16 with $T(q^{-1}) = C(q^{-1})$, the first term of the second member is zero, which corresponds to the minimization of the variance of $[y(t) - y^*(t)]$.

Note that Equation 4.2.30 allows a practical test for the optimal tuning of a digital controller to be defined, if the time delay d is known, since in this case the error must be a moving average of order d. Defining:

$$R(i) = \frac{1}{N}\sum_{t=1}^{N}\left[y(t) - y^*(t)\right] \cdot \left[y(t-i) - y^*(t-i)\right] \quad i = 0,1,2,... \quad (4.2.34)$$

and $RN(i) = R(i)/R(0)$ respectively, one theoretically must obtain for large N:

$$RN(i) \approx 0 \quad i \geq d+1 \quad (4.2.35)$$

In practice, based on finite length data, one considers as an acceptable value:

$$|RN(i)| \leq 2.17/\sqrt{N} \quad ; \quad i \geq d+1 \quad (4.2.36)$$

---

[5] The optimal prediction $\hat{y}(t+d+1)$ is given by the first two terms of Equation 4.2.32.

where N is the number of samples used for computing $RN(i)$ (for $N = 256$, $RN(i)$ ≤ 0.136, $i \geq d + 1$).

For more details on independence tests with finite length data see also Chapter 6, Section 6.2.

Finally, we remind that *this design method only applies to plants having discrete-time models with stable zeros since the controller cancels the plant zeros.* In the case of a discrete time plant model with unstable zeros one uses:

- either an approximation of the minimum variance tracking and regulation control law using the pole placement with a particular choice of the desired closed loop poles;
- or a control law based on a criterion that introduces a weight on the control signal energy.

Auxiliary Poles

In some applications the poles corresponding to $C(q^{-1})$ can be too fast with respect to the open loop system dynamics. This can lead to an unacceptable stress on the actuator or to unacceptable robustness margins. In this case, one can either use generalized minimum variance tracking and regulation design (see below Section 4.3), or add auxiliary poles.

If one chooses to add additional poles, the polynomial defining the desired closed loop poles becomes:

$$P(q^{-1}) = C(q^{-1}) \, P_F(q^{-1})$$

where $P_F(q^{-1})$ represents the polynomial corresponding to the additional poles. This corresponds to the polynomial $P(q^{-1})$ to be used in Equation 4.2.25 instead of $C(q^{-1})$ and in this case, consequently, $T(q^{-1}) = C(q^{-1}) \, P_F(q^{-1})$.

This modification corresponds to minimize the variance of the regulation error filtered by $P_F(q^{-1})$, i.e.:

$$min \; E \; \{[P_F(q^{-1}) \, [y(t+d+1) - y^*(t+d+1)]]^2\}$$

4.2.3 Minimum Variance Tracking and Regulation: Example

The considered plant model is the same as the one used for the tracking and regulation with independent objectives in Section 3.4.4 (the desired tracking performance is also the same). The results of the minimum variance tracking and regulation design are summarized in Table 4.1.

Table 4.1. Minimum Variance Tracking and regulation

Plant:
- $d = 0$
- $B(q^{-1}) = 0.2\ q^{-1} + 0.1\ q^{-2}$
- $A(q^{-1}) = 1 - 1.3\ q^{-1} + 0.42\ q^{-2}$

Tracking dynamics $\rightarrow T_S = 1s,\ \omega_0 = 0.5\ rad/s,\ \zeta = 0.9$

- $B_m = +0.0927 + 0.0687\ q^{-1}$
- $A_m = 1 - 1.2451\ q^{-1} + 0.4066\ q^{-2}$

Disturbance polynomial $\rightarrow C(q^{-1}) = 1 - 1.34\ q^{-1} + 0.49\ q^{-2}$

Pre-specifications: Integrator

*** CONTROL LAW ***

$$S(q^{-1})\ u(t) + R(q^{-1})\ y(t) = T(q^{-1})\ y^*(t+d+1)$$

$$y^*(t+d+1) = [(B_m q^{-1})/A_m(q^{-1})]\ .\ ref(t)$$

Controller:

- $R(q^{-1}) = 0.96 - 1.23\ q^{-1} + 0.42\ q^{-2}$
- $S(q^{-1}) = 0.2 - 0.1\ q^{-1} - 0.1\ q^{-2}$
- $T(q^{-1}) = C(q^{-1})$

Gain margin: 2.084	Phase margin: 61.8 deg
Modulus margin: 0.520 (- 5.68 dB)	Delay margin: 1.3 s

The controller design results for $R(q^{-1})$ and $S(q^{-1})$ are given in the lower part of Table 4.1. Before starting the simulation, the variance and mean value of the white noise generating the disturbance (through the filter $C(q^{-1})/A(q^{-1})$) must be specified.

The simulations results shown in Figure 4.16 illustrate the operation of the minimum variance controller in regulation and tracking. One can see that introduction of the controller effectively reduce the variance of the output.

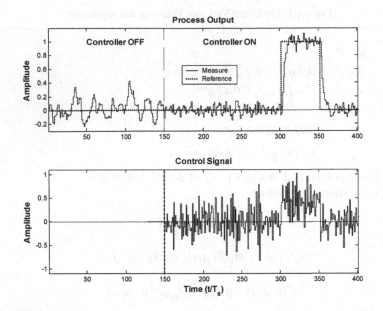

Figure 4.16. Minimum variance tracking and regulation

4.3 The Case of Unstable Zeros: Approximation of the Minimum Variance Tracking and Regulation by Means of Pole Placement

4.3.1 Controller Design

First remember that minimum variance tracking and regulation for the case where the zeros of the plant model are inside the unit circle (asymptotically stable zeros) is equivalent to a pole placement design with:

$$P(q^{-1}) = B^*(q^{-1})C(q^{-1}) \tag{4.3.1}$$

(see Equations 4.2.20 and 4.2.24).

In the case where $B^*(q^{-1})$ contains unstable zeros, we will apply pole placement design but replacing the unstable zeros of $B^*(q^{-1})$ by stable approximations.

For this, one first factorizes the polynomial $B^*(q^{-1})$ under the form:

$$B^*(q^{-1}) = B^+(q^{-1})B^-(q^{-1}) \tag{4.3.2}$$

where $B^+(q^{-1})$ contains all the zeros of $B^*(q^{-1})$ which lie inside the unit circle (asymptotically stable zeros) and $B^-(q^{-1})$ contains all the zeros of $B^*(q^{-1})$ located outside the unit circle (unstable zeros). The factorization is done such that the coefficient of the highest power q^{-n_B} of the polynomial $B^-(q^{-1})$ is equal to 1.

Let define the *reciprocal polynomial* of $B^-(q^{-1})$, the polynomial which is obtained by inverting the order of the coefficients and which will be denoted $B^{-'}(q^{-1})$. This monic polynomial will have all the zeros inside the unit circle.

Example:

$$B^-(q^{-1}) = 0.5 + q^{-1}; \qquad B^{-'}(q^{-1}) = 1 + 0.5 q^{-1}$$

$B^-(q^{-1})$ has a zero at $2+j0$ (outside the unit circle) and $B^{-'}(q^{-1})$ has a zero at $0.5+j0$ (inside the unit circle).

Observe that $B^{-'}(q^{-1})$ is a good approximation of $B^-(q^{-1})$ in the frequency domain since:

$$\left| \frac{B^-(e^{-j\omega})}{B^{-'}(e^{-j\omega})} \right| = 1 \quad \text{for all } \omega \qquad (4.3.3)$$

This can be easily verified for the example considered above. In this case one has:

$$\left| \frac{B^-(e^{-j\omega})}{B^{-'}(e^{-j\omega})} \right|^2 = \frac{1.25 + \cos\omega}{1.25 + \cos\omega} = 1 \quad \text{for all } \omega$$

The approximation of the minimum variance tracking and regulation is obtained by applying the pole placement design for:

$$\begin{aligned} P(q^{-1}) &= B^+(q^{-1}) B^{-'}(q^{-1}) C(q^{-1}) \\ &= A(q^{-1}) S(q^{-1}) + q^{-(d+1)} B^*(q^{-1}) R(q^{-1}) \end{aligned} \qquad (4.3.4)$$

Taking in account the structure of $B^*(q^{-1})$ (given in Equation 4.3.2) it results that $S(q^{-1})$ will have the form:

$$S(q^{-1}) = B^+(q^{-1}) S'(q^{-1})$$

and therefore $S'(q^{-1})$ and $R(q^{-1})$ are solutions of the equation:

$$B^{-1}(q^{-1})C(q^{-1}) = A(q^{-1})S'(q^{-1}) + q^{-(d+1)}B^{-}(q^{-1})R(q^{-1}) \qquad (4.3.5)$$

Polynomial $T(q^{-1})$ is given by:

$$T(q^{-1}) = B^{-1}(q^{-1})C(q^{-1})/B^{*}(1) \qquad (4.3.6)$$

For a proof of the optimality of this stochastic control law see (Astrom, Wittemmark 1997)[6].

4.3.2 An Example

Table 4.2 give the results of the design for the approximation of the minimum variance tracking and regulation by the pole placement.

Table 4.2. Approximation of minimum variance tracking and regulation design by means of the pole placement

> Plant:
> - $d = 0$
> - $B(q^{-1}) = 0.1 \, q^{-1} + 0.2 \, q^{-2}$
> - $A(q^{-1}) = 1 - 1.3 \, q^{-1} + 0.42 \, q^{-2}$
>
> Tracking dynamics → $T_s = 1s$, $\omega_0 = 0.5 \, rad/s$, $\zeta = 0.9$
> - $B_m = +0.0927 + 0.0687 \, q^{-1}$
> - $A_m = 1 - 1.2451 \, q^{-1} + 0.4066 \, q^{-2}$
>
> Regulation dynamics → $P(q^{-1}) = (1 + 0.5 \, q^{-1})(1 - 1.34 \, q^{-1} + 0.49 \, q^{-2})$
> Pre-specifications: Integrator
>
> ***** CONTROL LAW *****
>
> $$S(q^{-1}) u(t) + R(q^{-1}) y(t) = T(q^{-1}) y^{*}(t+d+1)$$
>
> $$y^{*}(t+d+1) = [B_m (q^{-1})/A_m (q^{-1})] r(t)$$
>
> Controller:
> - $R(q^{-1}) = 4.813 - 6.118 \, q^{-1} + 2.055 \, q^{-2}$
> - $S(q^{-1}) = 1 - 0.02139 \, q^{-1} - 0.9786 \, q^{-2}$
> - $T(q^{-1}) = -2.80 \, q^{-1} - 0.6 \, q^{-2} + 0.8166 \, q^{-3}$
>
> | Gain margin: 2.082 | Phase margin: 58.5 deg |
> | Modulus margin: 0.520 (- 5.69 dB) | Delay margin: 1.42 s |

---

[6] The idea of this method can be used also in a deterministic context if we would like a better approximation of the tracking and regulation with independent objectives by the pole placement when the plant model has unstable zeros.

The plant model has an unstable zero. For defining the desired closed loop poles one should first factorize $B^*(q^{-1})$:

$$B^*(q^{-1}) = 0.1 + 0.2q^{-1} = 0.2(0.5 + q^{-1}) = B^+(q^{-1})B^-(q^{-1})$$

So

$$B^-(q^{-1}) = 0.5 + q^{-1} \qquad B^-(q^{-1}) = 1 + 0.5q^{-1}$$

The disturbance model $C(q^{-1})$ is the same as the one used in the example 4.2.3 and the controller is computed using Equations 4.3.4 and 4.3.5.

4.4 Generalized Minimum Variance Tracking and Regulation

The minimum variance tracking and regulation strategy only applies to plants having a discrete-time model with stable zeros.

The *generalized minimum variance tracking and regulation* strategy is an extension of the minimum variance control strategy to plants having discrete-time models with unstable zeros. This method computes a control *u(t)* which minimizes the following criterion:

$$E\left\{ \left[y(t+d+1) - y^*(t+d+1) + \frac{Q(q^{-1})}{C(q^{-1})}u(t) \right]^2 \right\}$$

(4.4.1)

$$\approx \frac{1}{N} \sum_{i=1}^{N} \left[y(t+d+1) - y^*(t+d+1) + \frac{Q(q^{-1})}{C(q^{-1})}u(t) \right]^2 = \min$$

in which

$$Q(q^{-1}) = \frac{\lambda(1 - q^{-1})}{1 + \alpha q^{-1}}$$

(4.4.2)

and $C(q^{-1})$ characterizes the disturbance in the ARMAX model of the plant output.

In the case where $\alpha = 0$, the criterion of Equation 4.4.1 is written as:

$$E\left\{ \left[y(t+d+1) + \frac{\lambda}{C(q^{-1})}[u(t) - u(t-1)] - y^*(t+d+1) \right]^2 \right\} = \min \quad (4.4.3)$$

The quantity $y(t+d+1) + [\lambda/C(q^{-1})] [u(t) - u(t-1)]$ is interpreted as a *generalized output*, and its variations around $y^*(t+d+1)$ will be minimized.

For $\lambda = 0$, this strategy corresponds to the *minimum variance tracking and regulation*. For $\lambda > 0$, the variance of the difference between the *generalized output* and the reference trajectory y* will be minimized but this will no more assure the minimization of $E\{[y(t+d+1) - y^*(t+d+1)]^2\}$ and for this reason a small value of λ assuring however the stability of the closed loop is desirable in practice.

Note also that λ has a weighting effect upon the variations of $u(t)$ and therefore it can be used even in the case of plant models with stable zeros in order to reduce the control signal variation (and then the stress on the actuator).

4.4.1 Controller Design

Using Equation 4.2.33, the following expression is obtained for criterion 4.4.1:

$$E\left\{\left[y(t+d+1)-y^*(t+d+1)+\frac{Q(q^{-1})}{C(q^{-1})}u(t)\right]^2\right\}$$

$$= E\left\{\left[\frac{R(q^{-1})}{C(q^{-1})}y(t)+\frac{S(q^{-1})}{C(q^{-1})}u(t)+\frac{Q(q^{-1})}{C(q^{-1})}u(t)-y^*(t+d+1)\right]^2\right\} \quad (4.4.4)$$

$$+ E\left\{\left[S'(q^{-1})e(t+d+1)\right]^2\right\}$$

To minimize this criterion, $u(t)$ must be chosen such that the first term of the right hand side is zero. One thus obtains:

$$u(t) = \frac{C(q^{-1})y^*(t+d+1)-R(q^{-1})y(t)}{S(q^{-1})+Q(q^{-1})} \quad (4.4.5)$$

Controller design is carried out in two stages.

The first stage consists of computing the polynomials $R(q^{-1})$, $S(q^{-1})$ and $T(q^{-1})$ = $C(q^{-1})$ for $Q(q^{-1}) = 0$ by the minimum variance tracking and regulation method even if $B(q^{-1})$ has unstable zeros.

The second stage is to introduce the polynomial $Q(q^{-1})$ given by Equation 4.4.2 with $\lambda > 0$. Using Equation 4.4.5, the closed loop transfer operator from the reference trajectory to the output is given by:

$$H_{CL} = \frac{q^{-d}B(q^{-1})T(q^{-1})}{A(q^{-1})[S(q^{-1})+Q(q^{-1})]+q^{-(d+1)}B^*(q^{-1})R(q^{-1})}$$

$$= \frac{q^{-d}B(q^{-1})}{A(q^{-1})Q(q^{-1})+B^*(q^{-1})C(q^{-1})} \quad (4.4.6)$$

since from Equation 4.2.24, one has

$$A(q^{-1}) S(q^{-1}) + q^{-(d+1)} B^*(q^{-1}) R(q^{-1}) = B^*(q^{-1}) C(q^{-1}) \qquad (4.4.7)$$

One chooses λ (smaller than 1, as a general rule) and one checks that the polynomial $[(A(q^{-1}) Q(q^{-1}) + B^*(q^{-1}) C(q^{-1})]$ defining the closed loop poles is asymptotically stable. If this is not the case, the value of λ^7 is changed.

As initial choice, in the case of unstable roots of $B^*(q^{-1})$, one takes the lowest values of λ allowing to obtain a stable polynomial $S(q^{-1}) + Q(q^{-1})$ for $\alpha=0$ (since $S(q^{-1}) = B^*(q^{-1}) S'(q^{-1})$ is unstable). This makes the controller stable.

Note that the main limitation of this technique is that the existence of λ leading to a stable closed loop system is not guaranteed (especially in the case of several unstable zeros) whereas with the design illustrated in Section 4.3 one always obtains an asymptotically stable closed loop system.

4.5 Concluding Remarks

In this chapter the design of digital controllers in the presence of random disturbances has been considered. Many of the random disturbances encountered in practice may be modeled as a gaussian discrete-time white noise passed through a filter. The knowledge of this filter called the *disturbance model* allows the disturbance to be completely characterized (with the approximation of a scaling factor).

For the design of digital controllers in the presence of random disturbances one considers a joint plant and disturbance model called ARMAX (autoregressive moving average with exogenous input) model:

$$y(t) = \frac{q^{-d} B(q^{-1})}{A(q^{-1})} u(t) + \frac{C(q^{-1})}{A(q^{-1})} e(t)$$

where y is the plant output, u is the plant input and, and e is a discrete-time white noise sequence. The first term represents the effect of the plant input and the second term represents the effect of the disturbance upon the plant output.

The ARMAX model is often written in the form:

$$A(q^{-1}) y(t) = q^{-d} B(q^{-1}) u(t) + C(q^{-1}) e(t)$$

The *Minimum Variance Tracking and Regulation* control strategy minimizes the criterion:

---

[7] This technique can be also used in the deterministic case when the zeros of the plant model are unstable.

$$E\{[y(t+d+1) - y^*(t+d+1)]^2\} = min$$

where y^* is the desired output trajectory. This control strategy applies to plants models with stable zeros.

The poles of the closed loop system are the zeros of polynomial $C(q^{-1})$ characterizing the stochastic disturbance model, with the addition of the zeros of polynomial $B^*(q^{-1})$. The obtained controller is identical to the controller corresponding to *tracking and regulation with independent objectives* design used in a deterministic environment if the polynomial $P(q^{-1})$, defining the desired closed loop system poles, is chosen equal to $C(q^{-1})$. Note that, in some cases, in order to reduce the stress on the actuator or to improve the robustness margins, auxiliary poles should be added to those imposed by $C(q^{-1})$.

In the case of a plant model with unstable zeros one chooses:

- either an approximation of the minimum variance tracking and regulation control law using the pole placement ;
- or a control law based on a criterion that introduces a weight on the control signal energy.

The *Generalized Minimum Variance Tracking and Regulation* control strategy is an extension of the *Minimum Variance Tracking and Regulation* control strategy for the case of a plant model with unstable zeros. This strategy minimizes the criterion:

$$E\left\{\left\{y(t+d+1)+\frac{\lambda}{C(q^{-1})}[u(t)-u(t-1)]-y^*(t+d+1)\right\}^2\right\} = min$$

where λ (> 0) is chosen such that the resulting closed loop system poles are asymptotically stable. Note that λ has the effect to weight the variations of $u(t)$. It can also be used to smooth the variations of the control signal resulting from a minimum variance control design. However, since a value of λ that stabilizes the closed loop system may not always exist, the approximation of the *minimum variance tracking and regulation* by *pole placement* is recommended.

Regardless to the method used, in the final stage of the design it is required to verify the resulting sensitivity functions and the corresponding robustness margins. If these are not satisfactory, modifications should be done (see Chapter 3) because, in practice, one cannot use controllers leading to insufficient robustness margins or imposing an excessive stress on the actuator in the frequency regions where the gain of the model is low (leading to large peaks on the input sensitivity functions).

4.6 Notes and References

For the description of stochastic disturbances by means of ARMA(X) models and the minimum variance control see:

Box G.E.P., Jenkins G.M. (1970) Time Series Analysis, Forecasting and Control, Holden Day, S. Francisco.
Åström K.J. (1970) Introduction to Stochastic Control Theory, Academic Press, N.Y.

and also

Åström K.J., Wittenmark B. (1997) Computer Controlled Systems - Theory and Design, 3rd edition, Prentice-Hall, Englewood Cliffs, N.J.

The idea of the generalized minimum variance control has been introduced by

Clarke D.W., Gawthrop P.J., (1975) A self-tuning controller, Proc. IEEE, vol. 122, pp. 929-934.

and further developed in:

Clarke D.W., Gawthrop P.J. (1979) Self-tuning Control, Proc. IEEE, vol. 126, pp. 633-40.

For a unified approach to the control design in a stochastic and deterministic environment see:

Landau I.D. (1981) Model Reference Adaptive Controllers and Stochastic Self-tuning Regulators, A Unified Approach, Trans. A.S.M.E, J. of Dyn. Syst. Meas. and Control, vol. 103, n°4, pp. 404-416.

Control design techniques presented in Chapter 3 (with the time domain interpretation given in the Appendix B) and in Chapter 4 belong to the category of methods called « one step ahead predictive control » (more exactly $d+1$ steps). There also exist multi-step predictive control techniques (generalized predictive control). See Appendix B as well as:

Landau I.D., Lozano R., M'Saad M. (1997) Adaptive Control, (Chapter 7), Springer, London, UK.
Camacho, E.F., Bordons, C. (2004) Model Predictive Control, 2nd edition, Springer, London.

5

System Identification: The Bases

In this chapter the basic principles of identification of dynamic systems are first introduced. This is followed by a presentation of the main types of parameter adaptation algorithms used in recursive identification methods. The choice of input signals for identification and the influence of disturbances is also discussed. The last section presents the general structure of recursive identification methods.

5.1 System Model Identification Principles

Identification means the determination of the model of a dynamic system from input/output measurements. The knowledge of the model is necessary for the design and the implementation of a high performance control system.

Figure 5.1 sums up the general principles of controller design. In order to design and tune a controller correctly, one needs:

1. To specify the desired control loop performance and robustness
2. To know the dynamic model of the plant to be controlled (also known as the *control model*) which describes the relation between the control variations and the output variations
3. To possess of a suitable controller design method enabling to achieve the desired performance and robustness specifications for the corresponding plant model

The notion of the mathematical model of a system or phenomenon is a fundamental concept. In general, a multitude of model types exist, each one dedicated to a particular application.

For example, the *knowledge* type models (based on the laws of physics, chemistry, *etc...*) permit a fairly complete system description and are used for plant simulation and design. These models are in general extremely complex and can only rarely be directly used for the design of control systems. The dynamic control models that give the relation between the input and output variations of a system are, as indicated above, the type of model suitable for the design and tuning of control systems.

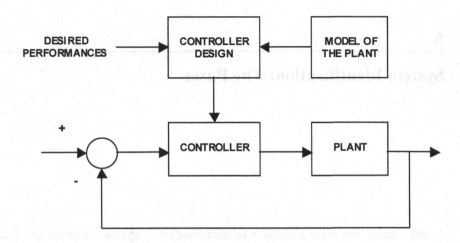

Figure 5.1. Controller design principles

Although indications concerning the structure of these control models can be obtained from the structure of the knowledge type model, it is in general very difficult to determine the significant parameter values from these models. This is why in the majority of practical situations, it is necessary to implement a methodology for direct identification of these dynamic (control) models from experimental data.

Note that there are two types of dynamic models:

1. *Non-parametric models* (example: frequency response, step response)
2. *Parametric models* (example: transfer function, differential or difference equation)

Henceforward we shall be concerned with the identification of sampled discrete-time parametric dynamic models, which are the most suitable for the design and tuning of digital control systems.

System identification is an experimental approach for determining the dynamic model of a system. It includes four steps:

1. Input/output data acquisition under an experimentation protocol
2. Selection or estimation of the "model" structure (complexity)
3. Estimation of the model parameters
4. Validation of the identified model (structure and values of the parameters)

A complete identification operation must necessarily comprise the four stages indicated above. The specific methods used at each stage depend on the type of model desired (parametric or non-parametric, continuous-time or discrete-time) and on the experimental conditions (for example: hypothesis on the noise, open loop or closed loop identification). The *validation* is the mandatory step to decide if the identified model is acceptable or not.

As there does not exist a unique parameter estimation algorithm and a unique experimental protocol that always lead to a good identified model, the models obtained may not always pass the validation test. In this case, it is necessary to reconsider the estimation algorithms, the model complexity or the experimental conditions. System identification should be then considered as an iterative procedure as illustrated in Figure 5.2.

We now present a brief sum-up of the essential elements that characterize the different steps of the identification.

Input/Output Data Acquisition with an Experimental Protocol
One should essentially select an excitation signal with a rich frequency spectrum in order to cover the bandwidth of the plant to be identified, but with small magnitude (since in practice, the accepted magnitude variations of the input signals are strongly constrained).

This aspect will be discussed in details in Section 5.3 and 7.2.

Selection or Estimation of the Model Complexity
The typical problem encountered is to define the orders of the polynomials (numerator, denominator) of the pulse transfer function that represents the plant model.

One often uses trial and error procedures, but valuable techniques for complexity estimation of models have been developed (see for example Duong and Landau 1996).

These aspects will be discussed in Section 6, Section 6.5 and in Chapter 7, Section 7.3.

Model Parameter Estimation
The "classic" identification methodology used to obtain parametric models based on non-parametric models of the type "step response" is illustrated in Figure 5.3. This methodology, initially used to obtain continuous time parametric models, has been extended to the identification of discrete time models.

From the shape of the plant step response, one selects a type of model and the parameters of this model are graphically determined. As the sampling frequency is known, one can obtain the corresponding discrete time model from conversion tables.

This methodology has several disadvantages:

- Test signals with large magnitude (seldom acceptable in the industrial systems)
- Reduced accuracy
- Bad influence of disturbances
- Models for disturbances are not available
- Lengthy procedure
- Absence of model validation

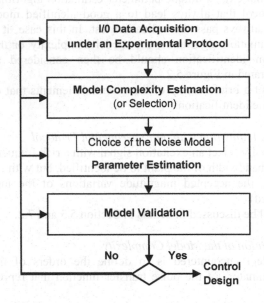

Figure 5.2. System identification methodology

The availability of a digital computer permits the implementation of algorithms that automatically estimate the parameters of the discrete time models. It should be emphasized that the identification of the parametric discrete time models allows to obtain (by simulation) non-parametric models of the *step-response* or *frequency-response* type, *with a far higher degree of accuracy with respect to a direct approach, and using extremely weak excitation signals.* The identification of parametric sampled data models leads to models of a very general use and offers several advantages over the other approaches.

High performance identification algorithms, which have a recursive formulation tailored to real-time identification problems and to their implementation on micro-computer, have been developed. The fact that these identification methods can operate with extremely weak excitation signals is a very much appreciated quality in practical situations.

The parameter estimation principle for discrete time models is illustrated in Figure 5.4. A sampled input sequence $u(t)$ (where t is the discrete time) is applied to the physical system (the cascade actuator-plant-transducer) by means of a digital-to-analog converter (DAC) followed by a zero order hold block (ZOH). The measured sampled plant output $y(t)$ is obtained by means of an analog-to-digital converter (ADC).

A discrete-time model with adjustable parameters is implemented on the computer. The error between the system output $y(t)$ at instant t, and the output $\hat{y}(t)$ predicted by the model (known as the *prediction error*) is used by a *parameter adaptation algorithm* that, at each sampling instant, will modify the model parameters in order to minimize this error on the basis of a chosen criterion.

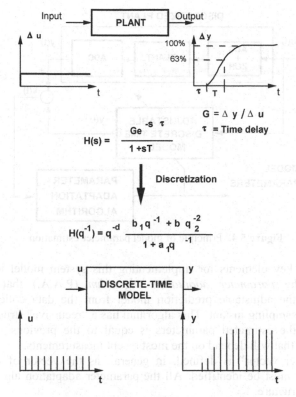

Figure 5.3. "Classic" identification methodology

The input is, in general, a very low level pseudo-random binary sequence generated by the computer (sequence of rectangular pulses with randomly variable duration). Once the model is obtained, an objective validation can be made by carrying out statistical tests on the prediction error $\varepsilon(t)$ and the predicted output $\hat{y}(t)$. The validation test enables the best model to be chosen (for a given plant), *i.e.* the best structure and the best algorithm for the estimation of the parameters.

Finally, by computing and graphically representing the step responses and the frequency response of the identified model, the characteristics of the continuous-time model (step response or frequency response) can be extracted.

This modern approach to system model identification avoids all the problems related to the previously mentioned "classical" methods and also offers other possibilities such as:

- Tracking of the variations of the system parameters in real time allowing retuning of controllers during operation
- Identification of disturbances models
- Modeling of the transducer noises in view of their elimination
- Detection and measurement of vibration frequencies
- Spectral analysis of the signals

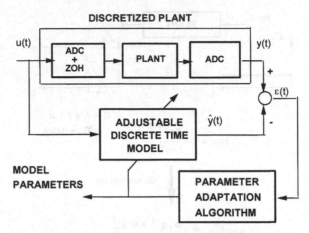

Figure 5.4. Principle of model parameter estimation

One of the key elements for implementing this system model identification approach is the *parameter adaptation algorithm* (P.A.A.) that drives the parameters of the adjustable prediction model from the data collected on the system at each sampling instant. This algorithm has a "recursive" structure, i.e. the new value of the estimated parameters is equal to the previous value plus a correction term that will depend on the most recent measurements.

A "parameter vector" is defined, in general, as the vector of the different parameters that must be identified. All the parameter adaptation algorithms have the following structure:

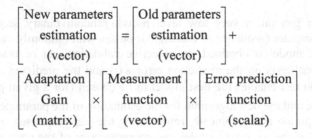

The measurement function vector is also known as the "observation vector".

Note that non-recursive parametric identification algorithms also exist (which process as a one block the input/output data files obtained over a certain time horizon). Recursive identification offers the following advantages with respect to these non-recursive techniques:

- Obtaining an estimated model as the system evolves
- Considerable data compression, since the recursive algorithms process at each instant only one input/output pair instead of the whole input/output data set
- Much lower requirements in terms of memory and CPU power
- Easy implementation on microcomputers

- Possibility to implement real-time identification systems
- Possibility to track the parameters of time variable systems

Section 5.2 introduces the main types of parameter estimation (identification) algorithms in their recursive form. The effect of the noise on the parameter estimation algorithms will be discussed in Section 5.4.

Model Validation

Different points of view can be considered for the choice of a model validation procedure. The goal is to verify that the output model excited by the same input applied to the plant reproduce the variations of the output caused by the variations of the input regardless the effect of the noise. Techniques for model validation will be presented in Chapter 6 (Section 6.2 and 6.4).

5.2 Algorithms for Parameter Estimation

5.2.1 Introduction

We will illustrate the principles of parametric identification presented in Figure 5.3 by an example.

Consider the discrete-time model of a plant described by

$$y(t+1) = -a_1 y(t) + b_1(t)u(t) = \theta^T \phi(t) \tag{5.2.1}$$

where a_1 and b_1 are the unknown parameters.

The model output can be also written under the form of a scalar product between the unknown *parameter vector*

$$\theta^T = [a_1, b_1] \tag{5.2.2}$$

and the vector of measures termed measurement vector or plant model regressor vector

$$\phi(t)^T = [-y(t), u(t)] \tag{5.2.3}$$

This vector representation is extremely useful since it allows easy consideration of models of any order.

Following the diagram given in Figure 5.4, one should construct an adjustable prediction model, which will have the same structure as the discrete-time model of the plant given in Equation 5.2.1:

$$\hat{y}^o(t+1) = \hat{y}(t+1|\hat{\theta}(t)) = -\hat{a}_1(t)y(t) + \hat{b}_1(t)u(t) = \hat{\theta}(t)^T \phi(t) \tag{5.2.4}$$

where $\hat{y}^o(t+1)$ is the predicted output at the instant t based on the knowledge of the parameters estimated at time t ($\hat{a}_1(t), \hat{b}_1(t)$). $\hat{y}^o(t+1)$ is called the *a priori* prediction. In Equation 5.2.4

$$\theta(t)^T = [\hat{a}_1(t), \hat{b}_1(t)] \tag{5.2.5}$$

is the *vector of estimated parameters* at time t.

One can define now the prediction error *(a priori)* as in Figure 5.4:

$$\varepsilon^o(t+1) = y(t+1) - \hat{y}^o(t+1) = \varepsilon^o(t+1, \hat{\theta}(t)) \tag{5.2.6}$$

The term $\hat{y}^o(t+1)$ is effectively computed between the sampling instants t and $t+1$ once $\hat{\theta}(t)$ is available, $\varepsilon^o(t+1)$ is computed at the instant $t+1$ after the acquisition of $y(t+1)$ (between $t+1$ and $t+2$). Note that $\varepsilon^o(t+1)$ depends on $\hat{\theta}(t)$.

Now it will be necessary to define a criterion in terms of the prediction error, which will be minimized by an appropriate evolution of the parameters of the adjustable prediction model, driven by the parameter adaptation algorithm. Since the objective is to minimize the magnitude of the prediction error independently of its sign, the choice of a quadratic criterion is natural. A first approach can be the synthesis of a parameter adaptation algorithm which at each instant minimizes the square of the *a priori* prediction error. This can be expressed as finding an expression for $\hat{\theta}(t)$ such that at each sampling one minimizes

$$J(t+1) = \left[\varepsilon^o(t+1)\right]^2 = \left[\varepsilon^o(t+1, \hat{\theta}(t))\right]^2 \tag{5.2.7}$$

The structure of the parameter adaptation algorithm will be of the form

$$\hat{\theta}(t+1) = \hat{\theta}(t) + \Delta\hat{\theta}(t+1) = \hat{\theta}(t) + f\left(\hat{\theta}(t), \phi(t), \varepsilon^o(t+1)\right) \tag{5.2.8}$$

The correction term $f\left(\hat{\theta}(t), \phi(t), \varepsilon^o(t+1)\right)$ should only depend upon the information available at the instant $t+1$ (last measurement $y(t+1)$, parameter vector $\hat{\theta}(t)$ and a finite number of measurements or information at $t, t-1,..., t-n$). The solution to this problem will be given in Section 5.2.2. A recursive adaptation algorithm will be derived enabling both on-line and off-line implementation.

The criterion of Equation 5.2.7 is not the only one step ahead criterion which can be considered and this aspect will also be discussed in Section 5.2.2.

When a set of input/output measurements over a time horizon t $(i=1, 2,..., t)$ is available, and we are looking for an off line identification, one may ask how to use this set of data optimally. The objective will be to search for a vector of parameters

$\hat\theta(t)$ using the available data up to instant t and that minimizes a criterion of the form

$$J(t+1) = \sum_{i=1}^{t} \left[\varepsilon^o(i,\hat\theta(t))\right]^2 \qquad (5.2.9)$$

that means the minimization of the sum of the squares of the prediction errors over the time horizon t. This point of view will lead to the least squares algorithm which will be presented in Section 5.2.3 (under the non-recursive and recursive form).

5.2.2 Gradient Algorithm

The aim of the gradient parameter adaptation algorithm is to minimize a one step quadratic criterion in terms of the prediction error (one-step ahead).

Consider the same example as in Section 5.2.1. The discrete time model of the plant is expressed by

$$y(t+1) = - a_1 y(t) + b_1 u(t) = \theta^T \phi(t) \qquad (5.2.10)$$

where

$$\theta^T = [a_1, b_1] \qquad (5.2.11)$$

is the parameter vector and

$$\phi(t)^T = [- y(t), u(t)] \qquad (5.2.12)$$

is the vector of measures (pant model regressor vector).

The adjustable prediction model (*a priori*) is described by

$$\hat{y}^o(t+1) = \hat{y}(t+1|\hat\theta(t)) = -\hat a_1(t)y(t) + \hat b_1(t)u(t) = \hat\theta(t)^T \phi(t) \qquad (5.2.13)$$

where $\hat{y}^o(t+1)$ represents the *a priori* prediction depending on the values of the parameters estimated at instant t and

$$\hat\theta(t)^T = [\hat a_1(t), \hat b_1(t)] \qquad (5.2.14)$$

is the estimated parameter vector[1].

The *a priori* prediction error is given by

---

[1] In this case the predictor regressor vector is identical to the measurement vector.

$$\varepsilon^o(t+1) = y(t+1) - \hat{y}^o(t+1) \tag{5.2.15}$$

To evaluate the quality of the new estimated parameter vector $\hat{\theta}(t+1)$, which will be provided by the parameter adaptation algorithm, it is useful to define the *a posteriori* output of the adjustable predictor, which corresponds to re-computing Equation 5.2.13 with the new values of the parameters estimated at *t+1*.

The *a posteriori* predictor output is defined by

$$\hat{y}(t+1) = \hat{y}(t+1|\hat{\theta}(t+1)) = -\hat{a}_1(t+1)y(t) + \hat{b}_1(t+1)u(t) = \hat{\theta}(t+1)^T \phi(t) \tag{5.2.16}$$

One also defines an *a posteriori* prediction error:

$$\varepsilon(t+1) = y(t+1) - \hat{y}(t+1) \tag{5.2.17}$$

A recursive parametric adaptation algorithm with memory is desired.
The structure of such an algorithm is[2]

$$\hat{\theta}(t+1) = \hat{\theta}(t) + \Delta\hat{\theta}(t+1) = \hat{\theta}(t) + f\left(\hat{\theta}(t), \phi(t), \varepsilon^o(t+1)\right) \tag{5.2.18}$$

The correction term $f\left(\hat{\theta}(t), \phi(t), \varepsilon^o(t+1)\right)$ must only depend upon the information available at instant *t+1* (last measure *y(t+1)*), parameters of $\hat{\theta}(t)$ and eventually a finite number of information at instants *t, t-1, t-2, ..., t-n*). The correction term should allow one to minimize at each step the *a priori* prediction error with respect to the criterion

$$\min_{\hat{\theta}(t)} J(t+1) = \left[\varepsilon^o(t+1)\right]^2 \tag{5.2.19}$$

If one represents the criterion *J* and the parameters $\hat{a}_1$ and $\hat{b}_1$ in three-dimensional space, one gets the form represented in Figure 5.5 (a reversed conic surface). The optimum of the criterion will correspond to the bottom of the cone and the projection of this point on the plane $\hat{a}_1, \hat{b}_1$ will give us the optimal values of the plant parameters: *a₁* and *b₁*. It is obvious that, in order to reach as quickly as possible this point (the optimum of the criterion), it will be advantageous to go down along the steepest descent. This solution is analytically given by the *gradient technique*.

---

[2] Effectively, if the correction term is null, one holds the previous value of the estimated parameters.

The horizontal sections of the surface correspond to curves along which the criterion has a constant value (isocriterion curves). If one represents the projection of the isocriterion curves ($J = const.$) in the plane of the parameters $\hat{a}_1, \hat{b}_1$, one obtains concentric closed curves around the point a_1, b_1 (the parameters of the plant model) which minimizes the criterion. As the value of the criterion J (= const.) increases, the isocriterion curves move further and further away from the minimum. This is illustrated in Figure 5.5.

In order to minimize the value of the criterion, one moves in the direction of the steepest descent that, in the plane $\hat{a}_1, \hat{b}_1$, corresponds to move in the opposite direction of the gradient associated to the isocriterion curve. This will lead us to a curve corresponding to $J = const$ of a smaller value, as shown in Figure 5.5.

The corresponding parametric adaptation algorithm will have the form

$$\hat{\theta}(t+1) = \hat{\theta}(t) - F\frac{\partial J(t+1)}{\partial \hat{\theta}(t)} \tag{5.2.20}$$

where $F = \alpha I$ ($\alpha > 0$) is the adaptation matrix gain (I – identity matrix) and $\partial J(t+1)/\partial \hat{\theta}(t)$ is the gradient of the criterion of Equation 5.2.19 with respect to $\hat{\theta}(t)$.

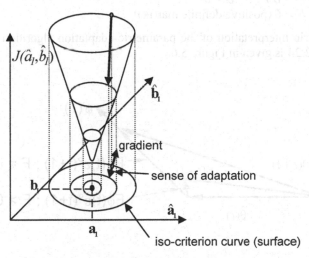

Figure 5.5. Principle of gradient method

From Equation 5.2.19, one gets

$$\frac{1}{2}\frac{\partial J(t+1)}{\partial \hat{\theta}(t)} = \frac{\partial \varepsilon^o(t+1)}{\partial \hat{\theta}(t)}\varepsilon^o(t+1) \qquad (5.2.21)$$

But

$$\varepsilon^o(t+1) = y(t+1) - \hat{y}^o(t+1) = y(t+1) - \hat{\theta}(t)^T \phi(t) \qquad (5.2.22)$$

and then

$$\frac{\partial \varepsilon^o(t+1)}{\partial \hat{\theta}(t)} = -\phi(t) \qquad (5.2.23)$$

By introducing Equation 5.2.23 into Equation 5.2.20, the parametric adaptation algorithm of Equation 5.2.20 becomes

$$\hat{\theta}(t+1) = \hat{\theta}(t) + F\phi(t)\varepsilon^o(t+1) \qquad (5.2.24)$$

where F is the adaptation matrix gain[3]. Two choices are possible:

1) $F = \alpha I$; $\alpha > 0$
2) $F > 0$ (positive definite matrix)[4]

The geometric interpretation of the parametric adaptation algorithm expressed by Equation 5.2.24 is given in Figure 5.6.

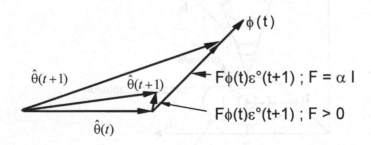

Figure 5.6. Geometric interpretation of the gradient adaptation algorithm

---

[3] In equations of the form of Equation 5.2.24 the vector ϕ is generally called the *observation vector*. In this particular case it corresponds to the measurement vector.

[4] A positive definite matrix is characterized by: (i) each diagonal term is positive; (ii) the matrix is symmetric ; (iii) the determinants of all principal matrix minors are positive.

The parametric adaptation algorithm given by Equation 5.2.24 presents some instability possibilities if the adaptation gain (respectively α) is large (this can be well understood with the support of Figure 5.5).

Let consider Equation 5.2.17 of the *a posteriori* error. By using Equations 5.2.13 and 5.2.14, it can be re-written as

$$\varepsilon(t+1) = y(t+1) - \hat{y}(t+1) = y(t+1) - \hat{\theta}(t)^T \phi(t) + \left[\hat{\theta}(t) - \hat{\theta}(t+1)\right]^T \phi(t) \quad (5.2.25)$$

From Equation 5.2.24 it results that

$$\hat{\theta}(t) - \hat{\theta}(t+1) = -F\phi(t)\varepsilon^o(t+1) \quad (5.2.26)$$

and by also taking into account Equation 5.2.15, Equation 5.2.25 becomes

$$\varepsilon(t+1) = \varepsilon^o(t+1) - \phi(t)^T F\phi(t)\varepsilon^o(t+1) \quad (5.2.27)$$

that for $F = \alpha I$ becomes:

$$\varepsilon(t+1) = \left(1 - \alpha\phi(t)^T \phi(t)\right)\varepsilon^o(t+1) \quad (5.2.28)$$

If $\hat{\theta}(t+1)$ is a better estimation than $\hat{\theta}(t)$ (which means that the estimation of the parameters goes in the good sense) one should get $\varepsilon(t+1)^2 < \varepsilon^o(t+1)^2$. Therefore it results from Equation 5.2.28 that the adaptation gain α should satisfy the (necessary) condition

$$\alpha < 2/\phi(t)^T \phi(t) \quad (5.2.29)$$

In this algorithm, in other words, the adaptation gain must be chosen as a function of the magnitude of the signals[5].

In order to avoid the possible instabilities, and the dependence of the adaptation gain with respect to the magnitude of the measured signals, one uses the same gradient approach but with a different criterion, which has as objective the minimization of the *a posteriori* prediction error at each step according to

$$\min_{\hat{\theta}(t+1)} J(t+1) = \left[\varepsilon(t+1)\right]^2 \quad (5.2.30)$$

Thus one gets:

---

[5] One can derives from Equation 5.2.28 that an optimal value for α is $\alpha \approx 1/\phi(t)^T \phi(t)$.

$$\frac{1}{2}\frac{\partial J(t+1)}{\partial \hat{\theta}(t+1)} = \frac{\partial \varepsilon(t+1)}{\partial \hat{\theta}(t+1)} \varepsilon(t+1) \qquad (5.2.31)$$

From Equations 5.2.16 and 5.2.17 it follows that

$$\varepsilon(t+1) = y(t+1) - \hat{y}(t+1) = y(t+1) - \hat{\theta}(t+1)^T \phi(t) \qquad (5.2.32)$$

and respectively that

$$\frac{\partial \varepsilon(t+1)}{\partial \hat{\theta}(t+1)} = -\phi(t) \qquad (5.2.33)$$

Introducing Equation 5.2.33 into Equation 5.2.31, the parameter adaptation algorithm of Equation 5.2.20 becomes

$$\hat{\theta}(t+1) = \hat{\theta}(t) + F\phi(t)\varepsilon(t+1) \qquad (5.2.34)$$

This algorithm depends on $\varepsilon(t+1)$, which is a function $\hat{\theta}(t+1)$. In order to implement this algorithm, it is necessary to express $\varepsilon(t+1)$ as a function of $\varepsilon^\circ(t+1)$: $(\varepsilon(t+1) = f(\hat{\theta}(t), \phi(t), \varepsilon^\circ(t+1)))$.
    Equation 5.2.32 can be rewritten as

$$\varepsilon(t+1) = y(t+1) - \hat{\theta}(t)^T \phi(t) - \left[\hat{\theta}(t+1) - \hat{\theta}(t)\right]^T \phi(t) \qquad (5.2.35)$$

The first two terms of the right side correspond to $\varepsilon^\circ(t+1)$ and, from Equation 5.2.34, one gets

$$\hat{\theta}(t+1) - \hat{\theta}(t) = F\phi(t)\varepsilon(t+1) \qquad (5.2.36)$$

which allows one to write Equation 5.2.35 in the form

$$\varepsilon(t+1) = \varepsilon^\circ(t+1) - \phi(t)^T F \phi(t) \varepsilon(t+1) \qquad (5.2.37)$$

from which one derives the desired relation between $\varepsilon(t+1)$ and $\varepsilon^\circ(t+1)$:

$$\varepsilon(t+1) = \frac{\varepsilon^\circ(t+1)}{1 + \phi(t)^T F\phi(t)} \qquad (5.2.38)$$

and the algorithm of Equation 5.2.34 becomes

$$\hat{\theta}(t+1) = \hat{\theta}(t) + \frac{F\phi(t)\varepsilon^{o}(t+1)}{1 + \phi(t)^{T} F\phi(t)} \tag{5.2.39}$$

that is a stable algorithm regardless of the gain F (positive definite matrix). The division by $1 + \phi(t)^{T} F \phi(t)$ introduces a normalization that reduces the sensitivity of the algorithm with respect to F and $\phi(t)$.

The sequence of operation corresponding to the recursive estimation algorithms can be summarized as follows:

1. Before $t+1$: $u(t), u(t-1),\ldots, y(t), y(t-1),\ldots,$ $\phi(t), \hat{\theta}(t), F$ are available

2. Before $t+1$ one computes: $\dfrac{F\phi(t)}{1 + \phi(t)^{T} F\phi(t)}$ and $y^{o}(t+1)$ (given by Equation 5.2.13)

3. At instant $t+1$ $y(t+1)$ is acquired and $u(t+1)$ is applied

4. The parametric adaptation algorithm is implemented

   a) One computes $\varepsilon^{0}(t+1)$ by using Equation 5.2.15
   b) One computes $\hat{\theta}(t+1)$ from Equation 5.2.39
   c) (Optionally) one computes $\varepsilon(t+1)$

5. Return to step 1

5.2.3 Least Squares Algorithm

By using the gradient algorithm, at each step $\varepsilon^{2}(t+1)$ is minimized or, more precisely, one moves in the steepest decreasing direction of the criterion, with a step update depending on F. The minimization of $\varepsilon^{2}(t+1)$ at each step does not necessarily lead to the minimization of

$$\sum_{i=1}^{t} \varepsilon^{2}(i)$$

on a t-steps time horizon, as illustrated in Figure 5.7. In fact, in the proximity of the optimum, if the gain is not small enough, oscillations may occur around the minimum. On the other hand, in order to obtain a satisfactory convergence speed at the beginning, when the current estimation is theoretically far from the optimum, a high adaptation gain is preferable. The least squares algorithm offers, in fact, such a variation profile for the adaptation gain.

The same equations, as in the gradient algorithm, are considered for the plant, the prediction model and the prediction errors, namely Equations 5.2.15 to 5.2.22.

The aim is to find a recursive algorithm of the form of Equation 5.2.18 that minimizes the "least squares" criterion

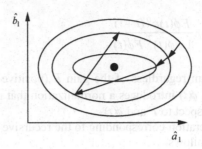

Figure 5.7. Evolution of an adaptation algorithm of the gradient type

$$\min_{\hat{\theta}(t)} J(t) = \frac{1}{t}\sum_{i=1}^{t}\left[y(i) - \hat{\theta}(t)^T \phi(i-1)\right]^2 = \frac{1}{t}\sum_{i=1}^{t}\varepsilon^2\left(i, \hat{\theta}(t)\right) \qquad (5.2.40)$$

The term $\hat{\theta}(t)^T \phi(i-1)$ corresponds to

$$\hat{\theta}(t)\phi(i-1) = -\hat{a}_1(t)y(i-1) + \hat{b}_1(t)u(i-1) = \hat{y}\left(i|\hat{\theta}(t)\right) \qquad (5.2.41)$$

This is the prediction of the output at instant i ($i \le t$) based on the parameter estimate at instant t obtained using t measurements. The objective is therefore the minimization of the sum of the squares of the prediction errors.

First, a parameter $\hat{\theta}$ must be estimated at instant t, so that it minimizes the sum of the squares of the differences between the output of the plant and the output of the prediction model over a horizon of t measurements. The value of $\hat{\theta}(t)$ that minimizes the criterion of Equation 5.2.40 is obtained by looking for the value that cancels $\partial J(t)/\partial \theta(t)$[6]:

$$\frac{\partial J(t)}{\partial \hat{\theta}(t)} = -2\sum_{i=1}^{t}\left[y(i) - \hat{\theta}(t)^T \phi(i-1)\right]\phi(i-1) = 0 \qquad (5.2.42)$$

From Equation 5.2.42, taking into account that

$$\left[\hat{\theta}(t)^T \phi(i-1)\right]\phi(i-1) = \phi(i-1)\phi(i-1)^T \hat{\theta}(t)$$

---

[6] This is the real minimum with the condition that the second derivative of the criterion, with respect to $\hat{\theta}(t)$ is positive, that is

$$\frac{\partial^2 J(t)}{\partial \hat{\theta}(t)^2} = 2\sum_{i=1}^{t}\phi(i-1)\phi(i-1)^T > 0 \text{, as it is in general the case for } t \ge dim\ \theta \text{ (see also Section}$$

5.3).

one obtains

$$\left[\sum_{i=1}^{t} \phi(i-1)\phi(i-1)^{T}\right]\hat{\theta}(t) = \sum_{i=1}^{t} y(i)\phi(i-1)$$

By left multiplying on the left both terms of this equation with

$$\left[\sum_{i=1}^{t} \phi(i-1)\phi(i-1)^{T}\right]^{-1}$$

it results in

$$\hat{\theta}(t) = \left[\sum_{i=1}^{t} \phi(i-1)\phi(i-1)^{T}\right]^{-1}\sum_{i=1}^{t} y(i)\phi(i-1) = F(t)\sum_{i=1}^{t} y(i)\phi(i-1) \quad (5.2.43)$$

where

$$F(t)^{-1} = \sum_{i=1}^{t} \phi(i-1)\phi(i-1)^{T} \quad (5.2.44)$$

This estimation algorithm is not recursive. In order to obtain a recursive algorithm, the estimation of $\hat{\theta}(t+1)$ is considered:

$$\hat{\theta}(t+1) = F(t+1)\sum_{i=1}^{t+1} y(i)\phi(i-1) \quad (5.2.45)$$

$$F(t+1)^{-1} = \sum_{i=1}^{t+1} \phi(i-1)\phi(i-1)^{T} = F(t)^{-1} + \phi(t)\phi(t)^{T} \quad (5.2.46)$$

And one should express it as a function of $\hat{\theta}(t)$:

$$\hat{\theta}(t+1) = \hat{\theta}(t) + \Delta\hat{\theta}(t+1) \quad (5.2.47)$$

From Equation 5.2.45 (adding and subtracting $\phi(t)\phi(t)^{T}\hat{\theta}(t)$) one gets

$$\sum_{i=1}^{t+1} y(i)\phi(i-1) = \sum_{i=1}^{t} y(i)\phi(i-1) + y(t+1)\phi(t) \pm \phi(t)\phi(t)^{T}\hat{\theta}(t) \quad (5.2.48)$$

Taking into account Equations 5.2.43, 5.2.45 and 5.2.46, Equation 5.2.48 can be rewritten as

$$\sum_{i=1}^{t+1} y(i)\phi(i-1) = F(t+1)^{-1}\hat{\theta}(t+1)$$

$$= F(t)^{-1}\hat{\theta}(t) + \phi(t)\phi(t)^T \hat{\theta}(t) + \phi(t)\left[y(t+1) - \hat{\theta}(t)^T \phi(t)\right]$$

(5.2.49)

But, on the basis of Equations 5.2.46 and 5.2.15, one gets

$$F(t+1)^{-1}\hat{\theta}(t+1) = F(t+1)^{-1}\hat{\theta}(t) + \phi(t)\varepsilon^o(t+1)$$

(5.2.50)

Multiplying on the left by $F(t+1)$, one gets

$$\hat{\theta}(t+1) = \hat{\theta}(t) + F(t+1)\phi(t)\varepsilon^o(t+1)$$

(5.2.51)

The adaptation algorithm of Equation 5.2.51 has a recursive form similar to the gradient algorithm given in Equation 5.2.24, with the difference that the gain matrix $F(t+1)$ is now time varying since it depends on the measurements (it automatically corrects the gradient direction and the step length). A recursive formula for $F(t+1)$ remains to be provided starting from the recursive formula for $F^{-1}(t+1)$ given in Equation 5.2.46. This is obtained by using the *matrix inversion lemma* (given below in a simplified form).

Lemma: Let F be a regular matrix of dimension ($n \times n$) and ϕ a vector of dimension n; then[7]

$$\left(F^{-1} + \phi\phi^T\right)^{-1} = F - \frac{F\phi\phi^T F}{1 + \phi^T F\phi}$$

(5.2.52)

From Equations 5.2.46 and 5.2.52 one gets

$$F(t+1) = F(t) - \frac{F(t)\phi(t)\phi(t)^T F(t)}{1 + \phi(t)^T F(t)\phi(t)}$$

(5.2.53)

and, regrouping the different equations, a first formulation of the recursive least squares (RLS) parameter adaptation algorithm (PAA) is given by

---

[7] One can simply multiply both terms by $F^{-1} + \phi\phi^T$ to verify the inversion formula.

$$\hat{\theta}(t+1) = \hat{\theta}(t) + F(t+1)\phi(t)\varepsilon^o(t+1) \tag{5.2.54}$$

$$F(t+1) = F(t) - \frac{F(t)\phi(t)\phi(t)^T F(t)}{1 + \phi(t)^T F(t)\phi(t)} \tag{5.2.55}$$

$$\varepsilon^o(t+1) = y(t+1) - \hat{\theta}(t)^T \phi(t) \tag{5.2.56}$$

An equivalent form of this algorithm is obtained by introducing the expression of $F(t+1)$ given by Equation 5.2.55 in Equation 5.2.54. Then it follows that

$$\left[\hat{\theta}(t+1) - \hat{\theta}(t)\right] = F(t+1)\phi(t)\varepsilon^o(t+1) = F(t)\phi(t)\frac{\varepsilon^o(t+1)}{1 + \phi(t)^T F(t)\phi(t)} \tag{5.2.57}$$

However from Equations 5.2.15, 5.2.16 and 5.2.17, also using Equation 5.2.57, one obtains:

$$\begin{aligned}
\varepsilon(t+1) &= y(t+1) - \hat{\theta}(t+1)\phi(t) = y(t+1) - \hat{\theta}(t)\phi(t) - \left[\hat{\theta}(t+1) - \hat{\theta}(t)\right]^T \phi(t) \\
&= \varepsilon^o(t+1) - \phi(t)^T F(t)\phi(t)\frac{\varepsilon^o(t+1)}{1 + \phi(t)^T F(t)\phi(t)} = \frac{\varepsilon^o(t+1)}{1 + \phi(t)^T F(t)\phi(t)}
\end{aligned} \tag{5.2.58}$$

which expresses the relation between the *a posteriori* prediction error and the *a priori* prediction error. Using this relation in Equation 5.2.57, an equivalent form of the parameter adaptation algorithm for the recursive least squares is obtained[8]:

$$\hat{\theta}(t+1) = \hat{\theta}(t) + F(t)\phi(t)\varepsilon(t+1) \tag{5.2.59}$$

$$F(t+1)^{-1} = F(t)^{-1} + \phi(t)\phi(t)^T \tag{5.2.60}$$

$$F(t+1) = F(t) - \frac{F(t)\phi(t)\phi(t)^T F(t)}{1 + \phi(t)^T F(t)\phi(t)} \tag{5.2.61}$$

---

[8] This equivalent form is especially used to analyze and understand the algorithm.

$$\varepsilon(t+1) = \frac{y(t+1) - \hat{\theta}(t)^T \phi(t)}{1 + \phi(t)^T F(t)\phi(t)} \qquad (5.2.62)$$

For the recursive least squares algorithm to be exactly equivalent to the non-recursive least squares algorithm, it must be started at instant $t_0 = dim\ \phi(t)$, since normally $F(t)^{-1}$ given by Equation 5.2.44 becomes non-singular for $t=t_0$. In practice, the algorithm is initialized at $t = 0$ by choosing

$$F(0) = \frac{1}{\delta}I = (GI)I \ ; \ 0 < \delta << 1 \qquad (5.2.63)$$

a typical value being $\delta = 0.001$ $(GI = 1000)$. It can be observed, from the expression of $F(t+1)^{-1}$ given by Equation 5.2.46 that the influence of this initial error decreases with time. A rigorous analysis (based on the stability theory - see Landau *et al.* 1997) shows nevertheless that for any positive definite matrix $F(0)$ $(F(0) > 0)$,

$$\lim_{t \to 0} \varepsilon(t+1) = 0$$

The recursive least squares algorithm is an algorithm with a decreasing adaptation gain. This is clearly seen if the estimation of a single parameter is considered. In this case $F(t)$ and $\phi(t)$ are scalars and Equation 5.2.61. becomes

$$F(t+1) = \frac{F(t)}{1 + \phi(t)^2 F(t)} \le F(t)$$

The recursive least squares algorithm gives, in fact, less and less weight to the new prediction errors, and thus to the new measurements.

As a consequence, this type of variation of the adaptation gain is not suitable for the estimation of time varying parameters, and other variation profiles must therefore be considered for the adaptation gain.

The least squares algorithm, presented up to now for $\theta(t)$ and $\phi(t)$ of dimension 2, may be generalized to the n-dimensional case on the basis of the description of discrete-time systems of the form

$$y(t) = \frac{q^{-d}B(q^{-1})}{A(q^{-1})} u(t) \qquad (5.2.64)$$

where

$$A(q^{-1}) = 1 + a_1 q^{-1} + \ldots + a_{n_A} q^{-n_A} \tag{5.2.65}$$

$$B(q^{-1}) = b_1 q^{-1} + \ldots + b_{n_B} q^{-n_B} \tag{5.2.66}$$

which can further be rewritten as

$$y(t+1) = -\sum_{i=1}^{n_A} a_i y(t+1-i) + \sum_{i=1}^{n_B} b_i u(t-d-i+1) = \theta^T \phi(t) \tag{5.2.67}$$

where

$$\theta^T = \left[a_1, \ldots a_{n_A}, b_1, \ldots, b_{n_B} \right] \tag{5.2.68}$$

$$\phi(t)^T = \left[-y(t) \ldots - y(t - n_A + 1), u(t-d) \ldots u(t-d-n_B+1) \right] \tag{5.2.69}$$

The *a priori* adjustable predictor is given in the general case by

$$\hat{y}^o(t+1) = -\sum_{i=1}^{n_A} \hat{a}_i y(t+1-i) + \sum_{i=1}^{n_B} \hat{b}_i u(t-d-i+1) = \hat{\theta}(t)^T \phi(t) \tag{5.2.70}$$

where

$$\hat{\theta}(t)^T = \left[\hat{a}_1(t), \ldots \hat{a}_{n_A}(t), \hat{b}_1(t), \ldots, \hat{b}_{n_B}(t) \right] \tag{5.2.71}$$

and, for the estimation of $\hat{\theta}(t)$, the algorithm given in Equations 5.2.54 to 5.2.56 is used with the appropriate dimension for $\hat{\theta}(t)$, $\phi(t)$ and $F(t)$.

5.2.4 Choice of the Adaptation Gain

The recursive formula for the inverse of the adaptation gain $F(t+1)^{-1}$ given by Equation 5.2.46 (or Equation 5.2.60) is generalized by introducing two weighting sequences $\lambda_1(t)$ and $\lambda_2(t)$, as indicated below:

$$F(t+1)^{-1} = \lambda_1(t) F(t)^{-1} + \lambda_2(t) \phi(t) \phi(t)^T$$
$$0 < \lambda_1(t) \leq 1 \; ; \; 0 \leq \lambda_2(t) < 2 \; ; \; F(0) > 0 \tag{5.2.72}$$

Note that $\lambda_1(t)$ and $\lambda_2(t)$ in Equation 5.2.72 have the opposite effect: $\lambda_1(t) < 1$ tends to increase the adaptation gain (the gain inverse decreases), $\lambda_2(t) > 0$ tends to decrease the adaptation gain (the gain inverse increases). For each choice of sequences $\lambda_1(t)$ and $\lambda_2(t)$ a different variation profile of the adaptation gain is found and, consequently, an interpretation in terms of the error criterion that is minimized by the PAA[9].

Using the matrix inversion lemma given by Equation 5.2.52, one obtains from Equation 5.2.72[10]

$$F(t+1) = \frac{1}{\lambda_1(t)}\left[F(t) - \frac{F(t)\phi(t)\phi(t)^T F(t)}{\dfrac{\lambda_1(t)}{\lambda_2(t)} + \phi(t)^T F(t)\phi(t)} \right] \qquad (5.2.73)$$

Next a selection of choices for $\lambda_1(t)$ and $\lambda_2(t)$ and their interpretations will be given.

A.1: Decreasing Gain (RLS)
In this case

$$\lambda_1(t) = \lambda_1 = 1 \quad ; \quad \lambda_2(t) = 1 \qquad (5.2.74)$$

and $F(t+1)^{-1}$ is given by Equation 5.2.60 which leads to a decreasing adaptation gain. The minimized criterion is expressed by Equation 5.2.40.

This type of profile is suited for the identification of stationary systems (with constant parameters).

A.2: Constant Forgetting Factor
In this case

$$\lambda_1(t) = \lambda_1 \quad ; \quad 0 < \lambda_1 < 1 \quad ; \quad \lambda_2(t) = \lambda_2 = 1 \qquad (5.2.75)$$

Typical values for λ_1 are: $\lambda_1 = 0.95,...,0.99$.

The criterion to be minimized will be

$$J(t) = \sum_{i=1}^{t} \lambda_1^{(t-i)}\left[y(i) - \hat{\theta}(t)^T \phi(i-1) \right]^2 \qquad (5.2.76)$$

---

[9] $F(t+1)^{-1}$ given by Equation 5.7.72 can be interpreted as the output of a filter characterized by the pulse transfer operator $H(q^{-1}) = \lambda_2(t)/(1-\lambda_1(t)q^{-1})$ whose input is $\phi\phi^T$.

[10] More numerically robust updating algorithms for the adaptation gain are available. See Appendix F.

The effect of $\lambda_1 < 1$ is to introduce a decreasing weighting on the past data ($i < t$). This is why λ_1 is known as the *forgetting factor*. The maximum weight is given to the most recent error.

This type of profile is suited for the identification of slowly time varying systems[11].

A.3: Variable Forgetting Factor
In this case

$$\lambda_2(t) = \lambda_2 = 1 \tag{5.2.77}$$

and the forgetting factor λ_1 is given by

$$\lambda_1(t) = \lambda_0 \lambda_1(t-1) + 1 - \lambda_0 \; ; \; 0 < \lambda_0 < 1 \tag{5.2.78}$$

typical values being: $\lambda_1(0) = 0.95,...,0.99$; $\lambda_0 = 0.95,...,0.99$.

Equation 5.2.78 leads to a forgetting factor that asymptotically tends towards 1. The criterion minimized will be

$$J(t) = \sum_{i=1}^{t} \left[\sum_{j=1}^{t-1} \lambda_1(j-i) \right] \left[y(i) - \hat{\theta}(t)^T \phi(i-1) \right]^2 \tag{5.2.79}$$

As $\lambda_1(t)$ tends towards 1 for large i, only the initial data are forgotten (the adaptation gain tends towards a decreasing gain).

This type of profile is *highly recommended for the identification of stationary systems*, since it avoids a too rapid decrease of the adaptation gain, thus generally resulting in an acceleration of the convergence (by maintaining a high gain at the beginning when the estimates are far from the optimum).

A.4: Constant Trace
In this case, $\lambda_1(t)$ and $\lambda_2(t)$ are automatically chosen at each step in order to ensure a constant trace of the gain matrix (constant sum of the diagonal terms)

$$tr F(t+1) = tr F(t) = tr F(0) = nGI \tag{5.2.80}$$

in which n is the number of parameters and GI the initial gain (typical values: $GI=0,1...,4$), the matrix $F(0)$ having the form

---

[11] When an excitation is not provided ($\phi(t)\phi(t)^T = 0$), $F(t+1)^{-1}$ goes towards zero (because in this case $F(t+1)^{-1} = \lambda_1 F(t)^{-1}$, $\lambda_1 < 1$), leading to very high adaptation gains, a situation that should be avoided.

$$F(0) = \begin{bmatrix} GI & & 0 \\ & \cdot & \\ & & \cdot \\ 0 & & GI \end{bmatrix}$$

(5.2.81)

The minimized criterion is of the form

$$J(t) = \sum_{i=1}^{t} f(t,i)\left[y(i) - \hat{\theta}(t)^T \phi(i-1)\right]^2$$

(5.2.82)

in which $f(t, i)$ represents the forgetting profile.

Using this technique, at each step there is a movement in the optimal direction of the RLS but the gain is maintained approximately constant (*reinflation* of the RLS gain).

The values of $\lambda_1(t)$ and $\lambda_2(t)$ are determined from the equation

$$trF(t+1) = \frac{1}{\lambda_1(t)} tr\left[F(t) - \frac{F(t)\phi(t)\phi(t)^T F(t)}{\alpha(t) + \phi(t)^T F(t)\phi(t)}\right] = trF(t)$$

(5.2.83)

by imposing the ratio $\alpha(t) = \lambda_1(t)/\lambda_2(t)$ (Equation 5.2.83 is obtained from Equation 5.2.73).

This type of profile is suited for the identification of systems with time varying parameters.

A.5: Decreasing Gain + Constant Trace
In this case, there is a switch from A1 to A4 when

$$trF(t) \le nG \quad ; \quad G = 0.1 \text{ to } 4$$

(5.2.84)

where G is fixed at the beginning.

This profile is suited for the identification of time variable systems in the absence of initial information on the parameters.

A.6: Variable Forgetting Factor +Constant Trace
In this case, there is a switch from A3 to A4 when

$$trF(t) \le nG$$

(5.2.85)

The domain of application is the same as for *A 5*.

A.7: Constant Gain (Improved Gradient Algorithm)
In this case

$$\lambda_1(t) = \lambda_1 = 1 \; ; \; \lambda_2(t) = \lambda_2 = 0 \tag{5.2.86}$$

and thus from Equation 5.2.72, it results that

$$F(t+1) = F(t) = F(0) \tag{5.2.87}$$

The improved gradient adaptation algorithm given by Equations 5.2.34 or 5.2.39 is then obtained.

This algorithm can be used to identify stationary or time varying systems with few parameters (≤ 3), and in the presence of a reduced noise level.

This type of adaptation gain results in performances which are inferior to those provided by the A1, A2, A3 and A4 profiles, but it is simpler to implement.

Choice of the Initial Adaptaion Gain F(0)
The initial adaptation gain $F(0)$ is of the form given by Equation 5.2.63, respectively Equation 5.2.81.

In the absence of initial information upon the parameters to be estimated (a typical choice is to set the initial estimation to zero), a high initial gain (*GI*) is chosen for reasons that have been explained in Section 5.2.3 (Equation 5.2.63). A typical value is *GI* = 1000.

On the other hand, if an initial parameter estimation is available (resulting for example from a previous identification), a low initial gain is chosen. In general, in this case *GI* ≤ *1*.

Since the adaptation gain decreases as the correct model parameter estimations are approached (a significant index is its trace), the adaptation gain may be interpreted as an index of the accuracy of the estimation (or prediction). This explains the choices of *F(0)* proposed above. Note that, under certain hypotheses, *F(t)* is effectively an index of the quality of the estimation because it represents the covariance of the parameter error vector $\tilde{\theta}(t) = \hat{\theta}(t) - \theta$ (see Landau 2001b). This property can give some information on the evolution of an estimation procedure. If the trace of $F(t)$ is not significantly decreasing, the parameter estimation, in general, is bad. This phenomenon occurs, for example, when the amplitude and the type of the input used are not suited for the identification. The importance of the nature of the identification signal will be discussed in the following section.

5.3 Choice of the Input Sequence for System Identification

5.3.1 The Problem

The convergence towards zero of the prediction error $\varepsilon(t)$ does not always imply that the estimated model parameters will converge towards the true parameters of the plant model.

This will be illustrated by an example. Let the discrete-time plant model be described by

$$y(t+1) = -a_1 y(t) + b_1 u(t) \tag{5.3.1}$$

and consider an estimated model described by

$$\hat{y}(t+1) = -\hat{a}_1 y(t) + \hat{b}_1 u(t) \tag{5.3.2}$$

where $\hat{y}(t+1)$ is the output predicted by the estimated model.

Now assume that $u(t) = constant$ and that the parameters a_1, b_1, $\hat{a}_1, \hat{b}_1$ verify the following relation:

$$\frac{b_1}{1+a_1} = \frac{\hat{b}_1}{1+\hat{a}_1} \tag{5.3.3}$$

i.e. that the steady state gains of the plant and of the estimated model are the same, even if the condition $\hat{b}_1 = b_1$ and $\hat{a}_1 = a_1$ don't hold.

Under the effect of the constant input $u(t) = u$, the plant output will be given by

$$y(t+1) = y(t) = \frac{b_1}{1+a_1} u \tag{5.3.4}$$

and the output of the estimated prediction model will be given by

$$\hat{y}(t+1) = \hat{y}(t) = \frac{\hat{b}_1}{1+\hat{a}_1} u \tag{5.3.5}$$

Therefore, considering Equation 5.3.3, it results that

$$\varepsilon(t+1) = y(t+1) - \hat{y}(t+1) = 0 \quad \text{for } u(t) = const; \hat{a}_1 \neq a_1 ; \hat{b}_1 \neq b_1 \tag{5.3.6}$$

It can thus be concluded from this example that the application of a constant input does not allow the two models to be distinguished, since they both have the same steady state gain.

If the frequency characteristics of both systems are plotted, we obtain the curves shown in Figure 5.8.

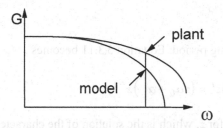

Figure 5.8. Frequency characteristics of two systems with the same steady state gain

It can be observed that, in order to put in evidence the differences between the two models (that is, between the parameters), one must apply a signal $u(t)=sin(\omega t)$ ($\omega \neq 0$) and not a signal $u(t) = constant$.

Let us analyze this phenomenon in greater detail. When the prediction error is null, from Equations 5.3.1 and 5.3.2 one obtains

$$\varepsilon(t+1) = y(t+1) - \hat{y}(t+1) = 0 \quad ; \quad -\left[a_1 - \hat{a}_1\right]y(t) + \left[b_1 - \hat{b}_1\right]u(t) = 0 \quad (5.3.7)$$

From Equation 5.3.1, $y(t)$ can be expressed as a function of $u(t)$:

$$y(t) = \frac{b_1 q^{-1}}{1 + a_1 q^{-1}} u(t) \quad (5.3.8)$$

Introducing the expression of $y(t)$ given by Equation 5.3.8 in Equation 5.3.7, the following equivalent relation follows:

$$\left\lfloor \left(\hat{a}_1 - a_1\right)b_1 q^{-1} + \left(b_1 - \hat{b}_1\right)\left(1 + a_1 q^{-1}\right)\right\rfloor u(t)$$
$$= \left[\left(b_1 - \hat{b}_1\right) + q^{-1}\left(b_1 \hat{a}_1 - a_1 \hat{b}_1\right)\right]u(t) = 0 \quad (5.3.9)$$

We are concerned with finding the characteristics of $u(t)$ so that verification of Equation 5.3.9. results in zero parametric errors.

Note that

$$b_1 - \hat{b}_1 = \alpha_0 \quad ; \quad b_1 \hat{a}_1 - a_1 \hat{b}_1 = \alpha_1 \quad (5.3.10)$$

Equation 5.3.9 can then be written as

$$\left(\alpha_0 + \alpha_1 q^{-1}\right)u(t) = 0 \tag{5.3.11}$$

that is a difference equation having a solution of the discretized exponential type.
Let

$$u(t) = z^t = e^{sT_s t} \tag{5.3.12}$$

where T_s is the sampling period. Equation 5.3.11 becomes

$$\left(\alpha_0 + z^{-1}\alpha_1\right)z^t = \left(z\alpha_0 + \alpha_1\right)z^{t-1} = 0 \tag{5.3.13}$$

and it will be verified for z, which is the solution of the characteristic equation

$$z\alpha_0 + \alpha_1 = 0 \tag{5.3.14}$$

One gets

$$z = -\frac{\alpha_1}{\alpha_0} = e^{\sigma T_s} \qquad \sigma = real \tag{5.3.15}$$

and the aperiodic solution

$$u(t) = e^{\sigma T_s t} \tag{5.3.16}$$

leads to the verification of Equations 5.3.11 and 5.3.7 respectively, without having $\hat{a}_1 = a_1$ and $\hat{b}_1 = b_1$. In practice, the signal $u(t) = constant$, previously considered, corresponds to $\sigma = 0$, that is - $\alpha_1 = \alpha_0$. However

$$-\alpha_1 = \alpha_0 \Rightarrow b_1 - \hat{b}_1 = a_1\hat{b}_1 - b_1\hat{a}_1 \Rightarrow \frac{b_1}{1+a_1} = \frac{\hat{b}_1}{1+\hat{a}_1}$$

In other words, if $u(t) = constant$, we can just identify the static gain.

Therefore it is necessary to apply a signal $u(t)$ such that $\hat{a}_1 = a_1$ and $\hat{b}_1 = b_1$. This can be obtained if $u(t)$ is not a possible solution of Equation 5.3.11.
Let

$$u(t) = e^{j\omega T_s t} \quad or \quad e^{-j\omega T_s t} \tag{5.3.17}$$

Equation 5.3.11 becomes (for $u(t) = e^{j\omega T_s t}$)

$$\left[e^{j\omega T_s}\alpha_0 + \alpha_1\right]e^{j\omega T_s(t-1)} = 0 \qquad (5.3.18)$$

As α_0 and α_1 are real, $e^{j\omega T_s}$ cannot be a root of the characteristic equation and it results that $\varepsilon(t) = 0$ will be obtained only if

$$\alpha_0 = \alpha_1 = 0 \Rightarrow \hat{b}_1 = b_1 \ , \ \hat{a}_1 = a_1 \qquad (5.3.19)$$

It is this type of input which was previously proposed when the frequency characteristics of the two models were examined ($sin\ \omega t = (e^{j\omega t}-e^{-j\omega t})/2j$). A non-zero frequency sinusoid is thus required in order to identify two parameters.

This approach for determining the input $u(t)$ allowing satisfactory model parameter identification may also be applied to systems of the general form

$$y(t) = -\sum_{i=1}^{n_A} a_i y(t-i) + \sum_{i=1}^{n_B} b_i u(t-d-i) \qquad (5.3.20)$$

for which the total number of parameters to be identified is:

$$number\ of\ parameters = n_A + n_B$$

In this case $u(t)$ can be chosen as a sum of p-sinusoids of different frequencies:

$$u(t) = -\sum_{i=1}^{p} sin\ \omega_i T_e t \qquad (5.3.21)$$

and the value p, allowing good parameter identification, is given by[12]

$$\left.\begin{array}{lll} n_A + n_B = & even & p \geq \dfrac{n_A + n_B}{2} \\[3mm] n_A + n_B = & odd & p \geq \dfrac{n_A + n_B + 1}{2} \end{array}\right\} \qquad (5.3.22)$$

In other words, *in order to identify a correct model, it is necessary to apply a frequency rich input. The standard solution in practice is provided by the use of "pseudo-random binary sequences".*

---

[12] The condition of Equation 5.3.22 also guarantees that $\sum_{i=1}^{t}\phi(i-1)\,\phi(i-1)^T$ is an invertible positive definite matrix for $t \geq n_A + n_B = dim\ \phi$ (see Equations 5.2.42 to 5.2.46).

5.3.2. Pseudo-Random Binary Sequences (PRBS)

Pseudo-random binary sequences are sequences of rectangular pulses, modulated in width, which approximate a discrete-time white noise and thus have a spectral content *rich* in frequencies.

They owe their name *pseudo-random* to the fact that they are characterized by a *sequence length* within which the variations in pulse width vary randomly, but that, over a large time horizon, they are periodic, the period being defined by the length of the sequence.

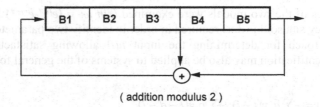

(addition modulus 2)

Figure 5.9. Generation of a PRBS of length $2^5-1 = 31$ sampling periods

The PRBS are generated by means of shift registers with feedback (implemented in hardware or software). The maximum length of a sequence is 2^N-1 in which N is the number of cells of the shift register. Figure 5.9 presents the generation of a PRBS of length $31 = 2^5-1$ obtained by means of a five cells shift register. Note that at least one of the N cells of the shift register should have an initial logic value different from zero (one generally takes all the initial values of the N cells equal to the logic value 1).

Table 5.1. Generation of maximum length PRBS

Number of cells N	Length of the sequence $L = 2^N - 1$	Bits added B_i and B_j
2	3	1 and 2
3	7	1 and 3
4	15	3 and 4
5	31	3 and 5
6	63	5 and 6
7	127	4 and 7
8	255	2, 3, 4 and 8
9	511	5 and 9
10	1023	7 and 10

Table 5.1 gives, for different numbers of cells, the structure enabling the generation of maximum length PRBS.

A C^{++} program as well as a MATLAB® function (*prbs.m*) for generation of pseudo-random binary sequences can be found on the book website.

Note also a very important characteristic element of the PRBS: *the maximum duration of a PRBS pulse (t_{im}) is equal to NT_s* (where N is, the number of cells and T_s is the sampling period). This property is to be considered when choosing a PRBS for system identification.

Sizing of a PRBS

In order to identify correctly the steady state gain of the plant dynamic model, the duration of, at least, one of the pulses (*e.g.* the maximum duration pulse) must be greater than the rise time t_R of the plant (including the time delay). The maximum duration of a pulse being $N.T_s$, the following condition results:

$$t_{im} = NT_s > t_R \tag{5.3.23}$$

that is illustrated in Figure 5.10.

From Equation 5.3.23, one derives N and thus the length of the sequence 2^N-1.

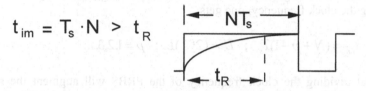

Figure 5.10. Choice of the maximum duration of a pulse in a PRBS

Furthermore, in order to cover the entire frequency spectrum generated by a particular PRBS, the length of a test must be at least equal to the length of the sequence. In a large number of cases, the duration of the test (L) is chosen equal to the length of the sequence. If the duration of the test is specified, it must therefore be ensured that

$$2^{N-1}T_s < L \; ; \; L = test\ duration \tag{5.3.24}$$

Note that the condition of Equation 5.3.23 can result in fairly large values of N, corresponding to sequence lengths of prohibitive duration.

This is why, in many practical situations, a sub-multiple of the sampling frequency is chosen as the clock frequency for the PRBS. If

$$f_{PRBS} = \frac{f_s}{p} \; ; \; p = 1,2,3,... \tag{5.3.25}$$

then Equation 5.2.23 becomes

$$t_{im} = pNT_s > t_R \tag{5.3.26}$$

This approach is more interesting than the extension of the sequence length (increase of N) in order to satisfy Equation 5.3.23. Indeed, if one passes from N to $N' = N + 1$, the maximum duration of a pulse passes from NT_s to $(N+1)T_s$, but the duration of the sequence is doubled $L' = 2L$. On the other hand, if $f_{PRBS} = f_s/2$ is chosen, the maximum duration of a pulse passes from NT_s to $2NT_s$ for a doubled sequence duration $L' = 2L$.

From a comparison of the two approaches, it results that the second approach (frequency division) enables a pulse of greater duration to be obtained for an identical duration of the sequence and thus of the test. If p is the integer frequency divider, one has in the case of clock frequency division (d_{max} = maximum pulse duration)

$$t_{im} = pNT_s \;\; ; \;\; L' = pL = p(2^{N-1})T_s \;\; ; \;\; p = 1,2,3,...$$

In the case of an augmentation of the number of registers N by p-1, without changing the clock frequency, one gets

$$t_{im} = (N+p-1)T_s \;\; ; \;\; L' = (2^{p-1})L \;\; ; \;\; p = 1,2,3,...$$

Note that dividing the clock frequency of the PRBS will augment the spectral density in low frequencies (which it is desired) but will reduce the frequency range corresponding to a constant spectral density.

As an example, the spectral densities of the PRBS sequences generated with $N=8$, for $p=1,2,3$ are represented in Figure 5.11.

One can observe that for $p>1$ the signal energy is augmented at low frequencies but it is reduced at high frequencies. Furthermore, for $p=3$ there is a hole at $f_s/3$ (the PRBS does not contain the sinusoid of frequency $f_s/3$).

In general this will not affect the quality of the identification either because the plant to be identified has a reduced band pass with respect to the sampling frequency, or because the effect of the reduction of the signal/noise ratio at high frequencies can be compensated by using appropriate parameter estimation algorithms. However it is recommended to use $p \le 4$ (see Landau et al. 1997; Landau 2001b).

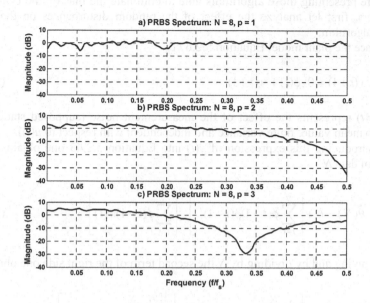

Figure 5.11a-c. Spectral density for a PRBS: **a** N=8, p=1 ; **b** N=8, p=2 ; **c** N=8, p=3

Choice of the Magnitude of the PRBS
The magnitude of the PRBS may be very small, but it must be larger than the amplitude of the residual noise. If the ratio signal/noise is too small, it is necessary to augment the length of the test in order to obtain a good parameter estimation.

Note that, in a large number of applications, the significant increase in the PRBS level may be undesirable in view of the non-linear character of the plants to be identified (we are concerned with the identification of a linear model around an operating point).

5.4 Effects of Random Disturbances upon Parameter Estimation

The plant measured output is in general contaminated by noise. This is due either to the effect of random disturbances acting at different points of the plant, or to measurement noises. These random disturbances are frequently modeled by ARMA models, the plant plus the disturbance being modeled by an ARMAX model (see Chapter 4, Section 4.1. for the description of random disturbances).

These disturbances introduce errors into the identification of the plant model parameters when the recursive (or non recursive) least squares algorithm is used. This type of error is called the *bias* of parameters.

Before presenting those algorithms able to eliminate the bias on the estimated parameters, first let analyze the effect of the random disturbances on the least squares algorithm.

Replace the plant model Equation 5.2.67 by

$$y(t+1) = \theta^T \phi(t) + w(t+1) \tag{5.4.1}$$

where $w(t)$ represents the effect of the measurement noise (supposed stationary, with zero mean value, finite variance and independent with respect to $u(t)$).

By introducing this expression of $y(t)$ into Equation 5.2.43 one obtains for a number of data N

$$\hat{\theta}(N) = \theta + \left[\sum_{t=1}^{N} \phi(t-1)\phi(t-1)^T \right]^{-1} \left[\sum_{t=1}^{N} \phi(t-1)w(t) \right] \tag{5.4.2}$$

By multiplying and by dividing by N the second term of the right side one obtains

$$\hat{\theta}(N) = \theta + \left[\frac{1}{N}\sum_{t=1}^{N} \phi(t-1)\phi(t-1)^T \right]^{-1} \left[\frac{1}{N}\sum_{t=1}^{N} \phi(t-1)w(t) \right] \tag{5.4.3}$$

We are interested in the properties of $\hat{\theta}(N)$ when $N \to \infty$ and we would like to find the conditions allowing one to obtain an asymptotically unbiased estimation ($\lim_{N\to\infty} \hat{\theta}(N) = \theta$). By examining Equation 5.4.3, as $\lim_{N\to\infty} \sum_{t=1}^{N} \phi(t-1)\phi(t-1)^T$ is supposed non-singular (the system is correctly excited – see Section 5.3), it results that the condition to obtain an asymptotically unbiased estimation is

$$\lim_{N\to\infty} \left[\frac{1}{N} \sum_{i=1}^{N-1} \phi(t-1)w(t) \right] = E\{\phi(t-1)w(t)\} = 0 \tag{5.4.4}$$

Thus, asymptotically unbiased parameter estimation will be obtained only if $\phi(t-1)$ and $w(t)$ are uncorrelated. Unfortunately, this will hold only in the case in which $w(t) = e(t) = white\ noise$. From Equation 5.4.1 it results that $y(t)$ depends on $w(t)$ and, as a consequence, $\phi(t-1)$, which contains $y(t-1),\ y(t-2), ...,\ y(t-n_A)$, is a function of $w(t-1),\ w(t-2),\ ...,\ w(t-n_A)$. Therefore

$$\phi(t-1)^T = f(w(t-1), w(t-2), ..., w(t-n_A), ...) \tag{5.4.5}$$

Equation 5.4.4 becomes

$$E\{w(t-i)w(i)\} = 0 \qquad \text{pour } i = 1,2,...n_A \tag{5.4.6}$$

and Equation 5.4.6 is satisfied only by white noise (see Section 4.1).

This clearly shows that the use of a least squares algorithm will lead to an unbiased estimation only in the very unlikely practical situation where $w(t) = e(t)$ = *white noise*.

In particular, the parameter estimation will be biased if the system "plant + disturbance" is modeled by an ARMAX model that is representative of many situations encountered in practice.

Let

$$y(t+1) = -a_1 y(t) + b_1 u(t) + c_1 e(t) + e(t+1)$$

One then gets:

$$w(t) = c_1 e(t-1) + e(t);$$
$$w(t-1) = c_1 e(t-2) + e(t-1)$$

and

$$E\{w(t)w(t-1)\} = c_1 E\{e(t-1)^2\} + E\{e(t)e(t-1)\} + c_1 E\{e(t-2)e(t)\}$$
$$+ c_1^2 E\{e(t-1)e(t-2)\} = c_1 E\{e(t-1)^2\} = c_1 \sigma^2 \neq 0$$

Suppose now that the exact value $\hat{\theta} = \theta$ is known, and we require that the estimation algorithm leaves unchanged (asymptotically) the estimated value. For $\hat{\theta} = \theta$ the least squares predictor equation is written as

$$\hat{y}(t+1|\theta) = \theta^T \phi(t) \tag{5.4.7}$$

and the prediction error becomes

$$\varepsilon(t+1|\theta) = y(t+1) - \hat{y}(t+1|\theta) = w(t+1) \tag{5.4.8}$$

It thus results from Equation 5.4.4 that in order to obtain $\hat{\theta}(N) = \theta$ when $t \to \infty$ there is a necessary condition to be satisfied for $\hat{\theta}(t) = \theta = const$ (by replacing $w(t)$ with $\varepsilon(t,\theta)$ in Equation 5.4.4)[13]:

---

[13] Notation $\phi(t-1, \theta)$ and $\varepsilon(t, \theta)$ indicates that these variables have been obtained with a fixed vector of estimated parameter θ.

$$\lim_{N\to\infty}\left[\frac{1}{N}\sum_{i=1}^{N-1}\phi(t-1,\theta)\varepsilon(t,\theta)\right] = E\{\phi(t-1,\theta)\varepsilon(t,\theta)\} = 0 \qquad (5.4.9)$$

To avoid *bias*, one needs to choose other observation vectors, other types of predictors and other adaptation errors in order that

$$E\{\phi(t)\varepsilon(t+1)\} = 0 \quad \text{for} \quad \hat{\theta} \equiv \theta \qquad (5.4.10)$$

Two criteria have been retained for the generation of algorithms fulfilling the condition of Equation 5.4.10 and asymptotically leading to unbiased parameter estimations:

1. $\varepsilon(t+1)$ (or $\nu(t+1) = adaptation\ error$) is a white noise for $\hat{\theta} \equiv \theta$
2. $\phi(t)$ and $\varepsilon(t+1)$ (or $\nu(t+1)$) are uncorrelated (or independent) for $\hat{\theta} \equiv \theta$

It is the elimination of the bias on the estimated parameters in presence of disturbances that is at the origin of the development of most of identification methods.

5.5 Structure of Recursive Identification Methods

All the recursive identification methods correspond to the basic scheme given in Figure 5.12.

They all use the same structure for the parametric adaptation algorithm (PAA) with the different possible choices for the "adaptation gain"

$$\hat{\theta}(t+1) = \hat{\theta}(t) + F(t)\Phi(t)\varepsilon(t+1) \qquad (5.5.1)$$

$$F^{-1}(t+1) = \lambda_1(t)F(t) + \lambda_2\Phi(t)\Phi(t)^T$$
$$0 < \lambda_1(t) \le 1 \ ; \ 0 \le \lambda_2(t) < 2 \ ; \ F(0) > 0 \qquad (5.5.2)$$

$$\varepsilon(t+1) = \frac{\varepsilon^0(t+1)}{1 + \Phi(t)^T F(t)\Phi(t)} \qquad (5.5.3)$$

where $\varepsilon(t+1)$ and $\varepsilon^0(t+1)$ are the *a posteriori* and *a priori* prediction errors, respectively, $\varPhi(t)$ is the *observation vector*, $F(t)$ is the *adaptation gain* and $\hat\theta(t)$ represents the vector of the *estimated parameters*.

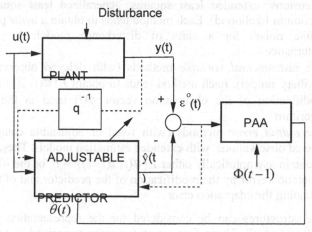

Figure 5.12. Structure of recursive identification methods

The different identification methods can be classified by:

- The structure of the predictor
- The origin of the elements of the observation vector ($\varPhi$)
- The dimension of the adjustable parameter vector $\hat\theta(t)$ and of the observation vector $\varPhi(t)$
- The way to compute the prediction error and respectively the adaptation errors

The convergence properties, in the presence of random disturbances, will depend on the different choices indicated above.

Table 5.2 gives a compact description of the various recursive identification methods which will be discussed in Chapter 6.

All these identification methods are classified in two categories, with respect to the criterion considered for their development (in order to obtain unbiased estimated parameters):

- Identification methods based on the whitening of the prediction error (ε)
- Identification methods based on the uncorrelation of the observation vector ($\varPhi$) and the prediction error (ε) ($E\{\varPhi(t)\,\varepsilon(t+1)\} \neq 0$)

Since the different methods have been developed in order to verify one of the two criteria mentioned above, the models identified with a particular method must be validated by using the corresponding criterion, used to define the objective of the method itself. It results that *two validation techniques* are available, allowing one to verify, according to the case, one of the two criteria.

If one considers in more detail the structure of the predictor and the choice of the observations vector, three types of methods can be distinguished:

- The *equation error* methods (recursive least squares and the different extensions: extended least squares, generalized least squares, recursive maximum likelihood). Each method tends to obtain a *white* prediction error (white noise) for a class of disturbance models by modeling the disturbance
- The *instrumental variable* methods (with delayed observations or with auxiliary model). Each method tends to obtain $E\{\Phi(t)\ \varepsilon(t+1)\} = 0$ by the modification of the observation vector $\Phi(t)$ used in the least squares algorithm
- The *output error* methods (with fixed or adjustable compensator, with filtered observations, with extended estimation model). These methods tend to obtain asymptotically either $E\{\Phi(t)\ \varepsilon(t+1)\} = 0$, or the whitening of the adaptation error by the modification of the predictor and of the method for obtaining the adaptation error

Four model structures can be considered for the representation of the system " plant + disturbance". These four structures are summarized in Figure 5.13 and they are briefly described below.

S1: $A(q^{-1})\ y(t) = q^{-d}\ B(q^{-1})\ u(t) + e(t)$
The method that can be used for this structure is:
1.   Recursive Least Squares (RLS)

S2: $A(q^{-1})\ y(t) = q^{-d}\ B(q^{-1})\ u(t) + A(q^{-1})\ w(t)$
This structure corresponds to a model of the form:

$$y(t) = \frac{q^{-d} B(q^{-1})}{A(q^{-1})} u(t) + w(t)$$

where $w(t)$ is a non-modeled disturbance for which it is assumed only that it is of zero mean value, finite power and *independent of the input*.
The methods that can be used for this structure are:

1.   Instrumental Variable with Auxiliary Model (IVAM)
2.   Output Error with Fixed Compensator (OEFC)
3.   Output Error with Filtered Observations (OEFO)
4.   Output Error with Adaptive Filtered Observations (OEAFO)

S3: $A(q^{-1})\ y(t) = q^{-d}\ B(q^{-1})\ u(t) + C(q^{-1})\ e(t)$
The methods that can be used for this structure are:

1.   Extended Least Squares (ELS)
2.   Recursive Maximum Likelihood (RML)
3.   Output Error with Extended Prediction Model (OEEPM)

S4: $A(q^{-1}) y(t) = q^{-d} B(q^{-1}) u(t) + [1/C(q^{-1})] e(t)$
The method that can be used for this structure is:

1.   Generalized Least Squares (GLS)

S1: $A(q^{-1}) y(t) = q^{-d} B(q^{-1}) u(t) + e(t)$

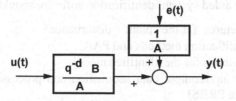

S2: $A(q^{-1}) y(t) = q^{-d} B(q^{-1}) u(t) + A(q^{-1}) w(t)$

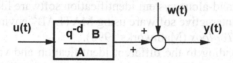

S3: $A(q^{-1}) y(t) = q^{-d} B(q^{-1}) u(t) + C(q^{-1}) e(t)$

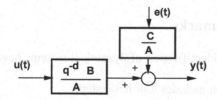

S4: $A(q^{-1}) y(t) = q^{-d} B(q^{-1}) u(t) + [1/C(q^{-1})]e(t)$

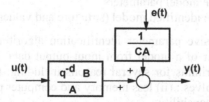

Figure 5.13. Structures of the "plant + disturbance" models
$(A(q^{-1}) = 1 + a_1 q^{-1} + ..., B(q^{-1}) = b_1 q^{-1} + ..., C(q^{-1}) = 1 + c_1 q^{-1} + ...)$

No single « plant + disturbance » structure exists that can describe all the situations encountered in practice

In practice, it can be pointed out that structure 3 corresponds to slightly less than two-thirds of situations, structure 2 corresponds to approximately one-third of the situations, and structures 1 and 4 correspond to the remaining situations.

Note that there does not exist a unique identification method which may be used with all the possible « plant + disturbance » structures such that the estimated parameters are always unbiased.

Since a unique "plant + disturbance" structure for describing all the situations that may be encountered in practice does not exist and since a unique identification method that always provides unbiased parameter estimations does not exist, it results that an interactive procedure for the system identification is required. Therefore a computer aided system identification software should provide:

- Different structures for the "plant + disturbance"
- Different identification methods and PAA
- Validation methods for the identified models
- A system for input/output data acquisition and processing (including the generation of a PRBS)
- Tools for the model analysis
- Tools for visualising different graphs and plots

There are dedicated stand-alone system identification software like *WinPIM⁺* (Adaptech 1996a) or interactive software using MATLAB® environment like the *System Identification Toolbox* (Mathworks 1998).

Routines corresponding to the different identification and validation methods, which will be presented in Chapter 6, are available in Scilab and MATLAB® environments (see Appendix H) and can be downloaded from the book website.

5.6 Concluding Remarks

Basic elements for the identification of dynamical systems have been laid down in this chapter.

System identification includes four basic steps:

1. Input/output data acquisition under an experimental protocol
2. Selection or estimation of *model* structure (complexity)
3. Estimation of the model parameters
4. Validation of the identified model (structure and values of parameters)

Recursive or non-recursive parameter identification algorithms can be used for estimating the parameter of a model from input-output data. Preference has been given to recursive algorithms for several reasons including: (i) estimation of the model as the system evolves ; (ii) less memory and computer power required ; (iii) real-time identification capability.

The core of the recursive identification methods is the *parameter adaptation algorithm* (PAA) which has the form

$$\hat{\theta}(t+1) = \hat{\theta}(t) + F(t+1)\Phi(t)\varepsilon^{\circ}(t+1)$$

where $\hat{\theta}$ is the estimated parameter vector and $F(t+1)$ $\Phi(t)$ $\varepsilon°(t+1)$ represents the correcting term at each step. F is the *adaptation gain* (constant or time varying), Φ is the observation vector and $\varepsilon°$ is the *prediction error* (or in general the *adaptation error*), *i.e.* the difference between the true output and the predicted one.

Different choices are possible for the adaptation gain sequence in relation with the type of identification problem (constant or time varying plant parameters, with or without initial information, *etc*...).

The uniqueness of the identified parameters depends upon the characteristics of the input signal. To obtain a unique set of identified parameters, in the case of the identification of a plant model characterized by an irreducible transfer function, the input signal should contain a number of distinct sinusoidal components superior to half of the number of parameters to be identified. In practice, one systematically uses as input for system identification the *pseudo-random binary sequences* (PRBS), which approximates a discrete-time white noise.

The stochastic disturbances which contaminate the measured output may cause *bias* in the parameter estimations. For specific types of disturbances appropriate recursive identification methods, providing asymptotically unbiased estimations, are available.

The different recursive identification methods available use the same structure for the PAA. They can be distinguished by:

1. The structure of the predictor
2. The origin of the elements of the observation vector (Φ)
3. The dimension of the adjustable parameter vector $\hat{\theta}(t)$ and of the observation vector Φ
4. The way in which the prediction errors (or the adaptation errors) are generated

These identification methods can be grouped in two categories:

1. Identification methods based on the whitening of the prediction error
2. Identification methods based on the uncorrelation of the observation vector and the prediction error

A unique *plant + disturbance* structure, which describes all the situations encountered in practice, does not exist, as also does not exist a unique identification method providing unbiased parameter estimates in all the situations.

Table 5.2. Recursive Identification Algorithms

	Recursive Least Squares (RLS)	Extended Least Squares (ELS)	Output Error with Extended Prediction Model (OEEPM)
Plant + Noise Model	$y = \dfrac{q^{-d}B}{A}u + \dfrac{1}{A}e$	$y = \dfrac{q^{-d}B}{A}u + \dfrac{C}{A}e$	$y = \dfrac{q^{-d}B}{A}u + \dfrac{C}{A}e$
Estimated Parameter Vector	$\hat{\theta}^T(t) = [\hat{a}^T(t), \hat{b}^T(t)]$ $\hat{a}^T(t) = [\hat{a}_1(t)...\hat{a}_{n_A}(t)]$ $\hat{b}^T(t) = [\hat{b}_1(t)...\hat{b}_{n_B}(t)]$	$\hat{\theta}^T(t) = [\hat{a}^T(t), \hat{b}^T(t), \hat{c}^T(t)]$ $\hat{a}^T(t) = [\hat{a}_1(t)...\hat{a}_{n_A}(t)]$ $\hat{b}^T(t) = [\hat{b}_1(t)...\hat{b}_{n_B}(t)]$ $\hat{c}^T(t) = [\hat{c}_1(t)...\hat{c}_{n_C}(t)]$	$\hat{\theta}^T(t) = [\hat{a}^T(t), \hat{b}^T(t), \hat{h}^T(t)]$ $\hat{a}^T(t) = [\hat{a}_1(t)...\hat{a}_{n_A}(t)]$ $\hat{b}^T(t) = [\hat{b}_1(t)...\hat{b}_{n_B}(t)]$ $\hat{h}^T(t) = [\hat{h}_1(t)...\hat{h}_{n_H}(t)], n_H = \max(n_A, n_C)$ $\hat{c}_1 = \hat{h}_1 + \hat{a}_1$
Predictor Regressor Vector	$\phi^T(t) = [-y(t)...-y(t-n_A+1),$ $u(t-d)...u(t-d-n_B+1)]$	$\phi^T(t) = [-y(t)...-y(t-n_A+1),$ $u(t-d)...u(t-d-n_B+1),$ $\varepsilon(t)...\varepsilon(t-n_C+1)]$	$\phi^T(t) = [-\hat{y}(t)...-\hat{y}(t-n_A+1),$ $u(t-d)...u(t-d-n_B+1),$ $\varepsilon(t)...\varepsilon(t-n_C+1)]\phi(t)$
a priori / *a posteriori* Predictor Output	$y^0(t+1) = \hat{\theta}^T(t)\phi(t) / \hat{y}(t+1) = \hat{\theta}^T(t+1)\phi(t)$		
a priori / *a posteriori* Prediction Error	$\varepsilon^0(t+1) = y(t+1) - \hat{y}^0(t+1) / \varepsilon(t+1) = y(t+1) - \hat{y}(t+1)$		
Adaptation Error	$v^0(t+1) = \varepsilon^0(t+1)$	$v^0(t+1) = \varepsilon^0(t+1)$	$v^0(t+1) = \varepsilon^0(t+1)$
Observation Vector	$\Phi(t) = \phi(t)$	$\Phi(t) = \phi(t)$	$\Phi(t) = \phi(t)$

Table 5.2. Continued

	Recursive Maximum Likelihood (RML)	Generalized Least Squares (GLS)	Output Error with Fixed Compensator (OEFC)
Plant + Noise Model	$y = \dfrac{q^{-d}B}{A}u + \dfrac{C}{A}e$	$y = \dfrac{q^{-d}B}{A}u + \dfrac{1}{AC}e$	$y = \dfrac{q^{-d}B}{A}u + v$
Estimated Parameter Vector	$\hat{\theta}^T(t) = [\hat{a}^T(t), \hat{b}^T(t), \hat{c}^T(t)]$ $\hat{a}^T(t) = [\hat{a}_1(t)\ldots\hat{a}_{n_A}(t)]$ $\hat{b}^T(t) = [\hat{b}_1(t)\ldots\hat{b}_{n_B}(t)]$ $\hat{c}^T(t) = [\hat{c}_1(t)\ldots\hat{c}_{n_C}(t)]$	$\hat{\theta}^T(t) = [\hat{a}^T(t), \hat{b}^T(t), \hat{c}^T(t)]$ $\hat{a}^T(t) = [\hat{a}_1(t)\ldots\hat{a}_{n_A}(t)]$ $\hat{b}^T(t) = [\hat{b}_1(t)\ldots\hat{b}_{n_B}(t)]$ $\hat{c}^T(t) = [\hat{c}_1(t)\ldots\hat{c}_{n_C}(t)]$	$\hat{\theta}^T(t) = [\hat{a}^T(t), \hat{b}^T(t)]$ $\hat{a}^T(t) = [\hat{a}_1(t)\ldots\hat{a}_{n_A}(t)]$ $\hat{b}^T(t) = [\hat{b}_1(t)\ldots\hat{b}_{n_B}(t)]$
Predictor Regressor Vector	$\phi^T(t) = [-y(t)\ldots -y(t-n_A+1),$ $u(t-d)\ldots u(t-d-n_B+1),$ $\varepsilon(t)\ldots\varepsilon(t-n_C+1)]$	$\phi^T(t) = [-y(t)\ldots -y(t-n_A+1),$ $u(t-d)\ldots u(t-d-n_B+1),$ $-\alpha(t)\ldots-\alpha(t-n_C+1)]$ $\alpha(t) = \hat{A}(t)y(t) - \hat{B}^*(t)u(t-d-1)$	$\phi^T(t) = [-\hat{y}(t)\ldots -\hat{y}(t-n_A+1),$ $u(t-d)\ldots u(t-d-n_B+1)]$
a priori / a posteriori Predictor Output		$y^0(t+1) = \hat{\theta}^T(t)\phi(t) \,/\, \hat{y}(t+1) = \hat{\theta}^T(t+1)\phi(t)$	
a priori / a posteriori Prediction Error		$\varepsilon^0(t+1) = y(t+1) - \hat{y}^0(t+1) \,/\, \varepsilon(t+1) = y(t+1) - \hat{y}(t+1)$	
Adaptation Error	$\nu^0(t+1) = \varepsilon^0(t+1)$	$\nu^0(t+1) = \varepsilon^0(t+1)$	$\nu^0(t+1) = \varepsilon^0(t+1) + \displaystyle\sum_{i=1}^{n_D} d_i\varepsilon(t+1-i), \quad n_D \le n_A$
Observation Vector	$\Phi(t) = \dfrac{1}{\hat{C}(t,q^{-1})}\phi(t)$	$\Phi(t) = \phi(t)$	$\Phi(t) = \phi(t)$

Table 5.2. Continued

	Instrumental Variable with Auxiliary Model (IVAM)	Output Error with Filtered Observations (OEFO)	Output Error with Adaptive Filtered Observations (OEAFO)
Plant + Noise Model	$y = \dfrac{q^{-d}B}{A}u + v$	$y = \dfrac{q^{-d}B}{A}u + v$	$y = \dfrac{q^{-d}B}{A}u + v$
Estimated Parameter Vector	$\hat{\theta}^T(t) = [\hat{a}^T(t), \hat{b}^T(t)]$ $\hat{a}^T(t) = [\hat{a}_1(t)...\hat{a}_{n_A}(t)]$ $\hat{b}^T(t) = [\hat{b}_1(t)...\hat{b}_{n_B}(t)]$	$\hat{\theta}^T(t) = [\hat{a}^T(t), \hat{b}^T(t)]$ $\hat{a}^T(t) = [\hat{a}_1(t)...\hat{a}_{n_A}(t)]$ $\hat{b}^T(t) = [\hat{b}_1(t)...\hat{b}_{n_B}(t)]$	$\hat{\theta}^T(t) = [\hat{a}^T(t), \hat{b}^T(t)]$ $\hat{a}^T(t) = [\hat{a}_1(t)...\hat{a}_{n_A}(t)]$ $\hat{b}^T(t) = [\hat{b}_1(t)...\hat{b}_{n_B}(t)]$
Predictor Regressor Vector	$\phi^T(t) = [-y(t)...-y(t-n_A+1),$ $u(t-d)...u(t-d-n_B+1)]$	$\phi^T(t) = [-y(t)...-y(t-n_A+1),$ $u(t-d)...u(t-d-n_B+1)]$	$\phi^T(t) = [-y(t)...-y(t-n_A+1),$ $u(t-d)...u(t-d-n_B+1)]$
a priori / a posteriori Prediction Output		$y^0(t+1) = \hat{\theta}^T(t)\phi(t) \,/\, \hat{y}(t+1) = \hat{\theta}^T(t+1)\phi(t)$	
a priori / a posteriori Prediction Error		$\varepsilon^0(t+1) = y(t+1) - \hat{y}^0(t+1) \,/\, \varepsilon(t+1) = y(t+1) - \hat{y}(t+1)$	
Adaptation Error	$v^0(t+1) = \varepsilon^0(t+1)$	$v^0(t+1) = \varepsilon^0(t+1)$	$v^0(t+1) = \varepsilon^0(t+1)$
Observation Vector	$\Phi^T(t) = [-y_{IV}(t)...-y_{IV}(t-n_A+1),$ $u(t-d)...u(t-n_B-d+1)]$ $y_{IV}(t) = \hat{\theta}^T(t)\Phi(t-1)$	$\Phi(t) = \dfrac{1}{L(q^{-1})}\phi(t)$	$\Phi(t) = \dfrac{1}{\hat{A}(t,q^{-1})}\phi(t)$

5.7 Notes and References

As basic references for system identification we cite:

Eykhoff P. (1974) System Identification: Parameter and State Estimation, Wiley, London.
Goodwin G.C., Payne R.L. (1977) Dynamic System Identification: Experiment Design and Data Analysis, Academic Press, N.Y.
Isermann R. (Ed.) (1981) Special Issue on System Identification, Automatica, vol 17, n° 1.
Ljung L. (1999) System Identification - Theory for the User, 2nd edition, Prentice Hall, Englewood Cliffs.
Söderström T., Stoica P. (1989) System Identification, Prentice Hall International, Hertfordshire.
Landau I.D. (2001b) Les bases de l'identification des systèmes, in Identification des Systèmes (I.D. Landau, A. Bensançon-Voda, ed), pp. 19-130, Hermes, Paris.

Many references are available for the parameter adaptation algorithms. For example:

Landau I.D. (1979) Adaptive Control - The model reference approach, Dekker, N.Y.
Ljung L., Söderström T. (1983) Theory and Practice of Recursive Identification, MIT Press, Cambridge, Mass.
Landau I.D., Lozano R., M'Saad M. (1997) Adaptive Control, Springer, London, UK.
Goodwin G.C., Sin K.S. (1984) Adaptive Filtering Prediction and Control, Prentice Hall, Englewood Cliffs, N.J.
Landau I.D. (1990) System Identification and Control Design, Prentice Hall, Englewood Cliffs, N.J.

The use of identification techniques in signal processing is discussed in several references among which we cite:

Landau I.D. (1984) A Feedback System Approach to Adaptive filtering, IEEE Trans. on Information Theory, vol. 30, n°2, pp. 251-262.
Landau I.D., M'Sirdi N., M'Saad M. (1986) Techniques de modélisation récursives pour l'analyse spectrale paramétrique adaptative, Traitement du Signal, vol.3, pp. 183-204.

and also Ljung and Söderström (1983).

The references for the software dedicated to system identification are:

Adaptech (1996a) WinPIM$^+$TR – System Identification Software, Adaptech, St. Martin d'Hères, France.
Mathworks (1998) System Identification toolbox for MATLAB®, The Mathworks Inc., Mass., U.S.A.

6

System Identification Methods

In this chapter two categories of identification methods: (1)methods based on the whitening of the prediction error; (2) methods based on the uncorrelation of the observation vector and the prediction error, are presented in their recursive form together with the corresponding model validation techniques.

Methods for the model order estimation are presented in the last part of the chapter.

6.1 Identification Methods Based on the Whitening of the Prediction Error   (Type I)

The following recursive identification methods given in Table 5.1 fall into this category:

- Recursive Least Squares (RLS)
- Extended Least Squares (ELS)
- Recursive Maximum Likelihood (RML)
- Output Error with Extended Prediction Model (OEEPM)
- Generalized Least Squares (GLS)

6.1.1 Recursive Least Squares (RLS)

The recursive least squares method has been presented in detail in Chapter 5, Section 5.2.3.

The analysis in the presence of random disturbances has been presented in Section 5.4. It should be remembered that the recursive least squares method gives unbiased estimations only for "plant + disturbance" models of the form (structure S1)

$$A(q^{-1})y(t) = q^{-d}B(q^{-1})u(t) + e(t) \tag{6.1.1}$$

i.e. for a disturbance model with $C(q^{-1}) = 1$ in the equation of the ARMAX models.

6.1.2 Extended Least Squares (ELS)

This method has been developed in order to identify without bias "plant + disturbance" models of the form (structure S3):

$$A(q^{-1})y(t) = q^{-d}B(q^{-1})u(t) + C(q^{-1})e(t) \tag{6.1.2}$$

The idea is to identify simultaneously the plant model and the disturbance model, in order to obtain a prediction (adaptation) error, which is asymptotically "white".

This method will be presented by means of an example. Let the "plant + disturbance" model be

$$y(t+1) = -a_1 y(t) + b_1 u(t) + c_1 e(t) + e(t+1) \tag{6.1.3}$$

Assume that the parameters are known and construct a predictor, which will give a white prediction error

$$\hat{y}(t+1) = -a_1 y(t) + b_1 u(t) + c_1 e(t) \tag{6.1.4}$$

Furthermore this predictor minimizes the variance of the prediction error $E\{[y(t+1) - \hat{y}(t+1)]^2\}$ (see Chapter 4, Section 4.1.4).

The prediction error is given by

$$\varepsilon(t+1) = y(t+1) - \hat{y}(t+1) = e(t+1) \tag{6.1.5}$$

This allows to rewrite Equation 6.1.4 in the form

$$\hat{y}(t+1) = -a_1 y(t) + b_1 u(t) + c_1 \varepsilon(t) \tag{6.1.6}$$

In the case of unknown parameters, the adjustable predictor will be given by Equation 6.1.6, in which the unknown parameters are replaced by their estimates.

The *a priori* adjustable predictor will therefore take the form

$$\hat{y}^o(t+1) = -\hat{a}_1(t)y(t) + \hat{b}_1(t)u(t) + \hat{c}_1(t)\varepsilon(t) = \hat{\theta}(t)^T \phi(t) \tag{6.1.7}$$

where

$$\hat{\theta}(t)^T = \left[\hat{a}_1(t), \hat{b}_1(t), \hat{c}_1(t)\right] \quad ; \quad \phi(t)^T = \left[-y(t), u(t), \varepsilon(t)\right] \qquad (6.1.8)$$

The *a posteriori* adjustable predictor will be given by

$$\hat{y}(t+1) = -\hat{a}_1(t+1)y(t) + \hat{b}_1(t+1)u(t) + \hat{c}_1(t+1)\varepsilon(t) = \hat{\theta}(t+1)^T \phi(t) \qquad (6.1.9)$$

The *a posteriori* prediction error $\varepsilon(t+1)$ is expressed as

$$\varepsilon(t+1) = y(t+1) - \hat{y}(t+1) \qquad (6.1.10)$$

and the *a priori* prediction error is defined as

$$\varepsilon^o(t+1) = y(t+1) - \hat{y}^o(t+1) \qquad (6.1.11)$$

Using the adjustable predictor given by Equation 6.1.7, the formulation of the problem of the simultaneous identification of the plant model and of the disturbance model has been reduced to a least squares type formulation as considered in Section 5.2.3.

The parameter adaptation algorithm (PAA) described in Section 5.5 (Equations 5.5.1 to 5.5.3) will be used with $\hat{\theta}(t)$ and $\Phi(t) = \phi(t)$ given by Equation 6.1.8. All "adaptation gain" policies may be used.

Compared to the simple least squares method, there is a larger number of parameters to be estimated. $\hat{\theta}(t)$ includes in addition the coefficients of $C(q^{-1})$. The observation vector will be obviously larger. It contains in addition the *a posteriori* errors $\varepsilon(t)$, $\varepsilon(t-1)...\varepsilon(t-n_C+1)$.

In the general case, the estimated parameters vector and the observation vector will be of the form

$$\hat{\theta}(t)^T = \left[\hat{a}_1(t)...\hat{a}_{n_A}, \hat{b}_1(t)...\hat{b}_{n_B}(t), \hat{c}_1(t)...\hat{c}_{n_C}(t)\right]$$

$$\Phi(t)^T = \left[-y(t)...-y(t-n_A+1), u(t-d)...u(t-d-n_B+1), \varepsilon(t)...\varepsilon(t-n_C+1)\right]$$

In the presence of a random disturbance corresponding to the ARMAX model and with an asymptotically decreasing adaptation gain, $\varepsilon(t)$ asymptotically tends towards a white noise, which guarantees an unbiased estimation of the parameters of $A(q^{-1})$, $B(q^{-1})$ (if the input is sufficiently rich)[1].

---

[1] The convergence of $\hat{\theta}$ is rigorously guaranteed within a convergence domain $D_C : \left\{\theta | (\theta^* - \theta)^T \phi(t, \theta) = 0\right\}$ in which θ^* is the vector of true parameters. For a rich input, this

Nevertheless, this convergence is subject to a sufficient (but not necessary) condition

$$\frac{1}{C(z^{-1})} - \frac{\lambda_2}{2} \quad ; \quad 2 > \lambda_2 \geq \max \lambda_2(t) \tag{6.1.12}$$

is a strictly positive real transfer function. A *strictly positive real* transfer function is characterized by the following two properties:

    1.   it is asymptotically stable
    2.   the real part of the transfer function is positive at all frequencies

This concept is illustrated for continuous time systems in the upper part of Figure 6.1 and for discrete time systems in the lower part of Figure 6.1.

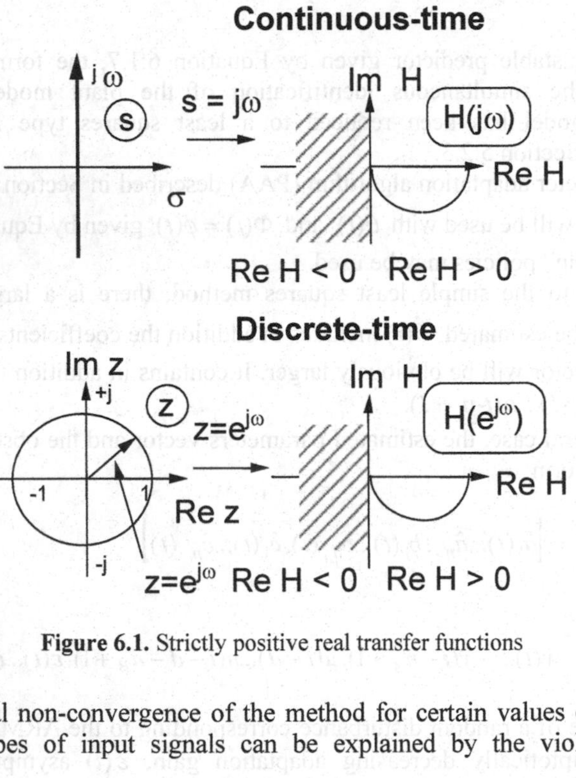

Continuous-time

Discrete-time

Figure 6.1. Strictly positive real transfer functions

An eventual non-convergence of the method for certain values of $C(q^{-1})$ and for certain types of input signals can be explained by the violation of this condition.

---

domain is reduced to a single point θ^*. The convergence of the parameters of $C(q^{-1})$ is slower than that of $A(q^{-1})$ and $B(q^{-1})$ parameters, and it depends on the realization of the stochastic process.

6.1.3 Recursive Maximum Likelihood (RML)

This is an improved version of the extended least squares method. Instead of using directly $\phi(t)$ given by Equation 6.1.8 in the parameter adaptation algorithm, it is first filtered by $1/\hat{C}\,(t,\,q^{-1})$ where $\hat{C}\,(t,\,q^{-1})$ is an estimation at instant t of $C(q^{-1})$. This modification has the effect of eliminating the positive real condition on $C(q^{-1})$ in the final convergence stage, and of accelerating the uncorrelation between the observation vector and the prediction error. An example will be used to present this method. The "plant + disturbance" model is given by Equation 6.1.3.

The *a priori* adjustable predictor is written as (similar to the ELS)

$$\hat{y}^{o}(t+1) = -\hat{a}_1(t)y(t) + \hat{b}_1(t)u(t) + \hat{c}_1(t)\varepsilon(t) = \hat{\theta}(t)^T \phi(t) \tag{6.1.13}$$

where

$$\hat{\theta}(t)^T = \left[\hat{a}_1(t), \hat{b}_1(t), \hat{c}_1(t)\right] \quad ; \quad \phi(t)^T = \left[-y(t), u(t), \varepsilon(t)\right] \tag{6.1.14}$$

The *a posteriori* adjustable predictor is written as

$$\hat{y}(t+1) = \hat{\theta}(t+1)^T \phi(t) \tag{6.1.15}$$

and the corresponding prediction errors are given by

$$\varepsilon^{o}(t+1) = y(t+1) - \hat{y}^{o}(t+1)$$
$$\varepsilon(t+1) = y(t+1) - \hat{y}(t+1) \tag{6.1.16}$$

The estimation of the polynomial $C(q^{-1})$ at instant t is written as

$$\hat{C}(t,q^{-1}) = 1 + \hat{c}_1(t)q^{-1} \tag{6.1.17}$$

The observation vector $\Phi(t)$ is defined by

$$\Phi(t)^T = \phi_f(t)^T = \frac{1}{\hat{C}(t,q^{-1})}\left[-y(t), u(t), \varepsilon(t)\right]$$
$$= \left[\frac{-y(t)}{\hat{C}(t,q^{-1})}, \frac{u(t)}{\hat{C}(t,q^{-1})}, \frac{\varepsilon(t)}{\hat{C}(t,q^{-1})}\right] \tag{6.1.18}$$

which corresponds to the filtering of the components of $\phi(t)$ by $1/\hat{C}\,(t,q^{-1})$.

The parameter adaptation algorithm is given by Equations 5.5.1, to 5.5.3, where $\Phi(t)$ is now given by Equation 6.1.18. In the general case

$$\hat{\theta}(t)^T = \left[\hat{a}_1(t)...\hat{a}_{n_A}(t), \hat{b}_1(t)...\hat{b}_{n_B}(t), \hat{c}_1(t)...\hat{c}_{n_C}(t)\right]$$

$$\Phi(t)^T = \phi_f(t)^T = \frac{1}{\hat{C}(t, q^{-1})}$$

$$\cdot \left[-y(t)... - y(t - n_A + 1), u(t - d)...u(t - d - n_B + 1), \varepsilon(t)...\varepsilon(t - n_C + 1)\right]$$

Nevertheless, this method cannot be implemented starting from the instant $t=0$ if a good estimation $C(q^{-1})$ is not available. First an initialization horizon must be considered, during which the extended least squares (ELS) method is used in order to obtain a first estimation. As a rule, the initialization horizon is taken equal to 5 (up to 8) times the number of the parameters to be identified.

On the other hand, filtering of the observations can only be carried out for stable estimations of $\hat{C}(t, q^{-1})$. A stability test must therefore be incorporated. Note that the transition ELS→ RML should only occur if $\hat{C}(t, q^{-1})$ is stable. If at the end of the initialization horizon $\hat{C}(t, q^{-1})$ is not stable, the commutation is delayed up to the instant for which $\hat{C}(t, q^{-1})$ becomes stable.

A gradual transition of the ELS towards the RML may also be used by introducing a "contraction factor" in $\hat{C}(t, q^{-1})$. This corresponds to a filter defined by $(1 + \alpha c_1 q^{-1})$ with: $0 \le \alpha \le 1$, instead of $(1 + c_1 q^{-1})$. This forces the polynomial roots into the unit circle. It is possible to use a variable contraction factor that asymptotically tends toward 1. This type of variation can be obtained using the formula

$$\alpha(t) = \alpha_0 \alpha(t - 1) + 1 - \alpha_0 \tag{6.1.19}$$

which corresponds to a first-order discrete stable filter with a unit static gain excited by a unit step (the final value is equal to 1). A possible choice for α_0 is

$$0.5 \le \alpha_0 = \alpha(0) \le 0.99$$

The recursive maximum likelihood method is used to improve, if necessary, the results of the extended least squares method. However, if the above mentioned precautions are not taken into account, the algorithm may diverge.

6.1.4 Output Error with Extended Prediction Model (OEEPM)

Originally, this is an extension of the output error method (see further on in Section 6.3). This method can be interpreted as a variant of the ELS. It offers asymptotic performances similar to the ELS, but with improved transient estimation (faster bias rejection). This method will be presented by means of an example. The "plant + disturbance" model is given by Equation 6.1.3.

The *a priori* adjustable predictor of the ELS (defined by Equation 6.1.7) can be rewritten in the form (adding and subtracting the term $\pm \hat{a}_1(t)\hat{y}(t)$)

$$\hat{y}^o(t+1) = -\hat{a}_1(t)y(t) + \hat{b}_1(t)u(t) + \hat{c}_1(t)\varepsilon(t) \pm \hat{a}_1(t)\hat{y}(t) \qquad (6.1.20)$$

and, by regrouping differently the terms of Equation 6.1.20, the structure of the *a priori* adjustable predictor used in the OEEPM is obtained:

$$\hat{y}^o(t+1) = -\hat{a}_1(t)\hat{y}(t) + \hat{b}_1(t)u(t) + \hat{h}_1(t)\varepsilon(t) = \hat{\theta}(t)^T \phi(t) \qquad (6.1.21)$$

where

$$\hat{h}_1(t) = \hat{c}_1(t) - \hat{a}_1(t) \qquad (6.1.22)$$

$$\hat{\theta}(t)^T = \left[\hat{a}_1(t), \hat{b}_1(t), \hat{c}_1(t)\right] \quad ; \quad \phi(t)^T = \left[-\hat{y}(t), u(t), \varepsilon(t)\right] \qquad (6.1.23)$$

The *a posteriori* adjustable predictor is given by

$$\hat{y}(t+1) = \hat{\theta}(t+1)^T \phi(t) \qquad (6.1.24)$$

and the prediction errors are given by

$$\begin{aligned} \varepsilon^o(t+1) &= y(t+1) - \hat{y}^o(t+1) \\ \varepsilon(t+1) &= y(t+1) - \hat{y}(t+1) \end{aligned} \qquad (6.1.25)$$

The parameter adaptation algorithm is the one given by Equations 5.5.1, 5.5.2 and 5.5.3 with $\hat{\theta}(t)$ and $\Phi(t) = \phi(t)$ as in Equation 6.1.23. In the general case:

$$\hat{\theta}(t)^T = \left[\hat{a}_1(t)..\hat{a}_{n_A}(t), \hat{b}_1(t)..\hat{b}_{n_B}(t), \hat{h}_1(t)..\hat{h}_{n_C}(t)\right]$$

$$\Phi(t)^T = \left[-\hat{y}(t)...-\hat{y}(t-n_A+1), u(t-d)..u(t-d-n_B+1), \varepsilon(t)..\varepsilon(t-n_C+1)\right]$$

As for the ELS, $\varepsilon(t)$ asymptotically tends towards a white noise, thereby guaranteeing an unbiased estimation of the parameters of $A(q^{-1})$ and $B(q^{-1})$ (if the input is sufficiently rich). The convergence is subject to the same sufficient condition as in the ELS, *i.e.* the transfer function given by Equation 6.1.12 must be strictly positive real.

Note that the values of the coefficients of $C(q^{-1})$ are obtained from the relation:

$$c_i = h_i + a_i \qquad (6.1.26)$$

The difference with the ELS lies essentially within the components of vector $\Phi(t)$ in which the measurements $y(t)$ directly affected by the disturbance are replaced by $\hat{y}(t)$, which only indirectly depends upon the disturbance. This explains why a better estimation is obtained with this method than with the ELS method over short horizons.

6.1.5 Generalized Least Squares (GLS)

The aim of this method is to obtain a "white" prediction error for a "plant + disturbance" model having the form (structure S4):

$$A(q^{-1})y(t) = q^{-d} B(q^{-1})u(t) + \frac{1}{C(q^{-1})} e(t) \tag{6.1.27}$$

(the term $C(q^{-1})e(t)$ of the ARMAX model has been replaced by $[1/C(q^{-1})]e(t)$). This method will be illustrated by means of an example. The "plant + disturbance" model is given by:

$$y(t+1) = -a_1 y(t) + b_1 u(t) + \frac{e(t+1)}{1 + c_1 q^{-1}} \tag{6.1.28}$$

Let us define

$$\alpha(t+1) = (1 + a_1 q^{-1})y(t+1) - b_1 u(t) = \frac{e(t+1)}{1 + c_1 q^{-1}} \tag{6.1.29}$$

The relation

$$(1 + c_1 q^{-1})\alpha(t+1) = e(t+1) \tag{6.1.30}$$

is then obtained ($\alpha(t)$ is a AR process – see Section 4.1.2).

Assuming that the parameters are known, a predictor can be constructed ensuring a white prediction error

$$\hat{y}(t+1) = -a_1 y(t) + b_1 u(t) - c_1 \alpha(t) \tag{6.1.31}$$

Indeed

$$y(t+1) - \hat{y}(t+1) = \frac{e(t+1)}{1 + c_1 q^{-1}} + c_1 \frac{e(t)}{1 + c_1 q^{-1}} = e(t+1) \tag{6.1.32}$$

If the parameters are not known, an adjustable predictor is built by replacing the known parameters in Equation 6.1.31 by their estimations. The a priori adjustable predictor will be given by

$$\hat{y}^o(t+1) = -\hat{a}_1(t)y(t) + \hat{b}_1(t)u(t) - \hat{c}_1(t)\alpha(t) = \hat{\theta}(t)^T \phi(t) \qquad (6.1.33)$$

in which

$$\hat{\theta}(t)^T = \left[\hat{a}_1(t), \hat{b}_1(t), \hat{c}_1(t)\right] \quad ; \quad \phi(t)^T = \left[-y(t), u(t), -\alpha(t)\right] \quad (6.1.34)$$

The quantity $\alpha(t)$ will be estimated using Equation 6.1.29, in which the unknown parameter values are replaced by their estimations:

$$\alpha(t) = \hat{A}(t, q^{-1})y(t) - q^{-d}\hat{B}(t, q^{-1})u(t)$$
$$= (1 + \hat{a}_1(t)q^{-1})y(t) - \hat{b}_1(t)u(t-1) \qquad (6.1.35)$$

The *a priori* prediction error is defined as

$$\varepsilon^o(t+1) = y(t+1) - \hat{y}^o(t+1) \qquad (6.1.36)$$

Since the adjustable predictor equation given in Equation 6.1.33 has the formulation permitting the use of the least squares algorithm, as in the case of the ELS, the parameter adaptation algorithm is given by Equations 5.5.1, 5.5.2 and 5.5.3. In this case $\Phi(t)$ is defined by Equation 6.1.34 and $\varepsilon^o(t+1)$ by Equation 6.1.36. In the general case

$$\hat{\theta}(t)^T = \left[\hat{a}_1(t)..\hat{a}_{n_A}(t), \hat{b}_1(t)..\hat{b}_{n_B}(t), \hat{c}_1(t)..\hat{c}_{n_C}(t)\right]$$
$$\Phi(t)^T = \left[-y(t)...-y(t-n_A+1), u(t-d)..u(t-d-n_B+1), -\alpha(t)...-\alpha(t-n_C+1)\right]$$

In the presence of a random disturbance corresponding to the S4 structure, and with an asymptotically decreasing adaptation gain, $\varepsilon(t)$ asymptotically tends towards white noise, thereby allowing an unbiased estimation of the model parameters.

The convergence of the algorithm is nevertheless linked to a sufficient (but not necessary) condition

$$C(z^{-1}) - \frac{\lambda_2}{2} \quad ; \quad 2 > \lambda_2 \geq \max \lambda_2(t) \qquad (6.1.37)$$

is a strictly positive real transfer function.

This method is used in particular if the disturbance has a narrow band frequency spectrum (for example a periodic disturbance close to a sinusoid). Indeed, this kind of disturbance can be fairly modeled by $[1/C(q^{-1})]e(t)$ with a few parameters whereas the modeling of this kind of disturbances by $C(q^{-1})e(t)$ requires a high order for $C(q^{-1})$.

6.2 Validation of the Models Identified with Type I Methods

This section is concerned with the validation of models identified using methods based on the "whitening" of the prediction error. These methods are

- Recursive Least Squares (RLS)
- Extended Least Squares (ELS)
- Recursive Maximum Likelihood (RML)
- Output Error with Extended Prediction Model (OEEPM)
- Generalized Least Squares (GLS)

The principle of the validation method is the following:

- If the "plant + disturbance" structure chosen is correct, *i.e.* representative of reality
- If the degrees of the polynomials $A(q^{-1})$, $B(q^{-1})$, $C(q^{-1})$ and the value of d (delay) have been correctly chosen

then the prediction error $\varepsilon(t)$ asymptotically tends towards white noise, that implies

$$\lim_{t \to \infty} E\{\varepsilon(t)\varepsilon(t-i)\} = 0 \qquad i = 1,2,3...;-1,-2,-3...$$

The validation method implements this principle. It is made up of several steps:

1. Creation of an I/O file for the identified model (using the same input sequence as for the system)
2. Creation of a prediction error file for the identified model (minimum 100 samples)
3. "Whiteness" (uncorellation) test on the prediction errors sequence (also known as residual prediction errors)

Whiteness test

Let $\{\varepsilon(t)\}$ be the centered sequence of the residual prediction errors (centered: the mean value is subtracted from the measured values).

One computes

$$R(0) = \frac{1}{N}\sum_{t=1}^{N} \varepsilon^2(t) \quad ; \quad RN(0) = \frac{R(0)}{R(0)} = 1 \tag{6.2.1}$$

$$R(i) = \frac{1}{N}\sum_{t=1}^{N} \varepsilon(t)\varepsilon(t-i) \quad ; \quad RN(i) = \frac{R(i)}{R(0)} \quad ; \quad i = 1,2,3...i_{max} \tag{6.2.2}$$

where

$$i_{max} = \max(n_A, n_B + d)$$

and the $RN(i)$ are estimations of the (normalized) autocorrelations.

If the residual prediction error sequence is *perfectly white* (theoretical situation) and the number of samples is very large ($N \rightarrow \infty$) one gets $RN(0) = 1$; $RN(i) = 0$ $i \geq 1$[2].

In real situations, however, this is never the case (*i.e.* $RN(i) \neq 0$; $i \geq 1$) since on one hand $\varepsilon(t)$ contains residual structural errors (order errors, non-linear effects, non-gaussian noises) and, on the other hand, the number of samples is generally relatively small (several hundreds). Also, it should be kept in mind that one always seeks to identify *good* simple models (with few parameters).

One considers as a practical validation criterion (extensively tested on applications)

$$RN(0) = 1 \quad ; \quad |RN(i)| \leq \frac{2.17}{\sqrt{N}} \quad ; \quad i \geq 1 \tag{6.2.3}$$

where N is the number of samples.

This test has been defined taking into account the fact that, for *white noise,* the sequence $RN(i)$ ($i \neq 0$) has an asymptotically gaussian (normal) distribution with zero mean and standard deviation: $\sigma = 1/\sqrt{N}$.

The confidence interval considered in Equation 6.2.3 corresponds to a 3% level of significance of the hypothesis test for a gaussian distribution.

Indeed, if $RN(i)$ obeys the gaussian distribution (0, $1/\sqrt{N}$), there is only a probability of 1.5 % that $RN(i)$ is larger than $2.17/\sqrt{N}$, or that $RN(i)$ is smaller than $-2.17/\sqrt{N}$. Therefore, if a computed value $RN(i)$ falls outside the range of the confidence interval, the hypothesis on the basis of which $\varepsilon(t)$ and $\varepsilon(t-i)$ are independent should be rejected, *i.e.* $\{\varepsilon(t)\}$ is not a white noise sequence.

Sharper confidence intervals can be defined. Table 6.1 gives the values of the validation criterion for various N and a 3% test level of significance as well as for 5 and 7% level of significance.

Table 6.1. Confidence intervals for independence (whiteness) tests

Level of significance	Criterion	N=128	N=256	N=512
3% (validation criterion)	$2.17/\sqrt{N}$	0.192	0.136	0.096
5%	$1.96/\sqrt{N}$	0.173	0.122	0.087
7%	$1.808/\sqrt{N}$	0.16	0.113	0.08

---

[2] For gaussian data, uncorrelation implies independence. In this case $RN(i) = 0$, $i \geq 1$ implies independence between $\varepsilon(t)$, $\varepsilon(t-1)$ i.e. the sequence of residuals $\{\varepsilon(t)\}$ is a gaussian white noise.

The following remarks are imperative:

- An *acceptable* identified model has in general

$$|RN(i)| \le \frac{1.8}{\sqrt{N}} \dots \frac{2.17}{\sqrt{N}} \qquad i = 1, \dots, i_{max}$$

- Taking into account the relative weight of various non-gaussian and modeling errors (which increase with the number of samples), the validation criterion may be slightly tightened for small N and slightly relaxed for large N. Therefore, for simplicity's sake, one can consider for the validation criterion the following practical value

$$|RN(i)| \le 0.15 \qquad i = 1, \dots, i_{max}$$

- A *too good* validation criterion indicates that model simplifications may be possible.
- If several identified models have the same complexity (number of parameters), one selects the model given by the methods which lead to the smallest $|RN(i)|$ and $R(0)$.

Note also that a complete model validation implies, after the validation done with the input/output sequence used for identification, a validation using a new plant input/output sequence.

There is another important point to consider: if the level of the residual prediction errors is very low compared to the output level (lets us say more than 60 dB) the whiteness test loses its significance. This occurs because on one hand the noise level is so low that the bias on the RLS is negligible and, on the other hand, because the residual noise may not be gaussian to a large extent (for example, noise caused by the propagation of round-off errors). This situation may occur when identifying simulated I/O data generated without disturbances.

6.3 Identification Methods Based on the Uncorrelation of the Observation Vector and the Prediction Error (Type II)

The following recursive identification methods fall into this category:

- Instrumental Variable with Auxiliary Model (IVAM)
- Output Error with Fixed Compensator (OEFC)
- Output Error with Filtered Observations (OEFO)
- Output Error with Adaptive Filtered Observations (OEAFO)

6.3.1 Instrumental Variable with Auxiliary Model

The general idea behind the instrumental variable methods consists in creating a new observation vector which is highly correlated with the uncontaminated variables (and therefore representative), but uncorrelated with the noise disturbance in order to obtain $E\{\Phi(t)\ \varepsilon(t+1)\} = 0$.

The "plant + disturbance" model considered in this case is that of structure S2 (see Section 5.5):

$$A(q^{-1})y(t) = q^{-d}B(q^{-1})u(t) + w'(t+1) \qquad (6.3.1)$$

in which

$$w'(t+1) = A(q^{-1})w(t) \qquad (6.3.2)$$

and corresponding to the diagram given in Figure 6.2.

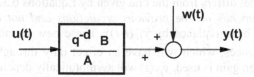

Figure 6.2. The "plant + disturbance" structure (type S2)

$w(t)$ is any disturbance, independent from $u(t)$, with zero mean value and a finite variance.

The final objective is to obtain unbiased estimations of the coefficients of $A(q^{-1})$ and $B(q^{-1})$ without taking into consideration a model for the disturbance. Thus, one builds an instrumental vector uncorrelated with the prediction error provided by the least squares predictor.

Let consider the following example, where the "plant + disturbance" is described by:

$$y(t+1) = -a_1 y(t) + b_1 u(t) + w'(t+1) \qquad (6.3.3)$$

where

$$w'(t+1) = (1 + a_1 q^{-1})w(t+1) \qquad (6.3.4)$$

Let consider the recursive least squares adjustable predictor

$$\hat{y}^0(t+1) = -\hat{a}_1(t)y(t) + \hat{b}_1(t)u(t) = \hat{\theta}(t)^T \phi(t) \qquad (6.3.5)$$

in which

$$\hat{\theta}(t)^T = \left[\hat{a}_1(t), \hat{b}_1(t)\right] \quad ; \quad \phi(t)^T = \left[-y(t), u(t)\right] \tag{6.3.6}$$

and respectively

$$\hat{y}(t+1) = \hat{\theta}(t+1)^T \phi(t) \tag{6.3.7}$$

the prediction errors being defined by:

$$\varepsilon^\circ(t+1) = y(t+1) - \hat{y}^\circ(t+1) \tag{6.3.8}$$

$$\varepsilon(t+1) = y(t+1) - \hat{y}(t+1) \tag{6.3.9}$$

Let us define an auxiliary prediction model that provides the instrumental variable:

$$y_{IV}(t) = -\hat{a}_1(t)y_{IV}(t-1) + \hat{b}_1(t)u(t-1) \tag{6.3.10}$$

This prediction model differs from the one given by Equations 6.3.5 or 6.3.7 as *the predicted output depends on the previous predictions and not on the previous outputs* ($y(t-1)$ has been replaced by $y_{IV}(t-1)$). These new variables will be less affected by disturbances, which will have an effect only through the PAA. If a decreasing adaptation gain is used, $y_{IV}(t)$ will asymptotically depend only on $u(t-1)$, that is not the case for $\hat{y}^\circ(t)$ or $\hat{y}(t)$ respectively given by Equations 6.3.5 and 6.3.7 (as $y(t)$ is a noisy signal).

The new observations vector is defined by

$$\Phi(t)^T = \phi_{IV}(t)^T = \left[-y_{IV}(t), u(t)\right] \tag{6.3.11}$$

One observes that $y(t)$, $y(t-1)$,..., which are used in recursive least squares, have been replaced by $y_{IV}(t)$, $y_{IV}(t-1)$...

In the general case, the observations vector has the structure:

$$\Phi(t)^T = \phi_{IV}(t)^T = \left[-y_{IV}(t), -y_{IV}(t-1),...,u(t-d),u(t-d-1),...\right] \tag{6.3.12}$$

and the instrumental variable y_{IV} is generated by the auxiliary model

$$\hat{A}(t,q^{-1})y_{IV}(t) = q^{-d}\hat{B}(t,q^{-1})u(t) \tag{6.3.13}$$

where

$$\hat{A}(t,q^{-1}) = 1 + \hat{a}_1(t)q^{-1} + ... \tag{6.3.14}$$

$$\hat{B}(t, q^{-1}) = \hat{b}_1(t)q^{-1} + \hat{b}_2(t)q^{-2} + \dots \qquad (6.3.15)$$

The parameter adaptation algorithm is given by Equations 5.5.1, 5.5.2 and 5.5.3, in which $\Phi(t)$ is now given by Equation 6.3.12. This method uses the same adjustable predictor of RLS for generating the prediction error but the observation vector is different. For $y_{IV}(t)$ to be representative of $y(t)$, an estimation of a_i and b_i is required. This is why this method must be initialized by the recursive least squares. In general, an initialization horizon is chosen equal to five (up to eight) times the number of parameters to be identified. If the initialization horizon is not enough long, divergence of the algorithm may occur.

6.3.2 Output Error with Fixed Compensator

The "plant + disturbance" model is the one given by Equations 6.3.1 and 6.3.2 and, for the first order example, by Equations 6.3.3 and 6.3.4 respectively.

The idea behind this method is the observation that, in the absence of disturbances, the output predicted by the RLS predictor $\hat{y}(t+1)$ tends towards $y(t+1)$. In these conditions one can consider replacing $y(t)$ by $\hat{y}(t)$ (the *a posteriori* prediction) in the predictor equation.

Indeed, consider, by means of an example, the RLS adjustable predictor for the "plant + disturbance" model of Equation 6.3.3

$$\hat{y}^\circ(t+1) = -\hat{a}_1(t)y(t) + \hat{b}_1(t)u(t) \qquad (6.3.16)$$

and replace in Equation 6.3.16 $y(t)$ by $\hat{y}(t)$ ("output error" type predictor). Then, we get

$$\hat{y}^\circ(t+1) = -\hat{a}_1(t)\hat{y}(t) + \hat{b}_1(t)u(t) = \hat{\theta}(t)^T \phi(t) \qquad (6.3.17)$$

in which

$$\hat{\theta}(t)^T = \left[\hat{a}_1(t), \hat{b}_1(t)\right] \quad ; \quad \phi(t)^T = \left[-\hat{y}(t), u(t)\right] \qquad (6.3.18)$$

where

$$\hat{y}(t+1) = \hat{\theta}(t+1)^T \phi(t) \Rightarrow \hat{y}(t) = \hat{\theta}(t)^T \phi(t-1) \qquad (6.3.19)$$

represents the new *a posteriori* prediction.

The difference between the RLS and the Output Error is illustrated in Figure 6.3.

The usefulness of this modification is clearly seen in the presence of disturbances. If $\hat{y}$ is used instead of $y(t)$ in the predictor equation and in the observations vector, one can see that, with an asymptotically decreasing adaptation gain, $\hat{y}(t)$ will only depend on $u(t)$ (which is not the case for the RLS predictor) and this will asymptotically lead to $E\{\phi(t)\ \varepsilon(t+1)\} = 0$. As a consequence a plant model with unbiased parameter estimates will be obtained.

In the general case, the observation vector used both in the predictor and in the parameter adaptation algorithm is of the form

$$\Phi(t)^T = \phi(t)^T = \left[-\hat{y}(t),-\hat{y}(t-1),...,-\hat{y}(t-n_A+1),u(t-d),...,u(t-d-n_B+1)\right] \qquad (6.3.20)$$

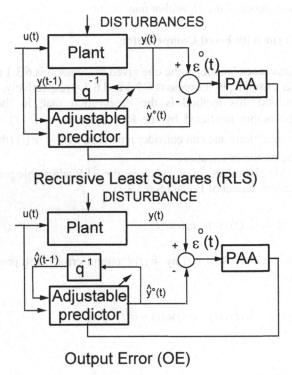

Figure 6.3. Comparison between the recursive least squares and the output error prediction structures

The prediction errors are defined from Equations 6.3.3, 6.3.17 and 6.3.19:

$$\varepsilon^o(t+1) = y(t+1) - \hat{y}^o(t+1)$$
$$\varepsilon(t+1) = y(t+1) - \hat{y}(t+1) \qquad (6.3.21)$$

and the adaptation algorithm given by Equations 5.5.1 through 5.1.3 is used.

One can also define, in the general case, an adaptation error (ν) obtained by filtering the prediction error:

$$v(t+1) = D(q^{-1})\varepsilon(t+1)$$
$$D(q^{-1}) = 1 + d_1 q^{-1} + \ldots + d_{n_D} q^{-n_D}$$
(6.3.22)

$$v^o(t+1) = \varepsilon^o(t+1) + \sum_{i=1}^{n_D} d_i \varepsilon(t+1-i)$$
(6.3.23)

In this case, we replace in the adaptation algorithm of Equations 5.5.1 through 5.5.3, $\varepsilon^o(t+1)$ and $\varepsilon(t+1)$ respectively by $v^o(t+1)$ and $v(t+1)$.

This is a method that assures therefore an unbiased identification of the coefficients of $A(q^{-1})$ and $B(q^{-1})$, without using a model for the disturbance to be estimated, and without requiring an initialization by another method.

This method is subject to a sufficient convergence condition (both in the case without and the case with disturbances):

$$\frac{D(z^{-1})}{A(z^{-1})} - \frac{\lambda_2}{2} \; ; \; 2 > \lambda_2 \geq \max \lambda_2(t)$$
(6.3.24)

should be a strictly positive real transfer function. This is why the filter $D(q^{-1})$ on the prediction error has been introduced.

For $n_A \leq 2$ ($n_A = deg\ A(q^{-1})$), the compensator $D(q^{-1})$ is not necessary (since $1/A(q^{-1}) - \lambda_2/2$ is strictly positive real almost everywhere $A(q^{-1})$ is asymptotically stable). For $n_A > 2$ and a decreasing adaptation gain, the compensator is generally unnecessary (it is proved that the prediction error remains bounded). If the compensator is introduced, one takes $n_d \leq n_A - 1$ or n_A ($n_d = degree\ D(q^{-1})$).

Since the condition on the transfer function of Equation 6.3.24 is only sufficient, $D(q^{-1}) = 1$ (no compensator) is always used to start, and a compensator is introduced only if the prediction error diverges.

It is also possible to estimate simultaneously model parameters and those of the filter on the prediction error (output error with adjustable compensator). For more details see Landau (2001b), Landau et al. (1997).

6.3.3 Output Error with (Adaptive) Filtered Observations

It is an extension of the basic output error method in which the filtering of the prediction error is replaced by the filtering of the observation vector (in order to satisfy the convergence conditions).

The equations for the adjustable predictor and the prediction errors are the same as for the output error with fixed compensator (Equations 6.3.16 to 6.3.19 and 6.3.21).

One uses the prediction error as adaptation error (it is not filtered). However, this algorithm uses a filtered version of the observation vector.

For the *output error with filtered observations* one defines a filter with constant parameters:

$$L(q^{-1}) = \hat{A}(q^{-1}) \qquad (6.3.25)$$

where $\hat{A}(q^{-1})$ is an estimation of the polynomial $A(q^{-1})$ obtained by another estimation algorithm (for example: recursive least squares or output error without filtering of the prediction error) and one defines the vector of filtered observations $\phi_f(t)$:

$$\Phi(t) = \phi_f(t) = \frac{1}{\hat{A}(q^{-1})} \phi(t) \qquad (6.3.26)$$

where $\phi(t)$ is given by Equation 6.3.20.

One can uses instead of the filter of Equation 6.3.25 a filter with time varying parameters:

$$L(q^{-1},t) = \hat{A}(t,q^{-1}) \qquad (6.3.27)$$

where $\hat{A}(t,q^{-1})$ is an estimation of the polynomial $A(q^{-1})$ at instant t (provided by the parameter adaptation algorithm) and the vector of filtered observations is defined by Equation 6.3.26 where $\hat{A}(q^{-1})$ is replaced by $\hat{A}(t,q^{-1})$. One obtains the output error with adaptive filtered observations.

As in the case of the recursive maximum likelihood algorithm, a stability test on $\hat{A}(t,q^{-1})$ should be done at each sampling.

The initialization of the algorithm can be done by the non-filtered output error algorithm (with no compensator) or the recursive least squares (in order to get a first estimation of $\hat{A}(q^{-1})$), but this is not mandatory since the algorithm can start with $\hat{A}(0,q^{-1}) = 1$).

The sufficient condition for the convergence of the algorithm is expressed for a fixed value of the filter $L(t,q^{-1}) = L(q^{-1},\hat{\theta}) = \hat{A}(q^{-1},\hat{\theta})$:

$$H'(z^{-1}) = \frac{\hat{A}(z^{-1},\hat{\theta})}{A(z^{-1})} - \frac{\lambda_2}{2} \ ; \ 2 > \lambda_2 \geq \max_t \lambda_2(t) \qquad (6.3.28)$$

should be a strictly positive real discrete time transfer function. In particular, this condition will always be satisfied around the correct value $\hat{\theta} = \theta$.

6.4 Validation of the models identified with Type II Methods

This section is concerned with the validation of models identified using the identification methods based on the uncorrelation between the observations and the prediction errors.

These methods are:

- Instrumental Variable with Auxiliary Model (IVAM)
- Output Error with Fixed Compensator (OEFC)
- Output Error with Filtered Observations (OEFO)
- Output Error with Adaptive Filtered Observations (OEAFO)

The principle of the validation method is as follows:

- If the disturbance is independent of the input ($E\{w(t)\,u(t)\} = 0$)
- If the "model + disturbance" structure chosen is correct, i.e., representative of the reality
- If an appropriate identification method has been used for the chosen structure
- If the degrees of the polynomials $A(q^{-1})$, $B(q^{-1})$ and the value of d (delay) have been correctly chosen

then the predicted outputs generated by a predictor model of "output error" type

$$\hat{A}\,(q^{-1})\,\hat{y}\,(t) = q^{-d}\,\hat{B}\,(q^{-1})\,u(t)$$

are asymptotically uncorrelated with respect to the output prediction error, that implies

$$E\{\varepsilon(t)\hat{y}(t-i)\} \approx \frac{1}{N}\sum_{t=1}^{N}\varepsilon(t)\hat{y}(t-i) = 0 \ ; \ i = 1,2,3,...$$

The validation method implements this principle. It is made up of several steps:

- Creation of an I/O file for the identified model (using the same input sequence as for the system)
- Creation of files for the sequences $\{y(t)\}$; $\{\hat{y}\,(t)\}$; $\{\varepsilon(t)\}$ (system output, model output, residual output prediction error). These files must contain at least 100 samples
- *Uncorrelation* test between the residual output prediction error sequence and the delayed prediction model output sequences

Uncorrelation test

Let $\{y(t)\}$ and $\{\hat{y}\,(t)\}$ be the centered output sequences of the plant and of the identified model respectively.

Let $\{\varepsilon(t)\}$ be the centered sequence of the residual output (prediction) errors (centered = measured values - mean values).

One computes

$$R(i) = \frac{1}{N} \sum_{t=1}^{N} \varepsilon(t)\hat{y}(t-i) \quad ; \quad i = 0,1,2,...,i_{max} \tag{6.4.1}$$

$$RN(i) = \frac{R(i)}{\left[\left(\frac{1}{N}\sum_{t=1}^{N}\hat{y}^2(t)\right)\left(\frac{1}{N}\sum_{t=1}^{N}\varepsilon^2(t)\right)\right]^{1/2}} \quad ; \quad i = 0,1,2,...,i_{max} \tag{6.4.2}$$

where

$$i_{max} = \max(n_A, n_B + d)$$

and the $RN(i)$ are estimations of the (normalized) cross-correlations. Note that $RN(0) \neq 1$.

If the uncorrelation is perfect between $\varepsilon(t)$ and $\hat{y}(t-i)$, $i \geq 1$ (theoretical situation) thus

$$RN(1) = RN(i) = 0 \quad ; \quad i = 1,2,...,i_{max}$$

In practice this is never the case, either because the duration of the identification was not long enough to approach the asymptotic character, or because the identification of *good* simple models is desired (with few parameters, *i.e.* under estimation of the orders of $A(q^{-1})$ and $B(q^{-1})$), which results in *structural* errors which are correlated with the input $u(t)$.

In practice it is desired that the values of $|RN(i)|$, $i = 1,..., i_{max}$ be as small as possible, but the difficulty lies in defining a reference (validation criterion).

The validation criterion for type II identification methods based on an uncorrelation test will be defined similarly to that for type I identification methods which also use an uncorrelation test, but on different sequences of variables (see Section 6.2). In this case one therefore considers as a practical validation criterion

$$|RN(i)| \leq \frac{2.17}{\sqrt{N}} \quad ; \quad i = 1,2,...,i_{max}$$

where N is the number of samples. Table 6.1 can be used for finding the numerical value of the validation criterion for various N and levels of signification of the test.

All the comments made in Section 6.2 upon the significance of the values of the validation criterion remain valid. In particular the basic practical numerical value for the validation criterion which is

$$|RN(i)| \leq 0.15 \quad ; \quad i = 1,2,...,i_{max}$$

is worth remembering.

One can say that the uncorrelation test allows one to compare the results provided by all identification methods since it only requires the parameters of the plant model $(B(q^{-1}),A(q^{-1}))$. Therefore, it is useful to perform this validation test even for models identified with type I identification methods in order to compare them with estimated models obtained with type II identification methods.

In short, one only takes the estimated plant model $(B(q^{-1}),A(q^{-1}))$ obtained with type I identification methods (the noise model is not considered) and one performs the uncorrelation test. The best model is that which will give the lowest cross-correlation terms[3].

6.5 Estimation of the Model Complexity [4]

6.5.1 An Example

In order to understand the principle of the model complexity estimation, let us consider a first-order discrete time model. It will be assumed that there is no noise.

Assume that the plant model is described by

$$y(t) = -a_1\,y(t\text{-}1) + b_1\,u(t\text{-}1) \tag{6.5.1}$$

and that the data are not corrupted by noise. The order of this model is $n = \max(n_A, n_B + d) = 1$

Question: Does there exist a test allowing one to verify if the hypothesis upon the order of the model is correct?

Construct the following matrix:

- On the first row one writes $y(t)$ as well as the variables represented in the right hand term of Equation6.5.1
- The other rows are formed by the delayed variables occurring in Equation 6.5.1.

---

[3] However, in comparing the models obtained by type I identification methods, priority will be given to the whiteness test.

[4] Can be omitted for a first reading.

$$\begin{bmatrix} y(t) & \vdots & -y(t-1) & u(t-1) \\ y(t-1) & \vdots & -y(t-2) & u(t-2) \\ y(t-2) & \vdots & -y(t-3) & u(t-3) \end{bmatrix} = \begin{bmatrix} Y(t) & R(1) \end{bmatrix}$$

(6.5.2)

$$\underbrace{\qquad\qquad\qquad\qquad}$$

$\uparrow$ $\uparrow$
$Y(t)$ $R(1) = R(\hat{n})$

The resulting matrix can be split into two parts: the first column denoted $Y(t)$, is a vector containing the current and previous outputs ; the remaining rows define a matrix denoted by $R(1)$.

If the structure of Equation 6.5.1 is true (which means that it can represent the relationship between u and y), one can replace the components of the first column of Equation 6.5.2 by their expression given by Equation 6.5.1. One obtains by evaluating the determinant of the matrix given in Equation 6.5.2:

$$rank[Y(t) \quad R(1)] = rank \begin{bmatrix} -a_1 y(t-1) + b_1 u(t-1) & \vdots & -y(t-1) & u(t-1) \\ -a_1 y(t-2) + b_1 u(t-2) & \vdots & -y(t-2) & u(t-2) \\ -a_1 y(t-3) + b_1 u(t-3) & \vdots & -y(t-3) & u(t-3) \end{bmatrix} = 2(<3)$$

(6.5.3)

since the first column is a linear combination of the two other columns. Effectively $Y(t) = R(1)\theta$ with $\theta^T = [a_1, b_1]$ and the rank of the matrix is 2 (instead of 3).

It results that the rank test on the matrix of Equation 6.5.2 (size of the largest non-null determinant) allows one to conclude upon the validity of the structure of the model considered for representing the plant.

Let assume that the plant model is second-order and that the data are not corrupted by noise. It can be described by the following equation:

$$y(t) = -\sum_{i=1}^{2} a_i y(t-i) + \sum_{i=1}^{2} b_i u(t-i)$$

(6.5.4)

If one computes the rank of the matrix of Equation 6.5.3, one gets:

$$rank \begin{bmatrix} y(t) & \vdots & -y(t-1) & u(t-1) \\ y(t-1) & \vdots & -y(t-2) & u(t-2) \\ y(t-2) & \vdots & -y(t-3) & u(t-3) \end{bmatrix} = 3 = full\ rank$$

(6.5.5)

$$\underbrace{\qquad\qquad\qquad\qquad}$$

$\uparrow$ $\uparrow$
$Y(t)$ $R(1) = R(\hat{n})$

The determinant of the matrix of Equation 6.5.5 is not null since $Y(t)$ cannot be expressed as a linear combination of columns 2 and 3.

However if one also uses the measurements from the instant *t-2,* one can construct the following matrix:

$$
\begin{bmatrix}
y(t) & \vdots & -y(t-1) & -y(t-2) & u(t-1) & u(t-2) \\
y(t-1) & \vdots & -y(t-2) & -y(t-3) & u(t-2) & u(t-3) \\
y(t-2) & \vdots & -y(t-3) & -y(t-4) & u(t-3) & u(t-4) \\
y(t-3) & \vdots & -y(t-4) & -y(t-5) & u(t-4) & u(t-5) \\
y(t-4) & \vdots & -y(t-5) & -y(t-6) & u(t-5) & u(t-6)
\end{bmatrix}
= \begin{bmatrix} Y(t) & R(2) \end{bmatrix}
\tag{6.5.6}
$$

$$
\underset{Y(t)}{\uparrow} \qquad\qquad\qquad \underset{R(2)=R(\hat{n})}{\uparrow}
$$

The rank of this matrix is strictly less than 5 (in fact it is equal to 4), since *Y(t)* is a linear combination of columns 2, 3, 4 and 5.

The objective of model order estimation will be to search for the number of columns that has to be added in order to be able to express Y(t) as a linear combination of the other columns (one adds the same number of delayed columns of *y* and *u*) . The algorithm will stop when the matrix of the form given in Equation 6.5.6 is no more of full rank. In this way $n = max\ (n_A,\ n_B + d)$.

6.5.2 The Ideal Case (No Noise)

In the ideal case, which means in the absence of noise, we are interested to estimate the order (complexity) of a model of the form

$$
y(t) = -\sum_{i=1}^{n} a_i y(t-i) + \sum_{i=1}^{n} b_i u(t-i)
\tag{6.5.7}
$$

by the *rank test.* To do this, one constructs the matrix

$$
R(\hat{n}) = \begin{bmatrix}
-y(t-1) & \cdots & -y(t-\hat{n}) & u(t-1) & \cdots & u(t-\hat{n}) \\
-y(t-2) & \cdots & -y(t-\hat{n}-1) & u(t-2) & \cdots & u(t-\hat{n}-1) \\
\vdots & \vdots & \vdots & \vdots & \vdots & \vdots \\
-y(t-N) & \cdots & -y(t-\hat{n}-N+1) & u(t-N) & \cdots & u(t-\hat{n}-N+1)
\end{bmatrix}
\tag{6.5.8}
$$

$$
= \begin{bmatrix}
\phi(t-1)^T \\
\phi(t-2)^T \\
\vdots \\
\phi(t-N)^T
\end{bmatrix}
$$

where

$$
\phi(t-j)^T = [-y(t-j),...,-y(t-\hat{n}-j+1), u(t-j),..,u(t-\hat{n}-j+1)];\ j=1,...,N \tag{6.5.9}
$$

and the vector

$$Y(t)^T = [y(t), y(t-1),..., y(t-N+1)] \tag{6.5.10}$$

and one tests the rank of the matrix

$$[Y(t) \; \vdots \; R(\hat{n})] \tag{6.5.11}$$

for increasing values of the estimated order $\hat{n}$.

Remark: the number of data N should be larger than $2n_{max}+1$ (where n_{max} is the supposed maximum value of the model order).

One has the following result:

- If $\hat{n} < n$, $[Y(t) \; \vdots \; R(\hat{n})]$ is a full rank matrix (of rank $2\hat{n}+1$).
- If $\hat{n} \geq n$, $[Y(t) \; \vdots \; R(\hat{n})]$ is not a full rank matrix (rank $2n$ instead of $2\hat{n}+1$).

The value used to test the rank can be the determinant, the singular values of the matrix, *etc.*.

6.5.3 The Noisy Case

In the presence of noise, Equation 6.5.7 becomes

$$y(t) = -\sum_{i=1}^{n} a_i y(t-i) + \sum_{i=1}^{n} b_i u(t-i) + w(t) \tag{6.5.12}$$

where *w(t)* represents the noise effect.

Unfortunately, the method presented earlier does not work when noise is present since the matrix $[Y(t) \; \vdots \; R(\hat{n})]$ never becomes singular.

White noise
It is important to remark that the rank test is equivalent to the search of a parameter vector $\hat{\theta}$ that minimizes the following criterion for an estimated value of the model order denoted by $\hat{n}$ [5]:

$$J_{LS}(\hat{n}) = \min_{\hat{\theta}} \frac{1}{N} \left\| Y(t) - R(\hat{n})\hat{\theta}_{\hat{n}} \right\|^2 \tag{6.5.13}$$

---

[5] $\|x\|^2 = x^T x$.

where $\hat{\theta}$ expresses the linear dependence of $Y(t)$ with respect to the columns of $R(\hat{n})$. However this criterion is the least squares criterion of Equation 5.2.40.

Taking in account Equations 6.5.8, 6.5.9 and 6.5.10, Equation 6.5.13 can be expressed as

$$J_{LS}(\hat{n}) == \min_{\theta} \frac{1}{N} \sum_{i=t-N+1}^{t} \left[y(i) - \hat{\theta}_{\hat{n}}^{T}(t)\phi(i-1) \right]^{2}$$

where

$$\hat{\theta}_{\hat{n}}^{T} = \left[\hat{a}_1, \ldots, \hat{a}_{\hat{n}}, \hat{b}_1, \ldots, \hat{b}_{\hat{n}} \right]$$

Therefore, for the case where the noise in Equation 6.5.12 is white, the use of the least squares algorithm (the least squares estimation algorithm is unbiased in this case) allows one to estimate the order of the model by evaluating the value of the criterion for increasing estimated orders $\hat{n}$. If $\hat{n} \geq n$ the criterion becomes practically null, or takes a constant value in the presence of white noise in Equation 6.5.12 (this is illustrated in Figure 6.4)

Non-white noise
However, the procedure for order estimation will not work when the noise in Equation 6.5.12 is not white since the difference $J_{LS}(\hat{n}) - J_{LS}(\hat{n}+1)$ does not really goes towards 0 when $\hat{n} \geq n$ (*i.e.* the procedure tends to overestimate the order of the model).

To overcome this problem one should use an algorithm allowing one to obtain an unbiased estimation of $\hat{\theta}_{\hat{n}}$. To achieve this one can use the technique of "instrumental variables" (used also in Section 6.3.1).

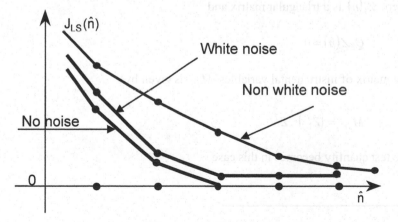

Figure 6.4. Least squares criterion evolution as a function of $\hat{n}$

In this approach one replaces the measurements by the instrumental variables that are correlated with the "good" measures but less correlated with the noise. The number of instrumental variables to be used ($2L$) should be larger than twice the order of the model to be estimated. One can use the delayed inputs as instrumental variables, since $u(t-L)$ is correlated with $y(t)$ for $L \geq \hat{n}$ (see Equation 6.5.12), but, on the other hand, $u(t)$ is not correlated with the noise.

In this case one replaces $R(\hat{n})$ and $Y(t)$ by $R_{IV}(\hat{n})$ and $Y_{IV}(t)$ where

$$R_{IV}(\hat{n}) = M_{IV} R(\hat{n}) \quad ; \quad Y_{IV}(t) = M_{IV} Y(t) \tag{6.5.14}$$

M_{IV} is the matrix of instrumental variables and it is constructed from the matrix

$$Z^T(\hat{n}) = [\phi_{IV}(t-1), \phi_{IV}(t-2), \ldots, \phi_{IV}(t-N)] \tag{6.5.15}$$

where

$$\phi_{IV}^T(t-j) = [-u(t-L-j), \ldots, u(t-L-\hat{n}-j+1), u(t-j), \ldots, u(t-\hat{n}-j+1)] \tag{6.5.16}$$

One searches first for a matrix

$$Q = \begin{bmatrix} Q_1 \\ Q_2 \end{bmatrix}$$

such that (triangularization of the matrix $Z(\hat{n})$)

$$Q_1 Z(\hat{n}) = Z_1(\hat{n})$$

where $Z_1(\hat{n})$ is a triangular matrix and

$$Q_2 Z(\hat{n}) = 0$$

The matrix of instrumental variables M_{IV} is given by[6]

$$M_{IV} = (Z_1^T)^{-1} Z^T$$

The test quantity becomes in this case

---

[6] The construction of the instrumental variables in this case is related to the orthogonal projection of the measurements on the space defined by $Z(\hat{n})$.

$$J_{IV}(\hat{n}) = \min_{\theta} \frac{1}{N} \left\| Y_{IV}(t) - R_{IV}(\hat{n})\hat{\theta} \right\|^2 \qquad (6.5.17)$$

A theoretical study shows that this quantity goes towards 0 as the estimated order approaches the true one (Duong and Landau 1996)[7]. As a consequence, a significant slope change will appears on the curve $J_{IV}(\hat{n})$, allowing, in general, a clear estimation of the order.

6.5.4 Criterion for Complexity Estimation

One of the objectives of system identification is to estimate models of reduced order (parsimony principle). It is therefore reasonable to add to the criterion of Equation 6.5.17 a term that penalizes the model complexity. Such a term can be of the form

$$S(\hat{n}, N) = \frac{2\hat{n} \log(N)}{N} \qquad (6.5.18)$$

but other choices are possible (Duong and Landau 1996; Söderström and Stoica 1989).

One defines the criterion for model order estimation as

$$CJ_{IV}(\hat{n}) = J_{IV}(\hat{n}) + S(\hat{n}, N) \qquad (6.5.19)$$

and therefore $\hat{n}$ will be given by the value that minimizes the criterion of Equation 6.5.19:

$$\hat{n} = \hat{n}^* \Rightarrow CJ_{IV}(\hat{n}) = min \qquad (6.5.20)$$

The complexity estimation obtained in this way is consistent, which means that one finds the exact order as the number of data tends toward infinity ($N \to \infty$).

A typical curve for the evolution of the criterion 6.5.18 as a function of $\hat{n}$ is given in Figure 6.5.

Once the order $\hat{n}$ is estimated, one can apply a similar procedure for estimating $\hat{n} - \hat{d}$, $\hat{n}_A$, $\hat{n}_B + \hat{d}$ from which $\hat{n}_A, \hat{n}_B$, and $\hat{d}$ are obtained.

The functions *estorderls.*sci, *estorderiv.sci* (Scilab) and *estorderls.m*, *estorderiv.m* (MATLAB®) implement the above described techniques for model order estimation[8].

---

[7] It is shown in (Duong, Landau 1996) that $J_{IV}(\hat{n})$ has a well defined statistical distribution allowing to define an "over estimation risk" for the estimation of the model order $\hat{n}$.

[8] Available on the book website.

Other aspects related to the estimation of n_A, n_B, and d will be presented in Chapter 7, Section 7.3.

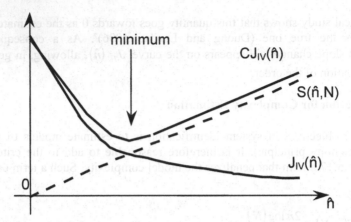

Figure 6.5. Evolution of the criterion for the estimation of the model order

6.6 Concluding Remarks

In this chapter recursive identification methods have been presented. They have been classified in two categories:

The first category includes the identification methods based on the *whitening of the prediction error* in order to get unbiased parameter estimates. The following identification methods belong to this category:

- Recursive Least Squares (RLS)
- Extended Least Squares (ELS)
- Recursive Maximum Likelihood (RML)
- Output Error with Extended Prediction Model (O.E.E.P.M)
- Generalized Least Squares (GLS)

The validation of the models identified using these methods consists in testing if the residual prediction error sequence is approaching the characteristics of a white noise sequence. This is done by carrying out a *whiteness test* on the sequence of residuals. An acceptable identified model should verify the condition

$$|RN(i)| \leq \frac{2.17}{\sqrt{N}} \quad ; \quad i = 1, 2, \dots, max(n_A, n_B + d)$$

where

$$RN(i) = \frac{1}{N} \sum_{t=1}^{N} \varepsilon(t)\varepsilon(t-i) / \left[\frac{1}{N} \sum_{t=1}^{N} \varepsilon^2(t) \right]$$

and N is the number of samples.

The second category includes the identification methods based on the *uncorrelation between the prediction error and the observations* in order to get unbiased parameter estimates. The following identification methods belong to this category:

- Instrumental Variable with Auxiliary Model (IVAM)
- Output Error with Fixed Compensator (OEFC)
- Output Error with Filtered Observations (OEFO)
- Output Error with Adaptive Filtered Observations (OEAFO)

The validation of models identified using these methods consist in testing if the residual prediction error sequence and the predicted outputs (generated by the model $A(q^{-1}) \hat{y}(t) = q^{-d} \hat{B}(q^{-1}) u(t)$) are uncorrelated. This is done by carrying out an *uncorrelation test*. An acceptable identified model should verify the condition:

$$|RN(i)| \le \frac{2.17}{\sqrt{N}} \quad ; \quad i = 1,2,\dots, \max(n_A, n_B + d)$$

where

$$RN(i) = \frac{1}{N} \sum_{t=1}^{N} \varepsilon(t)\hat{y}(t-i) / \left\{ \left[\frac{1}{N} \sum_{t=1}^{N} \hat{y}^2(t) \right] \left[\frac{1}{N} \sum_{t=1}^{N} \varepsilon^2(t) \right] \right\}^{1/2}$$

and N is the number of samples.

Taking into account the relative weight of various non-gaussian and modeling errors (which increases with the number of samples), the validation criterion can be slightly tightened for small N and slightly relaxed for large N. Therefore, for simplicity's sake, one can consider as a basic practical numerical value for the validation criterion (for both tests) the value

$$|RN(i)| \le 0.15 \quad ; \quad i = 1,2,\dots, \max(n_A, n_B + d)$$

Before doing a parameter estimation, it is obviously necessary either to estimate the order of the model from the data, or to choose a model order from the available *a priori* information. The techniques of order estimation from input/output data have been presented in Section 6.5

The problem of order estimation can be converted in a parameter identification problem whose objective is to minimize a criterion with two terms. One of the terms of the criterion corresponds to the variance of the prediction error, while the other term is a penalty terms which increases with the order of the model.

6.7 Notes and References

A complete presentation of identification methods in their recursive or non-recursive form, and also of methods for analysis and their properties, can be found in:

Landau I.D. (2001b) Les bases de l'identification des systèmes, in Identification des systèmes (I.D. Landau and A. Besançon-Voda Ed.), pp. 19-129, Hermes Paris.

Ljung L., (1999) System Identification. Theory for the User, 2nd edition, Prentice Hall, NJ.

Söderström T., Stoica P. (1989) System Identification, Prentice Hall, U.K.

For a detailed analysis of the convergence of recursive methods see:

Ljung L., Söderström T. (1983) Theory and Practice of Recursive Identification, MIT Press, Cambridge, U.S.A.

Landau I.D., Lozano R., M'Saad M., (1997) Adaptive Control, Springer, London, UK.

Solo V. (1979) The Convergence of A.M.L., IEEE Trans. Automatic Control, vol. AC-24, pp. 958-963.

Landau I.D. (1982) Near Supermartingales for Convergence Analysis or Recursive Identification and Adaptive Control Schemes, Int J. of Control, vol.35, pp. 197-226.

The method of "generalized least squares" presented in this book is based on:

Béthoux G. (1976) Approche unitaire des méthodes d'identification et de commande adaptative des procédés dynamiques, Ph.D. Thesis, Institut National Polytechnique de Grenoble, July.

The « output error » type identification methods find their origin in:

Landau I.D. (1976) Unbiased Recursive Identification Using Model Reference Adaptive Techniques, I.E.E.E., Trans. on Automatic Control, vol AC-20, n°2, pp. 194-202.

Landau I.D. (1979): Adaptive Control - the model reference approach, Dekker, N.Y., U.S.A.

For a detailed presentation of these methods and their properties see (Landau et al. 1997).

A detailed evaluation of output error identification methods can be found in:

Dugard L., Landau I.D. (1980) Recursive Output Error Identification Algorithms: Theory and Evaluation, Automatica, vol. 16, pp.443-462.

The methods of the instrumental variables are presented in details in:

Young P.C., Jakeman A.J. (1979) Refined Instrumental Variable Methods of Recursive Time Series Analysis, Part I, Single-input, Single-output Systems, Int. J. of Control, vol. 29, pp.1-30.

Söderström T., Stoica P. (1983) Instrumental variable methods for system identification, Lectures Notes in Control and Information Sciences, Springer, Berlin, Heidelberg, New York.

The methods of the instrumental variables presented in this book are based on:

Young P.C. (1969) An Instrumental Variable Method for Real Time Identification of a Noisy Process, Automatica, vol.6, pp. 271-288.

Duong H.N., Landau I.D. (1996) An I.V. based criterion for model order selection, Automatica, vol. 32, no. 6, pp. 909-914.

Validation techniques are also discussed in (Söderström and Stoica 1989, Ljung 1999).

Estimation of correlation functions and the independence tests are considered in:

Hogg R., Graig A. (1970) Introduction to Mathematical Statistics, MacMillan, N.Y.

Bendat J.S., Piersol A.G. (1971) Random Data: Analysis and Measurement Procedures, John Wiley.

Candy J.V., (1986) Signal Processing - The Model Based Approach, MacGraw-Hill, N.Y.

For model order estimation see (Söderström and Stoica 1989, Ljung 1999, Duong and Landau 1996).

7

Practical Aspects of System Identification

This chapter reviews a variety of practical aspects concerning system identification. Several identification examples are presented, using both simulated data and real data from several plants (air heater, distillation column, DC motor, flexible transmission).

7.1 Input/Output Data Acquisition

7.1.1 Acquisition Protocol

In Chapter 5, Section 5.3 it was shown that, in order to obtain a good identified model, the excitation signal (applied to the plant input) should contain a wide spectrum of frequencies. This signal will be superposed to the steady state control signal corresponding to the operating point around which we would like to identify the model of the plant.

As a general rule, one uses as an excitation signal a PRBS (pseudo random binary sequence) with a low magnitude. The properties of the PRBS have been examined in Chapter 5, Section 5.3.2.

We recall that, for correctly estimating the steady state gain of the dynamical model, it is necessary that at least the duration of one of the pulses of the PRBS be larger than the rising time of the plant to be identified. This leads to the following condition:

$$pNT_s \geq t_R$$

where

- T_s = sampling period
- p = frequency divider
- $p\,T_s$ = clock period of the PRBS generator (clock frequency = f_s/p)

279

- N = number of cells of the shift register
- t_M = rising time of the system

The magnitude of the PRBS should not exceed, as a general rule, a few percent of the steady state control signal amplitude.

Data acquisition is done either by dedicated data acquisition equipment or, very often, by using a computer equipped with a data acquisition board containing analog-to-digital and digital-to-analog converters. The board is emulated by the computer, which provides the PRBS and the input DC component corresponding to the operating point.

A C^{++} program as well as a MATLAB$^{®}$ function (*prbs.m*) for the generation of the PRBS can be downloaded from the book website.

Several options for connecting the data acquisition system to the plant are possible.

Plant Operated in Open Loop
This is the most typical case and the simplest one. It is illustrated in Figure 7.1.

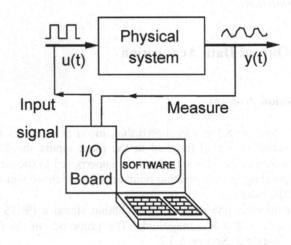

Figure 7.1. Data acquisition for system identification - plant operated in open loop

In this situation, the excitation signal superposed to the DC control signal corresponding to the operating point is directly applied to the process. *To start the data acquisition protocol, the plant should be in a steady state operation regime (i.e. the PRBS is sent once the transient related to the application of the DC control signal is over).*

Plant Operated in Closed Loop
In this situation two possibilities can be considered:

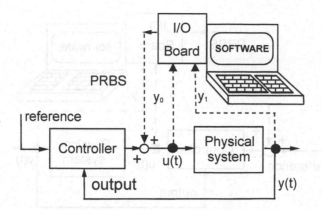

Figure 7.2. Data acquisition for identification – plant operated in closed loop with excitation signal added to the controller output

1. Excitation superposed to the controller output
    The PRBS is added to the controller output and one performs the acquisition of the PRBS and of the effective input applied to the plant together with the corresponding plant output. The transfer between y_0 and y_1 will be identified.
2. Excitation on the reference
    The PRBS is added to the reference defining the operating point and one performs the acquisition of the PRBS, the input and the output of the plant. This is shown in Figure 7.3.

If one would like to identify the plant model from data acquisition in closed loop operation, two options can be considered.

In the first case one tries to ignore the presence of the controller. For this, it is important to use what is called a "soft" controller. In the case of a PID one should suppress the derivative action and use a low value of proportional gain. However the integral action is important for maintaining in the average the operating point.

If an RST controller is used, the design based on an imprecise available model should be done such that the closed loop is slow but the presence of an integrator in the controller is important in order to maintain the operating point. The quality of the model obtained with this approach can vary considerably.

In the second case, one uses dedicated methods for plant model identification in closed loop operation. These methods, which take into account the presence of the controller, are presented in Chapter 9. They provide, in general, excellent identified models.

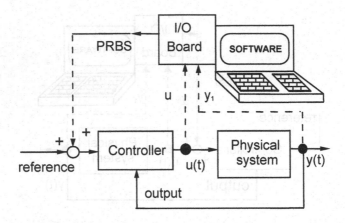

Figure 7.3. Data acquisition for identification – plant operated in closed loop with excitation added to the reference

7.1.2 Anti-Aliasing Filtering

It was shown in Chapter 2, Section 2.2, that the conversion of signals having a frequency spectrum exceeding $0.5\,f_s$ will introduce distortions on the discretized signal in the useful band between 0 and $0.5\,f_s$.

In order to avoid this situation, it is necessary to insert between the plant output and the analog to digital converter a continuous time anti-aliasing filter (low pass filter) allowing to significantly attenuate the components of the signal beyond $0.5f_s$.

The violation of this rule can produce important errors in identification. Even if one may be able to get discrete time models that can be validated, they do not represent, in practice, the real behavior of the system (*i.e.* the continuous time behavior of the plant).

7.1.3 Over Sampling

Different situations encountered in practice lead to the use of a data acquisition frequency that is a multiple of the sampling frequency used in the control loop (over – sampling). For example:

- In many digital control systems (DCS), the data acquisition frequency is significantly higher than the one used in the control loops.
- In systems using a sampling period for the control loop that is higher than 1s, it becomes difficult to implement a continuous time anti aliasing filter and it is necessary to use a digital filtering technique.

In both cases, one should do data acquisition at a frequency which is a multiple of the sampling frequency used for control (or conversely: the sampling frequency of the control loop should be a sub-multiple of the data acquisition frequency):

$$f_a = nf_s$$

where

- f_a = data acquisition frequency
- n = ratio of frequencies (integer)
- f_s = sampling frequency of the control loop

The discrete time signal obtained at the frequency f_a is passed through a digital anti-aliasing filter which will attenuate the signal components over $0.5 f_s = 0.5 f_a / n$. On the resulting sequence we will take data every n samples (decimation). The implementation of a data acquisition using over-sampling is illustrated in Figure 7.4.

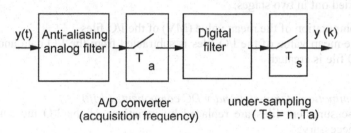

A/D converter   under-sampling
(acquisition frequency)   (Ts = n .Ta)

Figure 7.4. Implementation of the data acquisition using over-sampling

If n is sufficiently large ($n \geq 4$), a moving average type filter

$$y_f(t) = \frac{y(t) + y(t-1) + \ldots + y(t-n+1)}{n}$$

will be enough.

When using over-sampling for data acquisition, one should pay attention how the frequency divider (p) for the clock frequency of the PRBS is chosen. *The frequency divider used for the PRBS should be a multiple of n, in order that the final sampling frequency f_s is still an integer multiple of the clock frequency of the PRBS.*

The clock frequency of the PRBS is given by

$$f_{PRBS} = \frac{1}{p} f_a$$

The final sampling frequency is given by

$$f_s = \frac{1}{n} f_a = \frac{p}{n} f_{PRBS}$$

where (p/n) should be an integer.

7.2 Signal Conditioning

7.2.1 Elimination of the DC Component

The structures of the models used for identification correspond to dynamic models (that express output variations as a function of input variations around an operating point).

It is therefore necessary, for a correct identification, to eliminate from the input-output data either the DC components (corresponding to the operating point) or the slow drift of the operating point during identification.

Case 1. Elimination of stationary (or quasi-stationary) DC components.
This is carried out in two stages:

 a. Computation of the mean value (MV) of the I/O files
 b. The mean value of the I/O files is subtracted from the I/O data and a new I/O file is created

Case 2. Elimination of non-stationary DC components (drift)
The I/O measurement files are replaced by variations of the I/O measurements filtered if necessary:

$$y'(t) = \frac{y(t) - y(t-1)}{1 + f_1 q^{-1}} \quad ; \quad u'(t) = \frac{u(t) - u(t-1)}{1 + f_1 q^{-1}}$$

with

$$-0{,}5 \le f_1 \le 0$$

but other types of filters may be used.

7.2.2 Identification of a Plant Containing a Pure Integrator

A plant incorporating a pure integrator may be identified. However, if the existence of this integrator is known a priori, this a priori information can be used to reduce the complexity of the model to be identified and to improve the precision of the identification. Two methods can be used:

 1. The input is replaced by its integral, the output remaining the same:

$$y'(t) = y(t) \quad ; \quad u'(t) = \frac{u(t)}{1 - q^{-1}}$$

 2. The output is replaced by its variations, the input remaining the same:

$$y'(t) = y(t) - y(t-1) \qquad ; \qquad u'(t) = u(t)$$

$y'(t)$ and $u'(t)$ may also be filtered if necessary.

7.2.3 Identification of a Plant Containing a Pure Differentiator

If the presence of a differentiator is known, this a priori information can be used. In this case, the input is replaced by its variations, the output remaining the same:

$$y'(t) = y(t) \qquad ; \qquad u'(t) = u(t) - u(t-1)$$

7.2.4 Scaling of the Inputs and Outputs

The observation vector $\Phi(t)$ used in the parameter adaptation algorithm is built from the measurements $y(t)$, $y(t-1)$... (or by variables correlated with the output) and $u(t)$, $u(t-1)$...

$$\Phi(t)^T = \left[- y(t), -y(t-1), ..., u(t), u(t-1), ...\right]$$

On the other hand, the adaptation gain $F(t)$ is expressed as

$$F(t)^{-1} = \sum_{i=1}^{t} \Phi(i-1)\Phi^T(i-1) + \frac{1}{\delta}I \quad ; \quad \delta \ll 1$$

If the level of $u(t)$, $u(t-1)$... is very different from the level of $y(t)$, $y(t-1)$, the gain matrix will be extremely unbalanced, thus resulting in significantly different convergence speeds for $\hat{a}_i(t)$ and $\hat{b}_i(t)$.

It is therefore convenient to scale the I/O files (by multiplying, if required, either $u(t)$ or $y(t)$). This will result in the modification of the identified steady state gain. The values of the estimated parameters $\hat{b}_i$ must therefore be divided or multiplied accordingly in order to obtain a model with a static gain corresponding to the real one.

7.3 Selection (Estimation) of the Model Complexity

This section discusses the determination of the delay (d) and the degrees of the polynomials $A(q^{-1})$, $B(q^{-1})$ and $C(q^{-1})$.

The "plant + disturbance" model to be identified is of the form

$$A(q^{-1})y(t) = q^{-d}B(q^{-1})u(t) + w'(t)$$

where, according to the structures chosen, one has:

$$w'(t) = e(t) \tag{7.3.1}$$

$$w'(t) = A(q^{-1})w(t) \tag{7.3.2}$$

$$w'(t) = C(q^{-1})e(t) \tag{7.3.3}$$

$$w'(t) = \frac{1}{C(q^{-1})}e(t) \tag{7.3.4}$$

In order to start the parameter estimation methods, we need to specify

$$n_A = ? \quad ; \quad n_B = ? \quad ; \quad d = ?$$

and for structures [S3] and [S4]

$$n_C = ?$$

where n_A, n_B and n_C are the degrees of the polynomials $A(q^{-1})$, $B(q^{-1})$ and $C(q^{-1})$ respectively.

Remember that the order of a sampled-data system is given by $n = max\ (n_A, n_B+d)$.

Techniques for order estimation have been presented in Chapter 6, Section 6.5. If order estimation functions that allows the estimation of $n = max\ (n_A, n_B+d)$, n_A, n_B and d, are available (like *estimorderiv.sci* (Scilab) and *estimorderiv.m* (MATLAB®)[1], that estimate n), it is advisable to use them. Nevertheless the order estimation can also be done on the basis of "*a priori*" knowledge and a set of trials on the acquired data[2]. We give next details about this procedure.

"A priori" Choice for n_A
Two cases can be distinguished:

1.   Industrial plant (temperature control, flow rate, concentration, etc...).
      For this type of plant in general

$$n_A \le 3$$

---

[1] Available from the book website.
[2] This approach can be also used for a confirmation of the order estimation results obtained with the techniques described in Section 6.5.

and the value $n_A = 2$, which is very typical, is a good starting value to choose.

2. Electromechanical systems.

n_A results from the structural analysis of the system.

Example: flexible transmission with two resonant modes.
In this case, $n_A = 4$ is chosen, since a second-order is required to model each resonant mode.

Initial Choice of d and n_B

If no knowledge of the time delay is available, $d = 0$ is chosen as an initial value. If a minimum value is known, an initial value $d = d_{min}$ is chosen.

If the time delay has been underestimated during identification, the first coefficients of $B(q^{-1})$ will be very low. Thus n_B must then be chosen so that it can indicate the presence of the time delays and identify the transfer function numerator. $n_B = (d_{max} - d_{min}) + 2$ is then chosen as the initial value. At least two coefficients are required in $B(q^{-1})$ because of the fractional delay (see Section 2.3.7). If the time delay is known, $n_B \geq 2$ is chosen, however the value 2 is a good initial value.

Initial Choice of n_C

As a general rule $n_C = n_A$ is chosen.

Determination of the Time Delay d (First Approximation)
Method 1. One identifies the system using the RLS. The estimated numerator will be of the form

$$\hat{B}(q^{-1}) = \hat{b}_1 q^{-1} + \hat{b}_2 q^{-2} + \hat{b}_3 q^{-3} + \dots$$

If

$$\left| \hat{b}_1 \right| \leq 0.15 \left| \hat{b}_2 \right|$$

$b_1 \approx 0$ is considered, and time delay d is increased by 1: $d = d_{in} + 1$ (since if $b_1 = 0$, $B(q^{-1}) = q^{-1} (b_2 q^{-1} + b_3 q^{-2})$).

If

$$\left| \hat{b}_i \right| < 0.15 \left| \hat{b}_{d_i + 1} \right| \quad ; \quad i = 1, 2, \dots, d_i$$

time delay d is increased by d_i: $d = d_{in} + d_i$. After these modifications, a new identification session is performed.

Method 2. An identification of the system is performed with the RLS in order to obtain a model of the type "impulse response"

$$y(t) = \sum_{i=1}^{n_B} b_i u(t-i) \ ; \ n_B = Large \ (20,...,30)$$

If there is a time delay, then

$$\left| \hat{b_i} \right| < 0.15 \left| \hat{b}_{d+1} \right| \quad ; \quad i = 1,2,...,d$$

From this time delay estimation, a new identification is performed to find a pole-zero model.

Both these methods may of course be completed by a display of the step response.

A more accurate estimation of time delay d is carried out by performing a new identification, followed by a validation of the identified model. Note that if the system is contaminated by measurement system noise an accurate estimation of the delay will only be carried out with the method that enables the identified model to be validated.

Determination of $(n_A)_{max}$ and $(n_B)_{max}$

The aim is to obtain the simplest identified model that verifies the validation criteria. This is linked on the one hand to the complexity of the controller (which will depend on n_A and n_B), and, on the other hand, to the robustness of the identified model with respect to the operating conditions.

From the results presented in Section 6.5, a first approach to estimate the values of $(n_A)_{max}$ and $(n_B)_{max}$ is to use the RLS and to study the evolution of the variance of the residual prediction errors, *i.e.* the evolution of

$$R(0) = E\{\varepsilon^2(t)\} = \frac{1}{N} \sum_{t=1}^{N} \varepsilon^2(t)$$

as a function of the value of $n_A + n_B$. A typical curve is given in Figure 7.5.

In theory, if the example considered is simulated and noise free, the curve should present a neat elbow followed by a horizontal segment, which indicates that the increase in parameter number does not improve the performances. (see Section 6.5, Figure 6.5.1).

In practice, this elbow is not neat if non-white noise is present. The practical test used for determining $n_A + n_B$ is the following: consider first n_A, n_B and the corresponding variance of the residual errors $R(0)$. Consider now $n_A' = n_A + 1$, n_B and the corresponding variance of the residual errors $R'(0)$.

If

$$R'(0) \geq 0.8R(0)$$

it is unwise to increase the degree of n_A (same test with $n_B' = n_B + 1$).

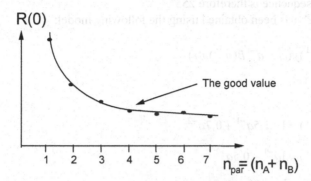

Figure 7.5. Evolution of the variance of residual errors as a function of the number of model parameters

It is advisable to increase the values of n_A and n_B in a non-simultaneous way. A simultaneous increase may induce the identification of a pair of very close poles and zeros that have a small influence on the criterion with respect to the increase of n_A or n_B but that could make the controller computation harder.

With the choice that results for n_A and n_B, the model identified by the RLS does not necessarily verify the validation criterion. Therefore, while keeping the values of n_A and n_B, other structures and methods must be tried out in order to obtain a "valid" model. If after all the methods have been tried none is able to give a model that satisfies the validation criterion, then n_A and n_B must be increased.

The estimated numerical values of the coefficients corresponding to the highest powers of polynomials $\hat{A}(q^{-1})$ and $\hat{B}(q^{-1})$ (after the estimation of the time delay d) also give a clue on the maximum order to be chosen. If these values are very small compared to the previous ones, it is often possible to reduce the order of $\hat{A}(q^{-1})$ and $\hat{B}(q^{-1})$ (one must obviously perform a new identification session with the updated orders).

The order estimation techniques presented in Section 6.5 rapidly provide good results. Moreover, it is recommended to compare these results with the available *a priori* information upon the system structure.

7.4 Identification of Simulated Models: Examples[3]

We consider now two files (T∅ and T1) containing I/O data generated by a known discrete-time model. Each file contains 256 input/output recordings. In each case, the input has been a PRBS generated by a register with 8 cells ($N = 8$). The length of a complete sequence is therefore 255[4].

The file T∅ has been obtained using the following model:

$$A(q^{-1})y(t) = q^{-1}B(q^{-1})u(t)$$

where

$$A(q^{-1}) = 1 - 1.5q^{-1} + 0.7q^{-2}$$

$$B(q^{-1}) = 1q^{-1} + 0.5q^{-2}$$

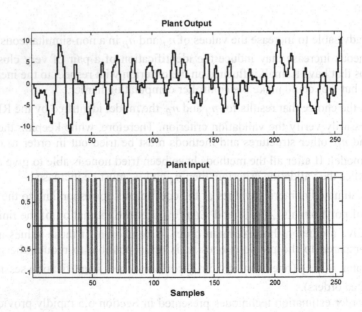

Figure 7.6. File T∅: I/O data set

---

[3] All the examples presented in this chapter have been worked out with WinPim (Adaptech) identification software and the MATLAB®/Scilab routines for model order estimation (see Section 6.5.4). Small numerical differences will result when using other identification routines.

[4] These files are available from the book web sites: *http://landau-bookic.lag.ensieg.inpg.fr*.

The file T1 has been generated with the same polynomials $A(q^{-1})$ and $B(q^{-1})$, but adding a stochastic disturbance, the ARMAX model being of the form

$$A(q^{-1})y(t) = q^{-1}B(q^{-1})u(t) + C(q^{-1})e(t)$$

in which $\{e(t)\}$ is an almost white noise generated by the computer and the degree of $C(q^{-1})$ is $n_C = 2$.

FILE T∅
We now come back to the file T∅[5]. Figure 7.6 shows the input and output sequences collected in this file. For the file T∅, by using the *recursive least squares* (RLS) method with the decreasing adaptation gain, the following results are obtained:

S = 1 M = 1(RLS) A = 1	FILE: T0 DELAY D = 1
INSTANT K	= 50
FORGETTING FACTOR	= 1
TRACE OF ADAPTATION GAIN	= 7.106747E-02
PROCESS OUTPUT	= -5.030791
MODEL OUTPUT	= -5.030789
PROCESS INPUT	= -1
ADAPTATION ERROR	= -1.430511E-06
A(1) = -1.49999	B(1) = 0.99998
A(2) = 0.69999	B(2) = 0.50000

where *S, M, and A* designate the structure, the method and the type of the adaptation gain[6] (see Chapter 5, Sections 5.5 and 5.2.4) respectively. $A(1)$, $A(2)$, $B(1)$, $B(2)$ designate the estimated coefficients of polynomials $A(q^{-1})$ and $B(q^{-1})$. It can be observed that the estimated parameters converge very fast towards the good values in the case of a noise free system (initial values $A(1) = A(2) = B(1) = B(2) = 0$).

Similar results are obtained with the *output error method, with fixed compensator,* decreasing adaptation gain and compensator degree $n_D=0$ (adaptation error = prediction error), as indicated below.

S = 2 M = 2 (OEFC) A = 1	FILE: T0 DELAY D = 1
INSTANT K	= 50
FORGETTING FACTOR	= 1
TRACE OF ADAPTATION GAIN	= 7.107091E-02
PROCESS OUTPUT	= -5.030791
MODEL OUTPUT	= -5.031748
PROCESS INPUT	= -1
ADAPTATION ERROR	= 9.570122E-04
A(1) = -1.50000	B(1) = 1.00002
A(2) = 0.700000	B(2) = 0.49993

---

[5] Before starting the identification, data must be centred.
[6] *S=1* designates the structure S1, *M=1* designates the first method used for the structure S1 (see Section 5.5) and *A=1* corresponds to the choice of the adaptation gain A.1 discussed in Section 5.2.4.

The estimation of the time delay for the file T∅ in the absence of *a priori* information will be illustrated next. Indeed with the recursive least squares method and a decreasing adaptation gain, for $n_A = 2$, $n_B = 4$ and $d = 0$ the following parameters are obtained:

S = 1	M = 1 (RLS)	A = 1	FILE: T0	DELAY D = 0
	INSTANT K		= 50	
	FORGETTING FACTOR		= 1	
	TRACE OF ADAPTATION GAIN		= .1623274	
	PROCESS OUTPUT		= -5.030791	
	MODEL OUTPUT		= -5.030764	
	PROCESS INPUT		= -1	
	ADAPTATION ERROR		= -2.717972E-05	
	A(1) = -1.49998	B(1) = 5.39346E-06		B(3) = .50001
	A(2) = .699984	B(2) = .99998		B(4) = 2.62828E-05

The results obtained ($B(1)$ and $B(4)$, very small compared to $B(2)$ and $B(3)$) clearly show that $d = 1$ and $n_B = 2$. However it must be stressed that this situation is without any ambiguity in this example, since there is no disturbing noise.

FILE T1
The file T1 is highly contaminated by noise and certain methods (in particular the recursive least squares) will give biased parameter estimates. The quality of the identification will be reflected in the validation results. The estimation has been carried out on 256 samples and the validation has been carried out on the same input/output data set.

The results obtained with the *recursive least squares* with decreasing gain are given below[7]:

S = 1 M = 1 (RLS) A = 1 FILE: T1 NS = 256 DELAY D = 1	
COEFFICIENTS OF POLYNOMIAL A:	A(1) = -1.403533
	A(2) = 0.6066453
COEFFICIENTS OF POLYNOMIAL B:	B(1) = 0.9831312
	B(2) = 0.6512049
VALIDATION TEST : Whiteness of the residual error	
System Variance: 18.3791	Model Variance: 17.9035
Error variance R(0): 0.4749	
NORMALIZED AUTOCORRELATION FUNCTIONS	
Validation Criterion: Theor. Val.: |RN(i)| ≤ 0.136, Pract. Val.: |RN(i)| ≤ 0.15	
RN(0) = 1.000000	→ RN(1) = -0.505234 ←
RN(2) = 0.115732	RN(3) = -0.054398
RN(4) = 0.016311	

The appearance of a bias on the estimated parameters is observed, which is also reflected in unsatisfactory validation results ($|RN(1)| > 0.15$). The residual prediction error is not close to white noise. One should thus consider another "plant

---

[7] In the tables shown in this section *the system variance* corresponds to the variance of the measured output, *the model variance* corresponds to the variance of the predicted output and *the error variance* corresponds to the variance of the residual prediction error ($R(0)$).

+ disturbance" structure such as, for example, the S3 structure, which replaces the disturbance model $e(t)$ in S1 by $C(q^{-1}) e(t)$.

We choose, among the identification methods applicable to structure S3, the *output error method with extended estimation model* (M3) and decreasing adaptation gain (A1). The results obtained are given in the following table.

```
S = 3   M = 3 (OEEPM)   A = 1   FILE:T1   NS = 256   DELAY D=1
COEFFICIENTS OF POLYNOMIAL A        A(1) = -1.50009
                                    A(2) =  0.69614
COEFFICIENTS OF POLYNOMIAL B        B(1) =  0.95782
                                    B(2) =  0.54005
COEFFICIENTS OF POLYNOMIAL C        C(1) = -0.83917
                                    C(2) =  0.05308
VALIDATION TEST: Whiteness of the residual error
        System Variance: 18.3791        Model Variance: 18.1894
        Error variance R(0): 0.2571
            NORMALIZED AUTOCORRELATION FUNCTIONS
Validation Criterion: Theor. Val.: |RN(i)| ≤ 0.136, Pract. Val.: |RN(i)| ≤ 0.15
        RN(0) = 1.000000              RN(1) = -0.141702
        RN(2) =  0.021206             RN(3) = 0.008497
        RN(4) = 0.051014
```

It can be observed that the estimated values of $A(1)$, $A(2)$ and $B(2)$ are better than in the case of the recursive least squares (the sum of the squared biases is lower in this case). On the other hand, the validation results are acceptable since all the normalized autocorrelation functions ($RN(1)$ to $RN(4)$) have a module less than 0.15. The residual prediction error becomes closer to white noise and its variance has been reduced compared to estimation using the recursive least squares.

The results obtained can be further improved if the output error with extended estimation model, and an adaptation gain with variable forgetting factor are used (A3 with $\lambda_1(0) = 0.97$). The following results are obtained:

```
S = 3   M = 3 (O.E.E.M.P.)   A = 3   FILE:T1   NS = 256   DELAY D = 1
COEFFICIENTS OF POLYNOMIAL A:       A(1) = -1.508445
                                    A(2) =  0.70574
COEFFICIENTS OF POLYNOMIAL B:       B(1) =  0.95120
                                    B(2) =  0.52940
COEFFICIENTS OF POLYNOMIAL C:       C(1) = -0.90610
                                    C(2) =  0.08344
VALIDATION TEST: Whiteness of the residual error
        System Variance: 18.3791        Model Variance: 18.2276
        Error variance R(0): 0.2531
            NORMALIZED AUTOCORRELATION FUNCTIONS
Validation Criterion: Theor. Val.: |RN(i)| ≤ 0.136, Pract. Val.: |RN(i)| ≤ 0.15
        RN(0) = 1.0000               RN(1) = -0.091665
        RN(2) =  0.032701            RN(3) = 0.025042
        RN(4) = 0.064717
```

which corresponds to an improvement of the results in terms of whiteness and variance of the residual prediction error, on the one hand, and in terms of the sum for the squared biases, on the other hand.

Better results than those obtained with the recursive least squares method can also be obtained with structure S2 by using the *output error method with fixed compensator*, decreasing adaptation gain and compensator degree $n_D=0$ (adaptation error = prediction error). The results obtained are given below:

```
S =2  M = 2 (OEFC)  A = 1   FILE:T1   NS = 256   DELAY  D=1
COEFFICIENTS OF POLYNOMIAL A         A(1) = -1.52885
                                     A(2) =  .73410
COEFFICIENTS OF POLYNOMIAL B         B(1) =  .93228
                                     B(2) =  .51900
VALIDATION TEST: Error / prediction uncorrelation
      System variance: 18.3791       Model variance: 18.5921
      Error variance R(0): 0.4860
         NORMALIZED AUTOCORRELATION FUNCTIONS
Validation Criterion: Theor. Val.: |RN(i)| ≤ 0.136, Pract. Val.: |RN(i)| ≤ 0.15
         RN(0) = -0.116266
         RN(1) = -0.028968
         RN(2) = 0.103195
```

One can see that the values obtained for the normalized cross-correlations satisfy the validation condition.

However, it is interesting to compare these results with those provided by the model identified with recursive least squares for the same validation test.

```
S =1  M = 1 (RLS)  A = 1   FILE:T1   NS = 256   DELAY  D=1
COEFFICIENTS OF POLYNOMIAL A         A(1) = -1.403533
                                     A(2) =  0.6066453
COEFFICIENTS OF POLYNOMIAL B         B(1) =  0.9831312
                                     B(2) =  0.6512049
VALIDATION TEST: Error / prediction uncorrelation
      System variance: 18.3791       Model variance: 18.5921
      Error variance R(0): 0.76355
         NORMALIZED AUTOCORRELATION FUNCTIONS
Validation Criterion: Theor. Val.: |RN(i)| ≤ 0.136, Pract. Val.: |RN(i)| ≤ 0.15
         RN(0) =  0.248184
      → RN(1) =  0.313691 ←
      → RN(2) =  0.284383 ←
```

It is observed that the parameters estimated by the output error method with fixed compensator are better than those obtained by the recursive least squares (the latter do not satisfy the validation criterion).

This is also confirmed by the validation results (the model identified with recursive least squares does not pass the uncorrelation test).

Finally it can be shown that, even in the presence of significant noise, the time delay can be determined from the relative values of the coefficients of the polynomial $B(q^{-1})$. The following results are obtained for $d = 0, n_B = 3, n_A = 2$, by using the recursive least squares method:

S = 1	M = 1 (RLS)	A = 1	File: T1	DELAY	D = 0
	INSTANT K				= 256
	FORGETTING FACTOR				= 1
	TRACE OF ADAPTATION GAIN				= 1.574461E-02
	PROCESS OUTPUT				= 11.39311
	MODEL OUTPUT				= 11.06473
	PROCESS INPUT				= -1
	PREDICTION ERROR				= 0.3283825
	A(1) = -1.4043				B(1) = -5.39615E-02
	A(2) = 0.60742				B(2) = .98315
					B(3) = .65067

One observes that $|B(1)| < 0.15 \, |B(2)|$. This leads to choose $d = 1$ and $n_B = 2$.

Exercise: Compare the model obtained using the output error with fixed compensator ($S = 2$, $M = 2$) with the model obtained using output error with extended estimation model ($S = 3$, $M = 3$). See Chapter 6, Section 6.4 for the comparison procedure.

If techniques for the model complexity estimation are used, (by using the error criterion of Equation 6.5.17 and the criterion for the complexity estimation of Equation 6.5.19) a minimum of the criterion is obtained for $n = max(n_A, n_B + d) = 3$, that is indeed the good value (see Figure 7.7). The function *estorderiv.m* has been used. A detailed estimation of the complexity leads to $n_A = 2$, $n_B = 2$ and $d = 1$.

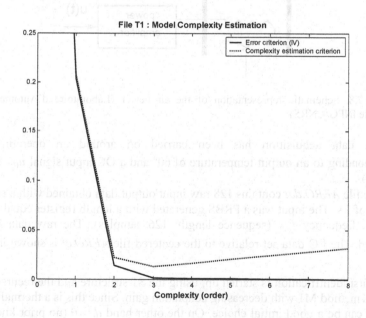

Figure 7.7. Estimation of the model complexity ($n=max(n_A,n_B+d)$) from the data collected in the file T1

7.5 Plant Identification Examples

7.5.1 Air Heater

The diagram of the system is represented in Figure 7.8. The air is heated at the pipe input by means of an electrical resistor supplied by a power amplifier. The temperature of the air at the output is measured by a thermocouple. This section is concerned with identifying the dynamic model linking the power amplifier control to the temperature of the output air, around a certain temperature. The steady state characteristic of the air heater + power amplifier is very non-linear. This results in the appearance of a DC component at the output even for centered input signals of low magnitude.

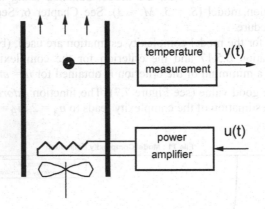

Figure 7.8. Schematic representation of the air heater (Laboratoire d'Automatique de Grenoble INPG/CNRS)

The data acquisition has been carried on around an operating point corresponding to an output temperature of $60°$ and a DC input signal $u_0=5V$ ($y_0(t) = 3.2V$).

The file *AERO.dat* contains 128 raw input/output data obtained with a sampling period of 5 s. The input was a PRBS generated with a length register equal to 6 and a clock frequency $f_s/2$ (sequence length: 126 samples). The raw data file was centered. The I/O data set relative to the centered file *AERO.c*[8] is shown in Figure 7.9.

A first identification is started up, using the S1 structure and the recursive least squares method M1 with decreasing adaptation gain. Since this is a thermal system, $n_A = 2$ can be a good initial choice. On the other hand $d = 0$ (no prior knowledge on the delay) and $n_B = 2$. The following results are obtained:

---

[8] Available from the web site: *http:/landau-bookic.lag.ensieg.inpg.fr*

S = 1 M = 1(RLS) A = 1 FILE: AERO.C NS = 128 DELAY D = 0
COEFFICIENTS OF POLYNOMIAL A: A(1) = -0.66872
                                     A(2) = 0.00115
COEFFICIENTS OF POLYNOMIAL B: B(1) = 0.17360
                                     B(2) = 0.05424

The system clearly has a time delay less than $0.5\ T_s$ (since $|b_1| > |b_2|$), thus the choice $d=0$ was a good one. As the coefficient $A(2)$ has a very small value (resulting from the product of poles) one can say that the value of one of the poles is very small and it can be neglected.

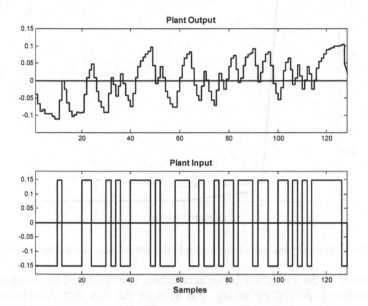

Figure 7.9. File AERO.c: I/O data set

This model structure is also compatible with the complexity estimation provided by the functions *estorderiv.m* (MATLAB®) or *estorderiv.sci* (Scilab) (estimation with the method of instrumental variable with delayed inputs) that gives $n = max\ (n_A, n_B + d) = 2$ and $n_A = 1$, $n_B = 2$, $d=0$ respectively (see Figure 7.10).

Another identification is thus carried out with the same structure, method and adaptation gain, but with $n_B = 2$, $d=0$, $n_A = 1$.

The results obtained are as follows:

S = 1 M = 1 (RLS) A = 1 FILE:AERO.C NS = 128 DELAY D=0
COEFFICIENTS OF POLYNOMIAL A: A(1) = -0.6672
COEFFICIENTS OF POLYNOMIAL B: B(1) = 0.1736
                                B(2) = 0.0545
VALIDATION TEST: Whiteness of the residual error
        System variance: 0.0035   Model variance: 0.0034
        Error variance R(0): 3.991 E-04

NORMALIZED AUTOCORRELATION FUNCTIONS
*Validation Criterion:*Theor. Val.:$|RN(i)| \leq 0.192$, Pract. Val.: $|RN(i)| \leq 0.15$
RN(0) = 1.000000   → RN(1) = -0.2284 ←
RN(2) = 0.1   RN(3) = 0.017
RN(4) = 0.0054

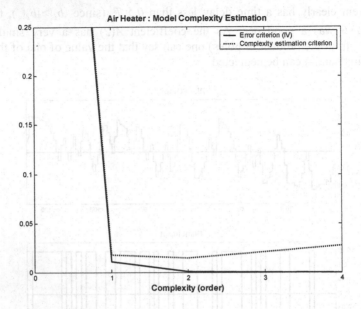

Figure 7.10. Complexity estimation for the air heater model based on the data file AERO.c

The result of the validation is not acceptable. The results of the identification can be further improved by taking into account the fact that disturbances are present. A first approach is to choose structure S3 with the method M3, the output error with extended prediction model, which gives also an estimation of the model of disturbances. The following results are obtained (with decreasing adaptation gain):

S=3 M=3 (O.E.E.M.P) A = 1 FILE:AERO.C NS = 128 DELAY D=0
COEFFICIENTS OF POLYNOMIAL A: A(1) = -0.6589
COEFFICIENTS OF POLYNOMIAL B: B(1) = 0.1724
                                                   B(2) = 0.0579
COEFFICIENTS OF POLYNOMIAL C: C(1) = -0.1248
VALIDATION TEST: Whiteness of the residual error
        System variance: 0.00351   Model variance: 0.00346
        Error variance R(0): 3.837E-05
            NORMALIZED AUTOCORRELATION FUNCTIONS
*Validation Criterion:*Theor. Val.:$|RN(i)| \leq 0.192$, Pract. Val.: $|RN(i)| \leq 0.15$
        RN(0) = 1.000000   RN(1) = 0.1303
        RN(2) = 0.1029   RN(3) = 0.0192
        RN(4) = 0.0326

The results of the validation are without doubt satisfactory.

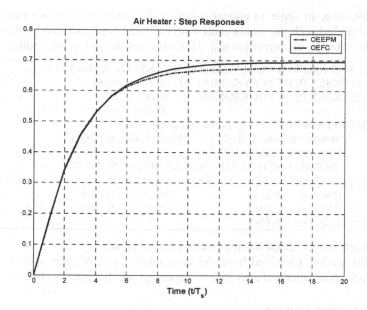

Figure 7.11. Step response for the models identified in the air heater example: OEEPM - output error with extended prediction model, variable forgetting factor, d = 0, n_A = 1, n_B = 2; OEFC - output error with fixed compensator, d = 0, n_A = 1, n_B = 2

The step response for this model is shown in Figure 7.11 (OEEPM). Structure S2 can also be used, which may lead to an asymptotically unbiased estimated model without modeling the disturbances. The output error with fixed compensator (method M2) is chosen with $n_D=0$ (because $n_A=1$), and with an adaptation gain with forgetting factor ($\lambda_1(0) = 0.97$). The results are summed up in the following table:

S=2 M=2 (OEFC) A=3 FILE: AERO.C NE=128 DELAY D=0
COEFFICIENTS OF POLYNOMIAL A: A(1) = -0.6837
COEFFICIENTS OF POLYNOMIAL B: B(1) = 0.1771 B(2) = 0.043
VALIDATION TEST: Error / prediction uncorrelation
System variance: 0.00351   Model variance: 0.00317
Error variance R(0): 9.77 E-05
NORMALIZED AUTOCORRELATION FUNCTIONS
*Validation Criterion:*Theor. Val.:|RN(i)| ≤ 0.192, Pract. Val.: |RN(i)| ≤ 0.15
RN(0) =   -0.2208   RN(1) =   0.1508
RN(2) =   0.0432   RN(3) =   0.0116
RN(4) =   0.0469

The step response for this model, which has passed the validation test, is presented in Figure 7.11 (OEFC). The model obtained with the output error with fixed compensator has a static gain and a rise time slightly larger than the corresponding values for the previous model (a larger rise time means that the identified model is *slower*).

In this case, in order to compare further the quality of the two models, the model identified with the output error with extended prediction model should also be validated by the uncorrelation test. This test gives the following results:

```
S=3  M=3 (OEEPM)  A=1  FILE: AERO.C  NS=128 DELAY D=0
COEFFICIENTS OF POLYNOMIAL A:   A(1) = -0.6589
COEFFICIENTS OF POLYNOMIAL B:   B(1) =  0.1724
                                B(2) =  0.0579
VALIDATION TEST: Error / prediction uncorrelation
      System variance: 0.00351       Model variance: 0.00317
      Error variance R(0): 1.02E-04
      NORMALIZED AUTOCORRELATION FUNCTIONS
Validation Criterion:Theor. Val.: |RN(i)| ≤ 0.192, Pract. Val.:|RN(i)| ≤ 0.15
      RN(0) =  - 0.2154             RN(1) =  - 0.1465
      RN(2) =  - 0.1033             RN(3) =  - 0.1094
      RN(4) =  - 0.1769
```

Both models are validated and the results are very close.

As the models identified have the same quality, one of them will be used to compute the controller and the result of the design will be tested on both models.

7.5.2 Distillation Column

The diagram of the binary column (water-methanol), which is the subject of the identification, is given in Figure 7.12a and a view of the plant is shown in Figure 7.12b.

The control variables are the heating power (QB) and the reflux rate (R). The flow rate (LF) and concentration (XF) of the input product are assumed constant (their variations produce disturbances in the system). The controlled variables are the flow rate of the top product (LD) and the concentration of the top product (XD). (T) designate the temperature and (LB) the flow rate of the product at the bottom. The corresponding block diagram is given in Figure 7.13.

The input and output variations will be related around an operating point by a linear model characterized by a 2×2 transfer matrix operator

$$\begin{bmatrix} LD \\ XD \end{bmatrix} = \begin{bmatrix} H_{11}(q^{-1}) & H_{12}(q^{-1}) \\ H_{21}(q^{-1}) & H_{22}(q^{-1}) \end{bmatrix} \begin{bmatrix} R \\ QB \end{bmatrix} + \begin{bmatrix} w_1 \\ w_2 \end{bmatrix}$$

where w_1 and w_2 represent the unmodeled disturbances. Each element of the transfer matrix operator is of the form

$$H_{ij}(q^{-1}) = \frac{q^{-d}\left(b_1 q^{-1} + b_2 q^{-2} + ...\right)}{1 + a_1 q^{-1} + a_2 q^{-2} + ...}$$

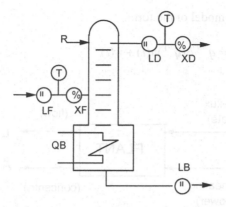

Figure 7.12a,b. Binary distillation column (Laboratoire d'Automatique de Grenoble, INPG/CNRS). **a** Functional diagram, **b** View

that corresponds to a model of the form:

$$A(q^{-1})y(t) = q^{-d}B(q^{-1})u(t) + w(t)$$

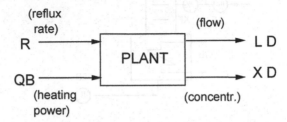

Figure 7.13. Block diagram of the distillation column (inputs-outputs)

Henceforward we shall be concerned with the identification of the transfer between the heating power (QB) and the concentration of the top product (XD). The input/output file is named QXD[9] file. It contains an I/O data set made of 256 samples. The input is a pseudo random binary sequence generated with a register with 8 cells ($N = 8, L = 255$). Figure 7.14 displays the inputs and outputs of the centered QXD file. The sampling period was 10 s.

The identification procedure starts after having centered the I/O sequences. As no prior knowledge of the system is available, first we choose structure S1, the recursive least squares method (M1) with a decreasing adaptation gain A1, delay $d=0$, degree for the polynomial B: $n_B = 4$ (in order to capture a possible time delay of the system) and degree for the polynomial A: $n_A = 2$ (since it is a chemical plant). The initial value of the parameters to be estimated is set to zero. The results obtained with this method are summarized below:

```
S=1   M=1 (RLS)  A=1   FILE: QXD   NS=256   DELAY  D=0
      INSTANT K                         = 256
      FORGETTING FACTOR                 = 1
COEFFICIENTS OF POLYNOMIAL A:   A(1) = -0.23644
                                A(2) = -0.21889
COEFFICIENTS OF POLYNOMIAL B:   B(1) =  0.01177
                                B(2) =  0.07982
                                B(3) =  0.18311
                                B(4) =  0.13271
```

Note that $B(1) < 0.15\ B(2)$. This shows that we have to set the delay d equal to 1, as a first approximation.

---

[9] Available from the book website *http://landau-bookic.lag.ensieg.inpg.fr*.

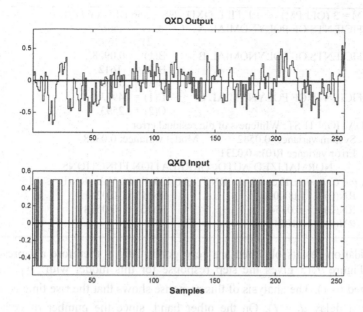

Figure 7.14. I/O data set for the distillation column QXD

Thus a new identification is performed with $d = 1$, $n_B = 3$, $n_A = 2$ and the results obtained are

```
S = 1  M = 1(RLS)  A = 1  FILE: QXD  NS = 256  DELAY D=1
COEFFICIENTS OF POLYNOMIAL A         A(1) = -.23806
                                     A(2) = -.18820
COEFFICIENTS OF POLYNOMIAL B         B(1) =  .07982
                                     B(2) =  .18303
                                     B(3) =  .13244
VALIDATION TEST: Whiteness of the residual error
      System variance: 0.0542       Model variance: 0.0274
      Error variance R(0):  0.0268
         NORMALIZED AUTOCORRELATION FUNCTIONS
Validation Criterion:Theor. Val.: |RN(i)| ≤ 0.136, Pract. Val.: |RN(i)| ≤ 0.15
         RN(0) =   1.0000            RN(1) = -0.1216
      → RN(2) =  -0.2278 ←           RN(3) =  0.0112
         RN(4) =   0.0718
```

It is observed that the validation is unsatisfactory since the residual error is not sufficiently white ($|RN(2)|) > 0.15$). We may consider that this situation arises from the fact that the disturbances are incorrectly modeled by the structure S1. A different structure (S3) and the output error method with extended estimation model (M3), still with decreasing adaptation gain, are then tried. In this way a disturbance model will be identified. The results obtained are

```
S = 3  M = 3 (OEEPM)  A = 1  FILE:QXD  NS = 256 DELAY D=1
COEFFICIENTS OF POLYNOMIAL A        A(1) = -.5065
                                    A(2) = -.1595
COEFFICIENTS OF POLYNOMIAL B        B(1) =  0.0908
                                    B(2) =  .1612
                                    B(3) =  .0776
COEFFICIENTS OF POLYNOMIAL C        C(1) = -.4973
                                    C(2) = -.2593
VALIDATION TEST: Whiteness of the residual error
      System variance: 0.0542          Model variance: 0.0308
      Error variance R(0):  0.0231
           NORMALIZED AUTOCORRELATION FUNCTIONS
Validation Criterion:Theor. Val.: |RN(i)| ≤0.136, Pract. Val.: |RN(i)| ≤0.15
      RN(0) =  1.000000              RN(1) = -0.0241
      RN(2) =  -0.0454              RN(3) =  0.0518
      RN(4) =  0.1014
```

The validation results obtained are very good. This is therefore a representative model. Figure 7.15 gives the step response for this model with $n_A = 2$ (gain normalized to 1). The analysis of this response shows that the rise time is $t_R \approx 10$ T_s (with a delay $d = 1$). On the other hand, since the number of cells of the generator register of the PRBS is $N = 8$, the duration of the largest pulse is less than t_R. The validation will thus not be significant for the steady state gain (several models may be validated without their steady state gain being the same). A direct verification of the steady state gain shows that the value obtained (0.987) is correct.

Note also that the polynomial $B(q^{-1})$ is unstable, $B(1) < B(2)$ revealing the presence of a fractional delay greater than $0.5T_s$ (which explains the presence of an unstable zero).

As the coefficient $A(2)$ is small if compared to $A(1)$ and the validation results are very good, one may think of identifying a new model with $n_A=1$. In this case the results are

```
S = 3  M = 3 (OEEPM)  A = 1  FILE:QXD  NS = 256  DELAY D=1
COEFFICIENTS OF POLYNOMIAL A        A(1) = -.7096
COEFFICIENTS OF POLYNOMIAL B        B(1) =  .0839
                                    B(2) =  0.1415
                                    B(3) =  0.0528
COEFFICIENTS OF POLYNOMIAL C        C(1) = -.688
VALIDATION TEST: Whiteness of the residual error
      System variance: 0.0542          Model variance: 0.02952
      Error variance R(0): 0.0227
        NORMALIZED AUTOCORRELATION FUNCTIONS
Validation Criterion:Theor. Val.: |RN(i)| ≤0.136, Pract. Val.: |RN(i)| ≤0.15
      RN(0) =  1.0000               RN(1) = -0.0035
      RN(2) =  -0.0748              RN(3) = -0.0131
      RN(4) =  0.0442
```

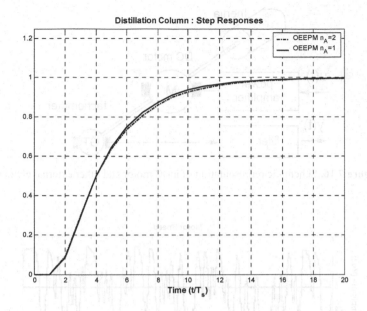

Figure 7.15. FIle QXD: normalized step responses for the two models identified (static gain normalized to 1)

The model obtained is validated. The static gain of the model is 0.958. It will be necessary to scale the values of the coefficients of the polynomial $B(q^{-1})$ in order to obtain the static gain previously obtained (nevertheless the difference found between the two static gain can be neglected).

The normalized step responses of both models are presented in Figure 7.15.

7.5.3 DC Motor

The identification of a DC motor model is examined in this section. The input of the system is the voltage applied to a power amplifier that feeds the motor, and the output is the speed measured by means of a tachometer. A short description of the global system is shown in Figure 7.16. From the identification point of view, let us consider the cascade power amplifier, motor, tachometer, filter on the measured output as the plant.

The file *MOT3.c*[10] contains 256 centered I/O data obtained with a sampling period of 15 ms. The input is a PRBS generated by a shift register with seven cells and a clock frequency $f_s/2$ (sequence length: 254).

---

[10] Available from the website: *http://landau-bookic.lag.ensieg.inpg.fr*.

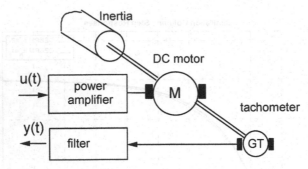

Figure 7.16. Schematic representation of a DC motor and other external elements

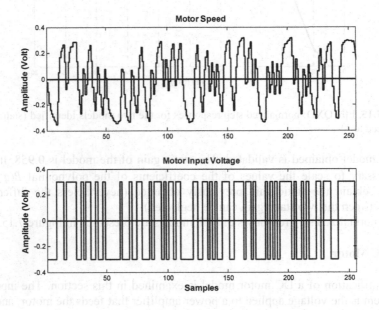

Figure 7.17. I/O data set for the DC motor (MOT3.c)

The magnitude of the PRBS is set to 0.3 V and applied when the operating point of the system is an input voltage of 3V (scales for u: 0-10V corresponding to a speed variation from 0 to +1500 rpm). This file is shown in Figure 7.17.

The result of the complexity estimation algorithm (instrumental variable with delayed inputs method using *estorderiv.sci* or *estorderiv.m*) is (see Figure 7.18)

$$n = max (n_A, n_B + d) = 2$$

The detailed complexity estimation gives the values $n_A = 1$, $n_B = 2$, $d = 0$. This complexity is coherent with the model obtained from physical equations that describe the DC motor, with at most two time constants, but the electrical time

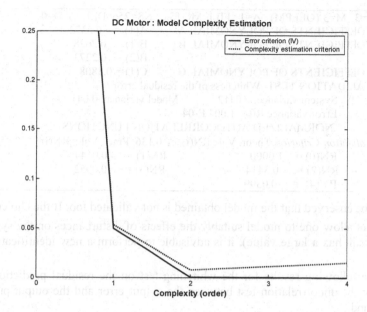

Figure 7.18. Model complexity estimation for the DC motor (input: voltage, output: speed) using the data file MOT3.c

constant in this case is very small if compared to the electro-mechanical one. The absence of a pure time delay is not surprising as well. The degree of the polynomial B , $n_B = 2$ reveals the presence of a fractional delay as a result of the filtering action on the measure.

A first identification is carried on with structure S1 and the recursive least squares (M1) with a decreasing adaptation gain (A1), the following results are obtained:

```
S=1   M=1(RLS)   A=1   FILE:MOT3.c   NS=256   DELAY D=0
COEFFICIENTS OF POLYNOMIAL A        A(1) = - 0.5402
COEFFICIENTS OF POLYNOMIAL B        B(1) =  0.2629
                                    B(2) =  0.2257
VALIDATION TEST: Whiteness of the residual error
      System variance: 0.0412        Model variance: 0.041
      Error variance R(0): 1.157 E-4
         NORMALIZED AUTOCORRELATION FUNCTIONS
Validation Criterion:Theor. Val.:|RN(i)| ≤ 0.136, Pract. Val.: |RN(i)| ≤ 0.15
         RN(0) = 1.0000          → RN(1) = - 0.4529 ←
         RN(2) = 0.3113             RN(3) = 0.0332
         RN(4) = - 0.0297
```

The model obtained is not validated. Next an identification with the structure S3 is performed by using the output error with extended prediction model method (M3) that simultaneously estimates the plant model and the disturbance model with the choice $n_C = 1$ (a decreasing adaptation gain is still used). In this case the results are

```
S=3  M=3 (OEEPM)  A=1  FILE:MOT3.c  NS=256  DELAY D=0
COEFFICIENTS OF POLYNOMIAL A        A(1) = - 0.5372
COEFFICIENTS OF POLYNOMIAL B        B(1) =  0.2628
                                    B(2) =  0.2273
COEFFICIENTS OF POLYNOMIAL C        C(1) = 0.3808
VALIDATION TEST: Whiteness of the residual error
      System variance: 0.0412       Model variance: 0.041
      Error variance R(0): 1.004 E-04
         NORMALIZED AUTOCORRELATION FUNCTIONS
Validation Criterion:Theor. Val.:|RN(i)| ≤ 0.136, Pract. Val.: |RN(i)| ≤ 0.15
      RN(0) =   1.0000              RN(1) =  - 0.0144
  --> RN(2) =   0.3414 <--          RN(3) =    0.0952
      RN(4) =   0.0399
```

It can be observed that the model obtained is not validated too. If the choice $n_C = 1$ does not allow one to model suitably the effects of disturbances on the system (as $RN(2)$ still has a large value), it is advisable to perform a new identification with $n_C = 2$.

The following results for the whitening test on the residual prediction errors and for the uncorrrelation test between the output error and the output prediction are found:

```
S=3  M=3 (OEEPM)  A=1  FILE:MOT3.c  NS=256  DELAY D=0
COEFFICIENTS OF POLYNOMIAL A      A(1) = - 0.535
COEFFICIENTS OF POLYNOMIAL B   B(1) =  0.2617      B(2) = 0.2288
COEFFICIENTS OF POLYNOMIAL C    C(1) = - 0.84848   C(2) = 0.19255
VALIDATION TEST: Whiteness of the residual error
System variance:0.0042  Model variance:0.041  Error variance R(0): 8.856 E-05
         NORMALIZED AUTOCORRELATION FUNCTIONS
Validation Criterion:Theor. Val.:|RN(i)| ≤ 0.136, Pract. Val.: |RN(i)| ≤ 0.15
      RN(0) =   1.0000      RN(1) =  - 0.0432      RN(2) =    0.0404
      RN(3) =   0.0453      RN(4) =    0.01179
VALIDATION TEST: Error / prediction uncorrelation
System variance:0.0412  Model variance:0.0408  Error variance R(0): 2.71 E-04
         NORMALIZED AUTOCORRELATION FUNCTIONS
Validation Criterion:Theor. Val.:|RN(i)| ≤ 0.136, Pract. Val.: |RN(i)| ≤ 0.15
      RN(0) = 0.0119       RN(1) = 0.0156          RN(2) = 0.0284
      RN(3) = 0.0356       RN(4) = 0.0186
```

Both the whitening test and the uncorrelation test give good values. Thus the model identified is validated. Another structure and method can be tested nevertheless with the same model complexity. If the structure S2 and the output error with fixed compensator method (M4) are chosen, in order to obtain an asymptotically unbiased estimation without estimating the disturbance model, one gets ($d = 0$)

```
S=2  M=4 (OEFC)  A=1  FILE:MOT2C  NS=256  DELAY D=0
COEFFICIENTS OF POLYNOMIAL A        A(1) = - 0.5347
COEFFICIENTS OF POLYNOMIAL B        B(1) =  0.2629
                                    B(2) =  0.2266
VALIDATION TEST: Error / prediction uncorrelation
      System variance: 0.0412       Model variance: 0.0406
      Error variance R(0): 2.723 E -04
```

NORMALIZED AUTOCORRELATION FUNCTIONS
Validation Criterion: Theor. Val.: $|RN(i)| \leq 0.136$, Pract. Val.: $|RN(i)| \leq 0.15$
$$RN(0) = 0.043 \qquad RN(1) = 0.0591 \qquad RN(2) = 0.0631$$
$$RN(3) = 0.0564 \qquad RN(4) = 0.031$$

Note that the results of the validation are very good, but slightly worse than those obtained with the output error with extended prediction model method. However, the parameters obtained in these two cases are very close. The step responses obtained using these models are extremely close each other and are shown in Figure 7.19.

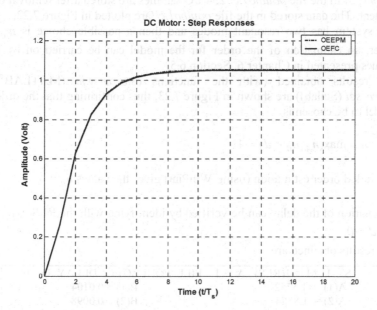

Figure 7.19. Step responses for the identified DC motor models (Output error with extended prediction model and output error with fixed predictor, $d = 0$, $n_A = 1$, $n_B = 2$)

7.5.4 Flexible Transmission

In this example the dynamic model of a flexible transmission, with low damped resonant modes, will be identified for control design purposes. A view of the flexible transmission is given in Figure 7.20 and the schematic representation of the control system is shown in Figure 7.21.

The flexible transmission is made of three pulleys linked by two elastic belts (see Figure 7.20). One of these pulleys is constrained to the axis of a DC motor. The motor position is controlled by a local servo (speed and position feedback). The dynamics of the local position control is very fast if compared to the mechanical system.

The control problem is to get the desired position of the third pulley, by modifying the input voltage of the position control of the motor that drives the first pulley. The output $y(t)$ of the system is the axis position of the third pulley, and the command signal $u(t)$ is the reference for the first pulley axis position. The mechanical loads that can be added on the third pulley modify the system inertia and, consequently, also the resonant modes of the mechanical system.

We are concerned in the following with the open loop identification of the model for this process (between $u(t)$ and $y(t)$) for the case without additional load (the controller in Figure 7.21 is not connected).

The sampling frequency is 20 Hz ($T_s = 50ms$). The excitation signal is a PRBS of small magnitude generated by a shift register with $N = 7$, and with a frequency divider $p = 2$. In the file *poulbo1.c* 254 I/O samples are stored after removal of DC component. The data stored in the file *poulbo1.c*[11]are plotted in Figure 7.22.

The system has two resonant modes and then a possible choice is $n_A = 4$. However, an estimation of the order for the model can be carried on by using techniques presented in Chapter 6, Section 6.5.

The results obtained (with the functions *estorderiv.m* (MATLAB®) or *estorderiv.sci* (Scilab)) are shown in Figure 7.23, thus confirming that the order for the model to be chosen is

$$n = \max(n_A, n_B + d) = 4$$

A detailed order estimation (using WinPim) gives the values $n_A = 4$, $n_B = 2$, $d = 2$.
The estimation of the delay can be verified by identifying with the RLS, with $n_A = n_B = 4$, $d = 0$.

The results obtained are

S = 1 M = 1(RLS) A = 1 FILE: POULBO1.C DELAY D = 0	
A(1) = -1.5752	B(1) = 0.0104
A(2) = 1.8384	B(2) = 0.0098
A(3) = -1.4788	B(3) = 0.3077
A(4) = 0.8896	B(4) = 0.4146

Note that $|B(1)|$, $|B(2)| << 0.15|B(3)|$, and that clearly implies $d = 2$.

---

[11] Available from the website: http://landau-bookic.lag.ensieg.inpg.fr.

Figure 7.20. View of the flexible transmission (Laboratoire d'Automatique de Grenoble INPG/CNRS/UJF)

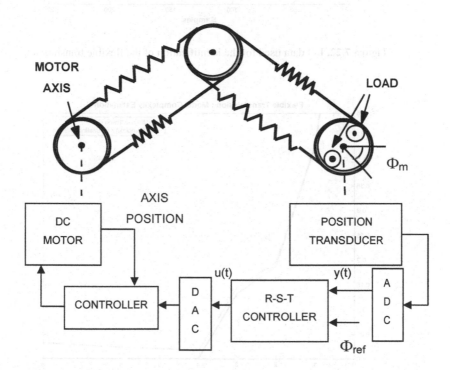

Figure 7.21. Control scheme for the flexible transmission

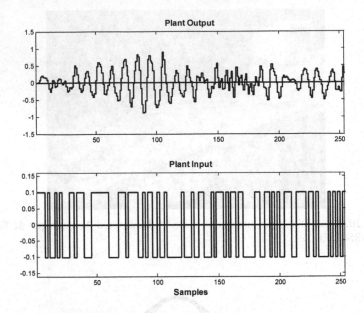

Figure 7.22. I/O data used for the identification of the flexible transmission

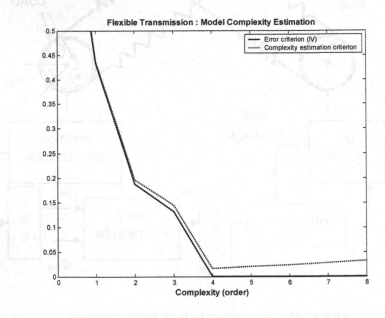

Figure 7.23. Complexity estimation for the flexible transmission model based on the file poulb01.c

If a new identification is performed with $d = 2$, $n_B = 2$, $n_A = 4$ by using structure S1 and the recursive least squares methods, one gets the following results:

```
S=1  M=1 (RLS)  A=1  FILE: POULBO1.C  NS=254  DELAY D=2
COEFFICIENTS OF POLYNOMIAL A:    A(1) = -1.5748
                                 A(2) =  1.8329
                                 A(3) = -1.4784
                                 A(4) =  0.8895
COEFFICIENTS OF POLYNOMIAL B:    B(1) = 0.3010
                                 B(2) = 0.4181
VALIDATION TEST: Whiteness of the residual error
     System variance: 0.3317  Model variance: 0.1053 Error variance R(0):
                              0.0007
           NORMALIZED AUTOCORRELATION FUNCTIONS
Validation Criterion: Theor. Val.: |RN(i)| ≤ 0.136, Pract. Val.: |RN(i)| ≤ 0.15
           RN(0) =  1.0000                → RN(1) =  -0.5727 ←
       → RN(2) =  0.2360 ←                  RN(3) =  -0.0475
           RN(4) =  -0.0158
```

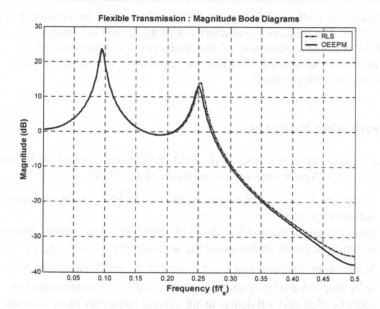

Figure 7.24. Frequency characteristics of the models identified for the flexible transmission (RLS – recursive least squares, OEEPM – output error with extended prediction model)

The validation is not satisfactory as $RN(1)$, $RN(2)$ are greater than 0.15. The frequency characteristics of this model is shown in Figure 7.25 (line RLS).

Thus one tries structure S3 which introduces a model for the disturbance (ARMAX model). The output error with extended prediction model is used (OEEPM) with a decreasing adaptation gain (A1). The results of identification and validation are given below:

```
S=3 M=3 (OEEPM) A=1 FILE:POULBO1.C NS=254 DELAY D=2
COEFFICIENTS OF POLYNOMIAL A:   A(1) = -1.60955
                                A(2) =  1.87644
                                A(3) = -1.49879
                                A(4) =  0.88574
COEFFICIENTS OF POLYNOMIAL B:   B(1) =  0.30530
                                B(2) =  0.39430
COEFFICIENTS OF POLYNOMIAL C:   C(1) = -0.67530
                                C(2) = 0.2283
                                C(3) = -0.0653
                                C(4) = -0.0585
VALIDATION TEST: Whiteness of the residual error
      System variance: 0.1061          Model variance:  0.1055
      Error variance R(0):  0.0004
         NORMALIZED AUTOCORRELATION FUNCTIONS
Validation Criterion: Theor. Val.: |RN(i)| ≤ 0.136, Pract. Val.: |RN(i)| ≤ 0.15
      RN(0) =  1.0000                  RN(1) = -0.0425
      RN(2) =  0.0959                  RN(3) =  -0.0563
      RN(4) = -0.0407
```

The validation is really satisfactory, as all the values $RN(i)$ are smaller than 0.136 for $i = 1,2,3,4$. The frequency characteristics of this model is shown in Figure 7.24 (line OEEPM). The comparison of the frequency characteristics of the models identified shows that a good validation corresponds to the identification of a less damped second resonant mode.

7.6 Concluding Remarks

This chapter has shown how effectively an identification of a plant model has to be carried out. The different steps can be summarized as follows:

- Input/output data acquisition using a PRBS (pseudo-random-binary sequence) as input
- Conditioning of the acquired data (DC removing, data scaling, filtering);
- selection or initial estimation of the system order $n = max\ (n_A,\ n_B + d)$
- Selection or estimation of $n_A,\ n_B,\ d$ either by order estimation techniques, or by inspection of the numerical values of the estimated parameters
- Identification and validation using several structures *plant + disturbance* and identification methods with the objective of obtaining the best acceptable model with lowest n_A and n_B
- Analysis of the model identified and validated both in the frequency and time domain

7.7 Notes and References

The proceedings of the IFAC Symposiums Identification and System Parameter Estimation, Pergamon Press, Oxford, are a good source of information about several types of system identification. See also:

Isermann R. (1980) Practical aspects of process identification, Automatica, vol.16, pp. 575-587.

Ljung L. (1999) System Identification - Theory for the User, 2nd edition, Prentice Hall, Englewood Cliffs.

For the identification of flexible structures see also:

Van den Bossche E., Dugard L., Landau I.D. (1986) Modelling and Identification of a Flexible Arm, Proceedings American Control Conference, Seattle, U.S.A.

Landau I.D., Langer J., Rey D., Barnier J. (1996) Robust control of a 360° flexible arm using the combined pole placement (sensitivity function shaping method), IEEE Trans. On Control Systems Technology, vol. 4, no. 4, Juillet 1996, pp. 369-383.

Landau I.D. (2001b) Identification des systèmes. Les bases, in Identification des systèmes, (I.D. Landau, A. Besançon-Voda éd.), Hermes, Paris.

For the identification of DC motors see also:

Landau I.D., Rolland F. (1993) Identification and digital control of electrical drives, Control Engineer Practice, vol.1, n°3.

8

Practical Aspects of Digital Control

Digital controller design methods and system identification techniques provide the basic tools for effectively solving a control problem. The final stages in the controller design are emphasized in this chapter. This involves implementation aspects, performance specifications and the interaction between plant identification and controller design.

The first part of the chapter reviews several topics related to the implementation of digital controllers: effect of the digital-to-analog conversion, effect of saturations, effect of the computational delay, manual to automatic switching, cascade control, performance evaluation, adaptation of controller parameters, the hardware for controllers implementation.

In the second part of the chapter, the joint use of system identification and controller design is illustrated through several examples:

- *Temperature control of an air heater*
- *Speed control of a DC motor*
- *Cascade position control of a DC motor*
- *Position control through a flexible transmission*
- *Position control of a 360° flexible robot arm*
- *Deposited zinc control in hot-dip galvanizing (SOLLAC-Florange)*

8.1 Implementation of Digital Controllers

8.1.1 Choice of the Desired Performances

The choice of desired performance, in terms of response time, is linked to the dynamics of the open loop system and to the power availability of the actuator during the transient. One should also take into account some "robustness" aspects.

Indeed, the acceleration of the natural response of a plant requires control "peaks" during transients that are greater than the steady state values.

317

If, for example, a system has the transfer function *1/(1+sT)* and it is desired to accelerate the closed loop system response twice, *i.e.* to obtain a transfer function *1/(1+sT/2)*, it is necessary to apply, for a certain time during the transient interval, a maximum input level which is twice the steady state value (*i.e.* $u_{max} = 2 \, u_{stat}$, (u_{stat} = value of control signal during the steady state).

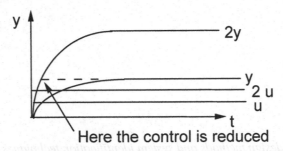

Figure 8.1. Acceleration of the natural response of a system

This is shown in Figure 8.1. By applying $2u_{stat}$, we expect to have the value *2y* at the end of the system natural time response, but the value *y* is approximately obtained after a period that is equal to half of the time response. The controller should then reduce the control signal to u_{stat} to get the desired steady state value *y*.

The following relation can thus be considered:

$$\frac{u_{max}}{u_{stat}} \approx \frac{desired \quad speed}{natural \quad speed} \approx \frac{desired \quad pass \quad band}{natural \quad pass \quad band}$$

It follows that the closed loop performances will depend on the actuator power availability. Or, *vice versa*, it will be necessary to choose the actuator according to the desired performances and the open loop response of the system.

With regard to the structure of the performance models for tracking $(A_m(q^{-1}))$ and for regulation $(P(q^{-1}))$, second-order models are preferred over first-order models for the same time response. This choice is a consequence of the fact that for a given time response the transient stress on the actuator will be weaker for a second-order model (as a result of a smoother response slope at the beginning of the transient).

It has been shown, in Chapter 2, Section 2.6 that the robustness of the closed loop system, with respect to uncertainties or variations of plant parameters, depends upon the ratio between the band pass (related to the *rise time*) of the regulation dynamics (mainly defined by $P(q^{-1})$), and the band pass (related to the *rise time*) of the plant in open loop (mainly defined by $A(q^{-1})$). The robustness of the closed loop systems is better as this ratio is approaching 1 (assuming that the open loop is stable).

In the situations where the desired dynamics in closed loop is significantly faster than the plant dynamics, a more careful design is necessary in order to assure the robustness margins. In such cases one should

1. Optimize the choice of the closed loop poles and of the fixed parts of the controller $S(q^{-1})$ and $R(q^{-1})$, in order to obtain the best possible robustness margins
2. Improve the quality of the plant model and reduce the uncertainties by:

    - Better identification in order to obtain a model which is relevant for the band pass desired in closed loop
    - Reduction of the size of the operating regions where only one controller is used
    - Use of "adaptive control" techniques

Note that the techniques for identification in closed loop described in Chapter 9 allow, in general, improvement of the quality of the models for the design of high performance controllers.

The achievable tracking dynamics is mainly limited by the power and the band pass of the actuators.

If the dynamics of the actuator is much faster than that of the plant, there is a risk that the high frequency dynamics of the plant be excited. It is therefore necessary, in these cases, on one hand to have a plant model which is relevant in the high frequencies and, on the other hand, to design a controller leading to an input sensitivity function with a low magnitude at high frequencies.

8.1.2 Effect of the Computational Time Delay

Two situations can be distinguished:

1. Computation time equal to or greater than $0.5\ T_s$.

    In this case, the values measured at instant t are used to compute the control u that will be sent at instant $t+1$. The computer thus introduces an additional time delay of 1 and the new delay to be considered for the design of the controller will be

    $$d' = d + 1$$

    This is illustrated in Figure 8.2a.

2. Computation time less than $0.5\ T_s$.

    In this case, the control is sent at the end of the computation. This is illustrated in Figure 8.2b. The computer introduces a fractional time delay which has the effect either of introducing a zero, or of modifying the existing zeros in the pulse transfer function of the plant. This effect is negligible if the computation time is much smaller than the sampling period.

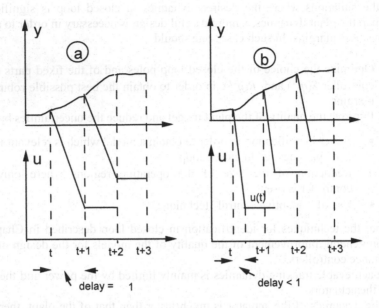

Figure 8.2a,b. Synchronization procedure for the measurement and the computed control: **a** computation time greater than $0.5\ T_s$; **b** computation time less than $0.5\ T_s$

8.1.3 Effect of the Digital-to-analog Conversion

The control signal generated by the digital controller is often computed with floating (or fixed) point arithmetic on 16, 32 or 64 bits. So the distinct values of this signal are larger than the distinct values of a digital to analog (D/A) converter which in general does not have more than 12 bits (4096 distinct values).

Figure 8.3 shows the characteristic of a digital to analog converter. The computed values of the control signal *(u)* are represented on the horizontal axis and the rounded values (u_r), effectively obtained at the output of the D/A converter, are represented on the vertical axis. Q represents the quantization step.

In a digital controller, the control signal generated at instant *t* depends upon the previous values of the control signal effectively applied on the plant input. It is therefore necessary to round the control *u(t)* inside the digital controller in order to correctly compute the future values.

The equation defining the controller has the form

$$S(q^{-1})u(t) + R(q^{-1})y(t) = T(q^{-1})y^*(t+d+1) \qquad (8.1.1)$$

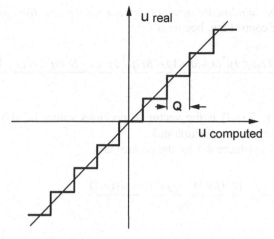

Figure 8.3. The characteristic of a digital to analog converter

and respectively

$$u(t) = \frac{1}{s_0}\left[T(q^{-1})y^*(t+d+1) - S^*(q^{-1})u(t-1) - R(q^{-1})y(t)\right] \qquad (8.1.2)$$

(where $s_0 = 1$ in the case of pole placement).

In order to take into account the effect of the D/A converter, the expression of the control $u(t)$ becomes

$$u(t) = \frac{1}{s_0}\left[T(q^{-1})y^*(t+d+1) - S^*(q^{-1})u_r(t-1) - R(q^{-1})y(t)\right] \qquad (8.1.3a)$$

where $u_r(t)$ is the rounded control signal sent to the D/A converter and satisfying

$$\left|u_r(t) - u(t)\right| \leq 1/2Q \qquad (8.1.3b)$$

If this rounding operation is not implemented, an equivalent noise is introduced at the plant input. This produces, in general, an increase of the variance of the plant output. In some cases low magnitude oscillations can be observed on the plant output.

8.1.4 Effect of the Saturation: Anti Windup Device

The effect of the saturation on the actuator can deteriorate the performance, in particular if the controller includes an integrator. Therefore, the effect of the actuator saturation has to be taken into account since for the computation of $u(t)$ the previous values of the control must be considered. The previous *computed values* are replaced by the *applied values*. *This is equivalent to the introduction of* a "copy" of the nonlinear actuator characteristic in the controller. As a consequence, the control law becomes

$$u(t) = \frac{T(q^{-1})y^*(t+d+1) - R(q^{-1})y(t) - S^*(q^{-1})\bar{u}(t-1)}{s_0} \quad (8.1.4)$$

In Equation 8.1.4, $\bar{u}\,(t\text{-}1)$ is the vector of previous values of $u(t)$ passed through the nonlinear characteristic (saturation).

This is shown in Figure 8.4 for the control law

$$u(t) = \frac{T(q^{-1})y^*(t+1) - r_0 y(t) - s_1 \bar{u}(t-1)}{s_0} \quad (8.1.5a)$$

$$\bar{u}(t) = \begin{cases} u(t) & \text{if } |u(t)| < u_{sat} \\ u_{sat} & \text{if } u(t) \geq u_{sat} \\ -u_{sat} & \text{if } u(t) \leq -u_{sat} \end{cases} \quad (8.1.5b)$$

corresponding to tracking and regulation with independent objectives for a discrete time model described by

$$y(t+1) = -a_1 y(t) + b_1 u(t) + b_2 u(t-1) \quad (8.1.6)$$

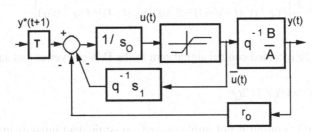

Figure 8.4. Digital control in the presence of actuator saturation (anti windup device)

The effects of the saturation of the control input and of the anti windup device are illustrated in Figure 8.5 for the case of a position control of a DC motor (described in Section 8.4). Both the plant and the controller contain an integrator.

One can observe that the saturation of the control slowdowns the system response and produces a significant overshoot, with respect to the case without saturation. The introduction of an anti windup device allows one to obtain a response without overshoot but which is obviously slower than the response in the linear operation case.

One can ask, looking at the evolution of the control signal, if it is possible to accelerate the system response in the presence of saturation, by maintaining the control at the saturation level for a longer duration, but still assuring a system response without or with a small overshoot.

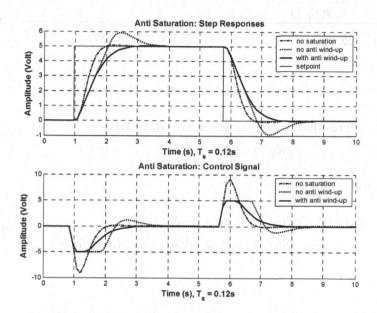

Figure 8.5. Response (in simulation) of a digital position control in the presence of saturation

In order to achieve this, it is necessary to impose dynamics on the trajectory of $u(t)$ when it leaves the saturation level (instead of a simple gain). Such a scheme is represented in Figure 8.6, where the dynamics of $u(t)$ when it leaves the saturation level is defined by the polynomial

$$P_S(q^{-1}) = 1 + q^{-1} P_S^*(q^{-1}) \tag{8.1.7}$$

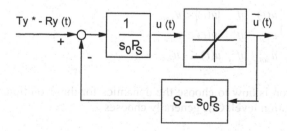

Figure 8.6. Anti windup device with specified dynamics when u(t) leaves the saturation level

In the linear domain ($|u(t)| < | u_{sat} | \rightarrow \bar{u}(t) = u(t)$), this system has the transfer function $1/S(q^{-1})$. Effectively from Figure 8.6 one gets

$$\frac{\dfrac{1}{s_0 P_s(q^{-1})}}{1 - \dfrac{S(q^{-1}) - s_0 P_s(q^{-1})}{s_0 P_s(q^{-1})}} = \frac{1}{s_0 P(q^{-1}) + S(q^{-1}) - s_0 P_S(q^{-1})} = \frac{1}{S(q^{-1})} \qquad (8.1.8)$$

The controller equation in the general case will take the form

$$s_0 P_s(q^{-1})u(t) = T(q^{-1})y^*(t+d+1) - R(q^{-1})y(t)$$
$$- \left[S(q^{-1}) - s_0 P_s(q^{-1}) \right] \bar{u}(t) \qquad (8.1.9)$$

and *u(t)* will be given by

$$u(t) = (1/s_0)\left\{ T(q^{-1})y^*(t+d+1) - R(q^{-1})y(t) \right.$$
$$\left. - \left[S(q^{-1}) - s_0 P_s(q^{-1}) \right] \bar{u}(t) - s_0 P_s^*(q^{-1})u(t-1) \right\} \qquad (8.1.10)$$

but

$$S(q^{-1}) - s_0 P(q^{-1}) = s_0 + q^{-1}S^*(q^{-1}) - \left[s_0 + q^{-1}s_0 P_s(q^{-1}) \right] =$$
$$= q^{-1}\left[S^*(q^{-1}) - s_0 P^*(q^{-1}) \right] \qquad (8.1.11)$$

One gets therefore the following equation for the control signal:

$$u(t) = (1/s_0)\left\{ T(q^{-1})y^*(t+d+1) - R(q^{-1})y(t) \right.$$
$$\left. - \left[S(q^{-1}) - s_0 P_s^*(q^{-1}) \right] \bar{u}(t-1) - s_0 P_s^*(q^{-1})u(t-1) \right\} \qquad (8.1.12a)$$

where

$$\bar{u}(t) = \begin{cases} u(t) & if \quad |u(t)| < u_{sat} \\ u_{sat} & if \quad u(t) \ge u_{sat} \\ u_{sat} & if \quad u(t) \le -u_{sat} \end{cases} \qquad (8.1.12b)$$

The next question is how to choose the dynamics for the evolution of *u(t)* when it leaves the saturation level. One generally chooses

$$P_S(q^{-1}) = 1 + p_{S_1}q^{-1} \qquad ; \qquad -0.8 < p_{S_1} < 0 \qquad (8.1.13)$$

This choice for p_{S_1} is easier to interpret if one considers Equation 8.1.13 as the denominator of a first-order discrete time filter resulting from the discretization of a first-order continuous time filter characterized by a time constant T_{sat} ($p_{S_1} = -e^{-T_s/T_{sat}}$). In this case the choices indicated in Equation 8.1.13 become

$$0 < T_{sat} \leq 4T_s$$

($T_{sat} = 4T_s$ corresponds to $p_1 = -0.778$).

Figure 8.7 illustrate the accelerating effect of this dynamics in the presence of saturation. One observes that for $T_{sat} = 2T_s$ ($p_1 = -0.602$) one gets an acceleration with respect to the case $T_{sat} = 0$ and still without overshoot.

Note that the value to be selected for p_{S_1} (T_{sat}), within the indicated bounds, is related to the specificity of each application.

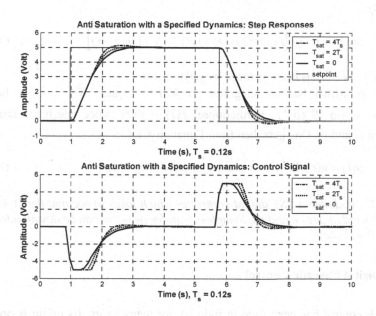

Figure 8.7. Digital position control (in simulation) with anti-wind-up device and dynamics when u(t) leaves the saturation level

8.1.5 Bumpless Transfer from Open Loop to Closed Loop Operation

To avoid large transients when one moves from the open loop operation to the closed loop operation, it is necessary to initialize the « memory » of the controller (i.e. to provide $y(t-1)$, $y(t-2)$ ..., $u(t-1)$, $u(t-2)$, ...$r(t-1)$, ...).

A method for controller initialization (when it is still in open loop operation) is described next:

1. One replaces the reference and the desired output by the measured output value ($y*(t+d+1) = y(t)$; $r(t) = y(t)$).
2. One stores the control $u(t)$, applied in open loop operation, in the controller memory.
3. One repeats step 1 and 2, for a number of times equal to $n = max (n_A + n_S, d + n_B + n_R)$.
4. One switches from open loop to closed loop operation.

Using this procedure, the control signal will have, at the switching instant, the value of the control signal applied at the previous instant in open loop operation.

To see this, suppose that $y(t) = constant$ during the initialization phase and that the controller has an integrator. The control signal at the switching instant will be given by

$$u(t) = (1/s_0)\left[T(1)y(t) - R(1)y(t) - S^*(1)u(t-1)\right]$$ (8.1.14)

(since $y*(t+d+1)$ has been replaced by $y(t)$).

The controller has an integrator so $S(1) = 0$ and $S^*(1) = -s_0$, because $S(q^{-1}) = s_0 + q^{-1}S^*(q^{-1})$. Furthermore, $T(1) = R(1)$ (since one has a unit gain between y^* and y). One gets then from Equation 8.1.14:

$$u(t) = u(t-1) \quad ; \quad t = max(n_A + n_S; d + n_B + n_R)$$ (8.1.15)

A C++ code for a RST digital controller which takes in account the D/A effect, the anti windup device and the bumpless transfer protocol can be downloaded from the book website[1].

8.1.6 Digital Cascade Control

Cascade control has been used in industry for many years. Its origin is probably the observation that between the actuator and the main controlled variable, there are intermediate physical variables that is interesting to control for various reasons. The block diagram of a cascade control is shown in Figure 8.8.

---

[1] http//: landau-bookic.lag.ensieg.inpg.fr

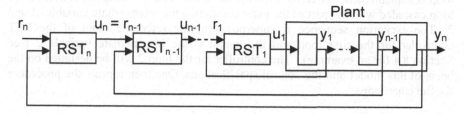

Figure 8.8. Block diagram of a cascade control

The controller for the main variable (n) provides the desired value for the intermediate variable (n-1). The controller for the intermediate variable (k) will provide the desired value for the variable (k-1). With cascade control, bounds on the intermediate variable (k-1) can be easily achieved by limiting the output of the (k) controller which provides the reference for the loop (k-1).

The logical order of the variables in a cascade control scheme results from the physical structure of the system. For example in a positioning system using a DC motor, the position will depend upon the speed, which in turn depends on the current which will be driven by the voltage applied to the motor. The physical constraints require that, during position transients (in tracking or regulation), the maximum rotational speed and the maximum current be bounded. These bounds will be obtained easily by bounding the output of the position controller and of the speed controller, respectively.

A similar situation is encountered in the level control of a liquid in a tank. The level of the liquid will depend upon the flow rate of the incoming liquid, which, itself, will depend upon the valve position.

Another important aspect in cascade control is related to the dynamics of the intermediate variables. These dynamics become more and more rapid as we approach the actuator. This allows decomposition of the transfer function from the control signal to the main output variable in several blocks, each being characterized by a different time scale.

This is particularly important in the case of digital control where the selection of the sampling frequency will depend for each loop upon the dynamics associated to the intermediate variable. It is this aspect that makes the difference between a continuous time cascade control and a digital cascade control.

In practice, we will start by selecting the sampling frequency for the first control loop characterized by the fastest dynamics. The sampling frequency for the other loops should be an integer under multiple of the sampling frequency for the first loop. In addition the sampling frequency for the loop (k) should be either an integer under multiple of the sampling frequency of the (k-1) loop or equal to it[2].

The implementation of a cascade control first requires one to identify the transfer function between the input to the actuator and the first intermediate variable followed by the design of the RST controller for this loop. Once the first

---

[2] This sampling frequency selection concerns the frequency at which the controller operates. This has not to be confounded with the data acquisition frequency that can be the same for all the loops.

loop is implemented, one has to identify the model corresponding to the first closed loop cascaded with the part of the plant connecting the intermediate variables 1 and 2. The excitation sequence is superposed to the reference input of the RST controller for the first loop and the output is the intermediate variable 2 (see Section 8.4 for an example). The controller for the loop 2 will be designed on the basis of this model and the control specifications. One then repeats the procedure for the other loops.

8.1.7 Hardware for Controller Implementation

There is a variety of hardware options for implementing a digital control scheme.

For a rapid test prior to final implementation on a specific hardware, one can use a "software in the loop" systems which include real time kernels (like Vissim (Visual solutions) or Simulink (Mathworks). Dedicated RST real time schemes operating on a PC can also be used (like Wintrac (Adaptech), which also provides the C^{++} code to be compiled and downloaded on the controller).

For the effective implementation on a micro-controller, a C++ code (Adaptech), available from the book website[3], can be used as a reference.

Many applications use a PC in which several controllers are implemented together with the logic related to the operation of a specific plant.

Another option is provided by the DCS (digital control systems). They have a high level programming language allowing to code the RST controller. See, for example, Roland and Landau (1991).

The PLC (programmable logic controller) is probably the device on which RST controllers are mostly implemented. The PLC produced by various manufacturers (see Figure 8.9) incorporate the function "RST controller" (like LTI 160 (Leroy), Alspa (Alstom), Quantum (Schneider Modicon)). A guide for the integration of RST controllers on a programmable system is available (Adaptech 2001a)[4].

A number of programmable controllers (like T640 (Eurotherm), μ - pilot (Soléa)) also allow the implementation of RST controllers.

The micro-controller boards, the VME boards under real time operating system OS9 and DSP boards also allow one to implement RST digital controllers operating at high sampling frequency.

To summarize: any digital computer with a data acquisition system and a real operating system can be used for the implementation of a digital control.

8.1.8 Measuring the Quality of a Control Loop

An important aspect of controller design and implementation is the measure of the achieved performance. This is easy, in general, for the tracking aspects (some step changes may be enough) but less obvious for the regulation aspects since, in practice, one cannot apply the disturbances. Evaluating the regulation performance for continuous type production processes is extremely important since any

---

[3] *http://landau-bookic.lag.ensieg.inpg.fr*
[4] To be downloaded from the website: *http://www.adaptech.com*

improvement in the regulation performance has an economic impact (improvement of the product quality, reduction of raw materials and energy consumption, reduction of machine stress).

For continuous time production processes, evaluation of the regulation performance is done using histograms of the measurements of the controlled variable over a certain time horizon (the horizon should be long enough in order that the results have a statistical meaning).

Figure 8.10 illustrates the histogram of a controlled variable for a "poor" control and for a "good" control.

If the control performance is "poor", one observes that a significant number of measurements are far from the average value (corresponding to the reference value). This will require to move the reference value towards higher values in order to guarantee a minimum acceptable value for the controlled variable (for example: minimum depth coating on a steel strip, minimum humidity in drying processes, etc.).

Figure 8.9. PLC Leroy LT160 embedding several RST controllers and I/O modules (courtesy of Leroy Automatisme)

If one has a "good" control, the dispersion of the measurements around the mean value will be significantly reduced. This corresponds to the reduction of the variance of the controlled variable (and of the standard deviation). As a consequence, on one hand a better quality (uniformity) of the products will be obtained and, on the other hand, the set point can be moved close to the tolerance limits. This in general will induce raw material and energy savings.

From a histogram one makes the following computations:

- Mean value:

$$y_M = \frac{1}{N} \sum_{i=1}^{N} y(i)$$

If the controller has an integrator, the mean value is often very close to the reference value (assuming that the disturbances over the measurement horizon have an almost zero mean value – this is often the case in continuous time production processes if the measurement horizon is enough large).

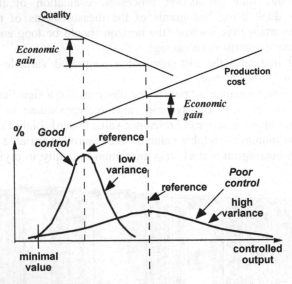

Figure 8.10. Histograms of the controlled variable

- Standard deviation:

$$\sigma = \left\{ \frac{1}{N} \sum_{i=1}^{N} \left(y(i) - y_M \right)^2 \right\}^{1/2}$$

If the histogram has a form close to that of a Gauss bell (Gaussian distribution - see Appendix A), 63% of the measurements will be in the interval $\pm$ σ around the mean value and more than 95% of the measurements will be in the interval $\pm$ $2\ \sigma$ around the mean value. In addition, from the histogram one can directly determine the percentage of measurements which are inside a certain tolerance zone.

It is important to mention that for high capacity production processes an improvement of the regulation performance (i.e. the reduction of the standard deviation) of 1%, or less, may have an important economic impact. This impact may be significantly higher than the investment for improving the control performance (in many cases a plant model identification and replacement of a PID controller by a digital RST controller designed on the basis of the identified model may achieve this).

8.1.9 Adaptive Control

"Adaptive Control" is a set of techniques used for on-line automatic tuning of controllers in order to achieve (or to maintain) a certain level of performance when

- The plant parameters are unknown or vary during plant operation
- The plant parameters are known but the characteristics of the disturbances are unknown or vary during operation

Adaptive control techniques should be considered only when the large variability of the plant parameters, or of the disturbance characteristics, does not allow to design a robust linear controller assuring satisfactory performance. We distinguish two categories:

1. "Closed loop" adaptive control
2. "Open loop" adaptive control

A "closed loop" adaptive control combines in general a real time (recursive) identification algorithm with the computation in real time of the controller parameters based on the current values of the estimated model parameters and of the desired performance. The tuning of the controller can be done at each sampling instant or at a lower frequency. In this second case one has a time horizon for the estimation of the plant model followed by the computation of the new controller parameters.

If this estimation horizon is significantly larger than the number of parameters to be estimated and, in addition, an external excitation is added during the parameter estimation stage (the controller being constant over this horizon), one has an iterative scheme which combines plant model identification in closed loop[5] with the redesign of the controller.

A "closed loop" adaptive control system should incorporate a supervision function, which monitors the correct operation conditions for the adaptive loop (ex: richness of the excitation signal used for identification, compatibility of the identified model with the control design, *etc.*). A block diagram of a "closed loop adaptive" system is shown in Figure 8.11a. It includes a control loop, an adaptation loop, and a supervision device. Closed loop adaptive control schemes, featuring explicitly a real time model estimation block followed by a real time controller design block, are termed "indirect adaptive control".

In a number of cases, one can directly tune (estimate) the controller parameters without an explicit use of the current model parameters for controller redesign. This type of scheme, shown in Figure 8.11b, is termed "direct adaptive control" (examples: model reference adaptive control, adaptive tracking and regulation, adaptive minimum variance control, adaptive rejection of unknown disturbances).

---

[5] The techniques for plant model identification in closed loop operation are presented in Chapter 9.

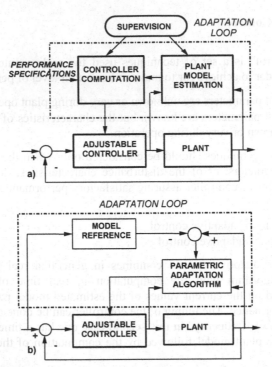

Figure 8.11a,b. Closed loop adaptive control: **a** indirect adaptive control; **b** direct adaptive control

Closed loop adaptive control schemes have basically two modes of operation: *self-tuning* operation and *adaptive* operation. In the self-tuning operation, the adaptation starts when a degradation of the performance is detected and stops when the desired performance is achieved. In the adaptive operation the algorithm operates all the time.

The characteristics of a plant dynamic model often depend upon a set of measurable variables which define an *operating point* (for example in hot dip galvanizing, the dynamic characteristics of the process depend upon the steel strip speed and the position of the air knives with respect to the steel strip – see Section 8.7 for details). In such cases one can use an "open loop" adaptive control scheme, as shown in Figure 8.12.

The range of operating points is divided in a number of operation regions. For each region, a relevant operating point is selected and a corresponding controller is designed based on an identified model. This controller assures the desired performances for all the operating points located within this region. The corresponding controllers are stored in a table. When the plant will operate in one of the operation regions, the corresponding values of the controller parameters will be used according to the table.

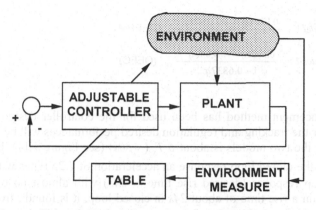

Figure 8.12. Open loop adaptive control scheme

For many applications characterized by a large variability of the plant dynamics, resulting from the change of the operating point, an "open loop" adaptive control is enough for assuring the desired performance over the range of possible operating points.

It is this approach which is illustrated in Section 8.6 (hot-dip galvanizing at SOLLAC-Florange).

For a detailed presentation of adaptive control techniques see: Landau (1979), Landau (1986), Åström and Wittenmark (1995), Landau *et al.* (1997), Landau (1993) and Landau *et al* (2005).

8.2 Digital Control of an Air Heater[6]

This section aims at illustrating the implementation on a real system (air heater) of an RST digital controller, designed according to one of the methods presented in Chapter 3 and based on an identified discrete-time model of the plant (for details concerning the identification of the air heater see Chapter 7, Section 7.5.1).

The diagram of the system and of the digital control loop is represented in Figure 8.13. The air is heated by means of a resistor supplied by a computer controlled thyristor power amplifier. The controlled variable is the air temperature at the output which is measured by a thermocouple.

Two models have been identified and validated for this air heater with $T_s = 5s$ (see Chapter 7, Section 7.5.1):

---

[6] The examples presented in this chapter have been worked out with WinPim (Adaptech) identification software and WinReg (Adaptech) control design software. Small numerical differences will result when using other identification and control design routines.

$$Model\ 1: \quad \frac{0,1724q^{-1}+0,0579q^{-2}}{1-0,6589q^{-1}} \quad (OEEPM)$$

$$Model\ 2: \quad \frac{0,1771q^{-1}+0,043q^{-2}}{1-0,6837q^{-1}} \quad (OEFC)$$

The pole placement method has been used for the controller design[7]. The same dynamics for the tracking and regulation desired performances will be chosen. The rise time for the two models is about $6\ T_s$ ($\approx 30s$) (see Figure 7.11). First we will consider for the closed loop response an acceleration of 1.25 times with respect to the open loop response (desired rise time $\approx 24s$) with almost no overshoot. In order to obtain a rise time of about $24s$ in closed loop, it is found , from diagrams of Figure 1.10 (or 1.11b), that for a second-order with damping $\zeta = 0.9$, it is necessary to set $\omega_0 = 0.136\ rad/s$. The desired closed loop poles will result from the discretization of a second order continuous time system with $\omega_0 = 0.136\ rad/s$ and $\zeta = 0.9$ (the response has almost no overshoot) for $T_s = 5s$.

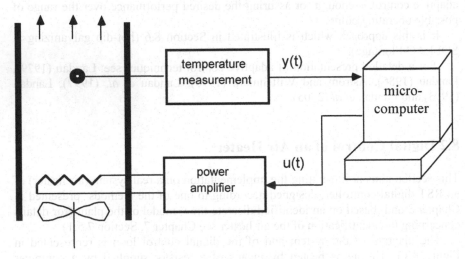

Figure 8.13. Digital control of an air heater

Table 8.1 summarizes the results of the digital controller design based on the model 1 and using the *pole placement*. Figure 8.14 shows the magnitude of the frequency response of the output sensitivity function S_{yp}. Note that the resulting curve respects the robustness template.

The designed controller has been tested first in simulation, both on model 1 and model 2. Figure 8.15 illustrates the closed loop response with both models

---

[7] In this case one can also use the tracking and regulation with independent objectives, as the zeros of the plant model are stable ($b_2 < b_1$).

(simulation). The rise time is about five sampling periods ($\approx 25s$) matching the desired performance. The differences between the responses obtained with the two models are negligible (one can note a difference in the input signal applied to the system, caused by the different identified static gains).

Table 8.1. Controller designed by "pole-placement" for the air heater, with $\omega_0 = 0.136$ rad/s, and $\zeta = 0.9$ (both tracking and regulation)

Plant:

- $d = 0$
- $B(q^{-1}) = 0.1724 \ q^{-1} + 0.0579 \ q^{-2}$
- $A(q^{-1}) = 1 - 0.6589 \ q^{-2}$

Tracking dynamics $\rightarrow T_s = 5s$, $\omega_0 = 0.136 \ rad/s$, $\zeta = 0.9$

- $B_m \ (q^{-1}) = 0.1543 \ + 0.1024 \ q^{-1}$
- $A_m \ (q^{-1}) = -1.0372 \ q^{-1} + 0.2940 \ q^{-2}$

Regulation dynamics $\rightarrow T_s = 5s$, $\omega_0 = 0.136$, $\zeta = 0.9$

$$P(q^{-1}) = 1 - 1.0372 \ q^{-1} + 0.2940 \ q^{-2}$$

Pre-specifications: Integrator

*** CONTROL LAW ***

$$S(q^{-1}) \ u(t) + R(q^{-1}) \ y(t) = T(q^{-1}) . y^*(t+d+1)$$

$$y^*(t+d+1) = (B_m(q^{-1}) / A_m(q^{-1})) . r(t)$$

Controller:

- $R(q^{-1}) = 2.7637 - 1.6491 \ q^{-1}$
- $S(q^{-1}) = 1 - 0.8549 \ q^{-1} - 0.1451 \ q^{-2}$
- $T(q^{-1}) = 4.34 - 4.5017 \ q^{-1} + 1.2762 \ q^{-2}$

Gain Margin: 5.61	Phase margin: 63.6 deg
Modulus margin: 0.732 (- 2.71 dB)	Delay margin: 9.6 s

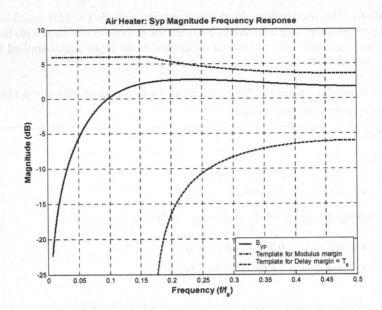

Figure 8.14. Output sensitivity function with the controller based on the specifications $\omega_0 = 0.136$ rad/s, and $\zeta = 0.9$

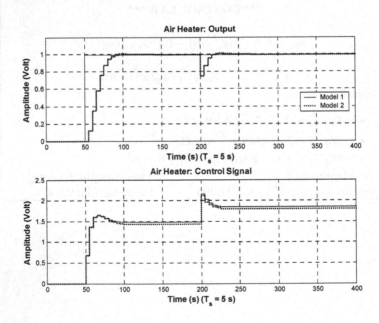

Figure 8.15. Simulated closed loop response for the air heater (model 1 and model 2) with the controller based on model 1 and the desired performances: $\omega_0 = 0.136$ rad/s and $\zeta = 0.9$

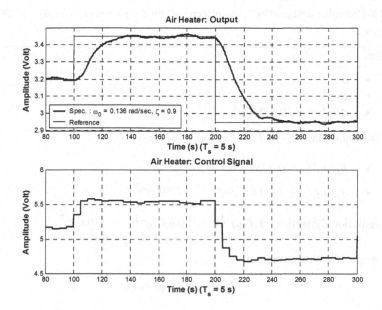

Figure 8.16. Real time responses for the controller designed from the specifications $\omega_0 = 0.136$ rad/s, $\zeta = 0.9$

The experimental results for temperature variations around 60°C (3.2V) are given in Figure 8.16. One can see that the rise time is indeed about 25s.

We now consider the case in which a stronger acceleration is required for the closed loop system if compared to the natural system response (open loop response). For a rise time about *15s* (rise time reduced to half of open loop rise time), it results, from the diagrams of Figure 1.11, that the desired dynamics for the closed loop corresponds to that of a second-order system characterized by $\omega_0 = 0.226\ rad/s$; $\zeta = 0.9$. The new controller must guarantee, nevertheless, a modulus margin greater or equal to 0.5 and a delay margin at least equal to 5s (*1 T_s*).

On the basis of these new specifications, the results of the controller design are given in Table 8.2. Note that the robustness margins are reduced with respect to the margins found in the previous case, as a consequence of the stronger acceleration required (rise time reduction). The delay margin obtained, in particular, is very close to the limits (T_s=5s). However the robustness conditions for the modulus and the delay margins are still satisfied.

The magnitude of the output sensitivity function is presented in Figure 8.17. We observe that the sensitivity function reaches the upper template. The system response cannot be further accelerated while preserving the robustness (without specific calibration of the sensitivity functions).

338 Digital Control Systems

Table 8.2. Controller designed by "pole-placement" for the air heater, with $\omega_0 = 0.226$ rad/s and $\zeta = 0.9$

Plant:

- $d = 0$
- $B(q^{-1}) = 0.1724\ q^{-1} + 0.0579\ q^{-2}$
- $A(q^{-1}) = 1 - 0.6589\ q^{-2}$

Tracking dynamics → $T_s = 5s$, $\omega_0 = 0.226$, $\zeta = 0.9$

- $B_m\ (q^{-1}) = 0.3281 + 0.1652\ q^{-1}$
- $A_m\ (q^{-1}) = 1 - 0.6373\ q^{-1} + 0.1308\ q^{-2}$

Regulation dynamics → $P\ (q^{-1}) = 1 - 0.637\ q^{-1} + 0.1308\ q^{-2}$
$$T_s = 5s, \ \omega_0 = 0.226, \ \zeta = 0.9$$

Pre-specifications: Integrator

*** CONTOL LAW ***

$$S(q^{-1})\ u(t) + R(q^{-1})\ y(t) = T(q^{-1})\ y^*(t+d+1)$$

$$y^*(t+d+1) = (B_m(q^{-1}) / A_m(q^{-1}))\ .\ r(t)$$

Controller:

- $R(q^{-1}) = 4.6461 - 2.5045\ q^{-1} + 0.1147\ q^{-2}$
- $S(q^{-1}) = 1 - 0.7797\ q^{-1} - 0.2202\ q^{-2}$
- $T(q^{-1}) = 4.34 - 2.7662\ q^{-1} + 0.5677\ q^{-2}$

Gain Margin: 3.159	Phase margin: 51.9 deg
Modulus margin: 0.613 (- 4.25 dB)	Delay margin: 5.05 s

The simulated responses for both models of the air heater are presented in Figure 8.18. The responses obtained in simulation with the two models are again very close. The only difference one can see is the steady state value of the input signal (as the model gain is different). The rise time is about 15s. One can note an increase of the actuator effort needed during the transient as the rise time has been reduced (to be compared with Figure 8.15).

The experimental results obtained for temperature variations around 60°C (3.2V) are given in Figure 8.19. The controller designed with the specifications $\omega_0 = 0.226\ rad/s$ and $\zeta = 0.9$ allows one to obtain the desired performances (rise time $\approx 15s$, no overshoot).

Figure 8.20 shows the behavior during regulation (around 65°C) for a disturbance introduced by a step variation of the fan speed (see Figure 8.13).

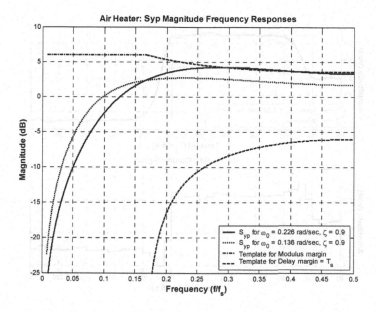

Figure 8.17. Output sensitivity functions for the controllers designed from the specifications $\omega_0 = 0.226$ rad/s, $\zeta = 0.9$ and $\omega_0 = 0.136$ rad/s, $\zeta = 0.9$ respectively

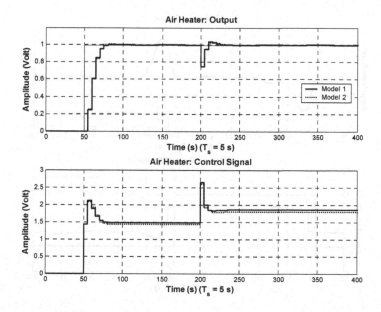

Figure 8.18. Simulated closed loop response for the air heater (model 1 and model 2) for the controller designed from $\omega_0 = 0.226$ rad/s, $\zeta = 0.9$

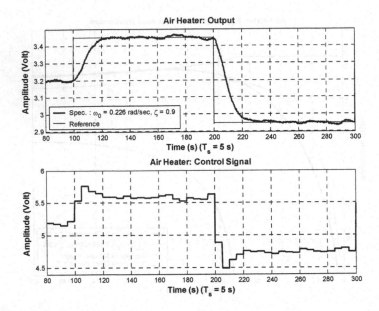

Figure 8.19. Real time response for the air heater with the controller designed from the specifications: ω_0=0.226 rad/s and $\zeta = 0.9$

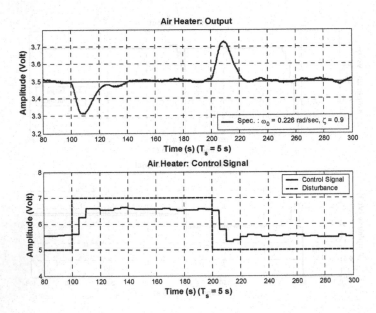

Figure 8.20. Air heater output in regulation with the controller based on the specifications: ω_0=0.226 rad/s and $\zeta = 0.9$

8.3 DC Motor Speed Control

The block diagram of the DC motor digital speed control is shown in Figure 8.21. For controller design, the model of the motor previously identified (see Chapter 7, Section 7.5.3) has been used. The specific real time implementation has been done using Wintrac (Adaptech 2004).

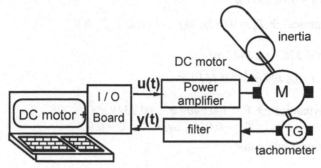

Figure 8.21. Digital speed control of a DC motor

The model identified with OEEPM (T_s = *15 ms*) has been used (see Chapter 7, Section 7.5.3).

$$\text{Motor model:} \quad \frac{0.2617q^{-1} + 0.2288q^{-2}}{1 - 0.535q^{-1}}$$

The model has a stable zero since $b_2 < b_1$, but very close to the unit circle. It is therefore not recommended to use the *tracking and regulation with independent objectives*. The *pole placement* method will be used. The model rise time is around 4.5 T_s ($\approx$ *67 ms*) (see Figure 7.19). An acceleration by a factor of 1.33 is desired for the closed loop time response (desired rise time: $\approx$ *50 ms*) together with a small overshoot. In order to obtain a rise time $t_R \approx$ *50 ms,* from Figure 1.10 it results that a second-order continuous time system with ζ = *0.9* o, $\omega_0 \approx$ *75 rad/s* has to be chosen as desired closed loop dynamics. Therefore the desired closed loop poles will result from the discretization of a second-order continuous time system with ω_0 =*75 rad/s*, ζ = *0.9* and T_s = *15 ms*. The same dynamics have been chosen for tracking.

Table 8.3 summarizes the controller design for the identified model of the motor. The robustness margins are satisfactory. Figure 8.22 illustrates the simulation results using the identified model both in tracking and regulation. The rising time is $\approx$ *55 ms*. This is normal, since in pole placement there is the additional dynamics of $B(q^{-1})$ which will influence the results. In this case it is a fractional delay of about $\approx$ *5 ms*.

Table 8.3. "Pole placement" speed controller with $\omega_0 = 75$ rad/s and $\zeta = 0.9$ (DC motor)

Plant:

- $d = 0$
- $B(q^{-1}) = 0.2617\,q^{-1} + 0.2288\,q^{-2}$
- $A(q^{-1}) = 1 - 0.5351\,q^{-1}$

Tracking dynamics $\rightarrow T_s = 0.015s$, $\omega_0 = 75$ rad/s, $\zeta = 0.9$

- $B_m = 0.3262 + 0.1647\,q^{-1}$
- $A_m = 1 - 0.6409\,q^{-1} + 0.1319\,q^{-2}$

Regulation dynamics $\rightarrow P = 1 - 0.6409\,q^{-1} + 0.1319\,q^{-2}$
$$T_s = 0.015s, \quad \omega_0 = 75, \quad \zeta = 0.9$$

Pre-specifications: Integrator

*** CONTROL LAW ***

$$S(q^{-1}) \cdot u(t) + R(q^{-1}) \cdot y(t)\, T(q^{-1})\, y^*(t+d+1)$$

$$y^*(t+d+1) = (B_m(q^{-1}) / A_m(q^{-1})) \cdot ref(t)$$

Controller:

- $R(q^{-1}) = 1.9176 - 0.9168\,q^{-1}$
- $S(q^{-1}) = 1 - 0.6078\,q^{-1} - 0.3921\,q^{-2}$
- $T(q^{-1}) = 2.0382 - 1.3065\,q^{-1} - 0.269\,q^{-2}$

Gain margin: 3.205	Phase margin: 57.8 deg
Modulus margin: 0.625 (- 4.08 dB)	Delay margin: 22 ms (*1.46 T_s*)

Real time results are shown in Figures 8.23 and 8.24. Figure 8.23 shows the time response of the motor speed for a step change in the desired speed. The rising time is close to the simulated one (≈ 55 *ms*). Figure 8.24 illustrates the behavior of the closed loop system for a load disturbance. It is an almost step disturbance followed by an almost instantaneous take off of the load at *393 T_s*. The disturbance rejection time is ≈ 105 *ms*.

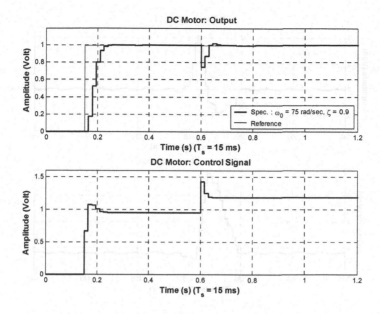

Figure 8.22. Simulation results for the digital speed control with ω_0 = 75 rad/s and ζ = 0.9

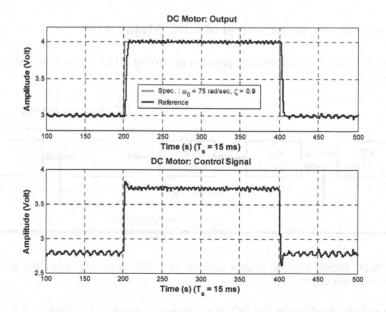

Figure 8.23. Closed loop speed response of a DC motor for a step on the reference (tracking dynamics ω_0 = 75 rad/s, ζ = 0.9)

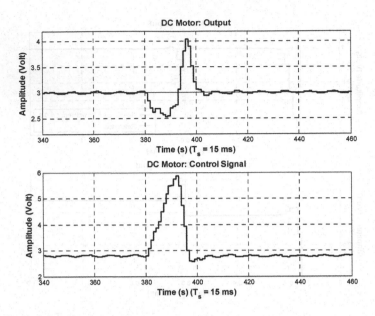

Figure 8.24. Motor speed response for a load disturbance (regulation dynamics: $\omega_0 = 75$ rad/s, $\zeta = 0.9$)

8.4 Cascade Position Control of a DC Motor Axis

The block diagram of the cascade position control of a DC motor axis is shown in Figure 8.25.

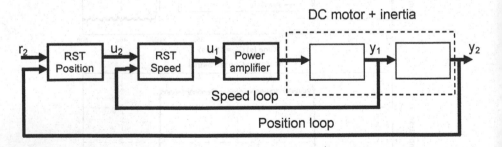

Figure 8.25. Digital cascade control position of a DC motor axis (the D/A and A/D converters are not shown)

In this specific application the DC motor already considered in Section 8.3 (speed control) will be used. We will discuss now the integration of the speed control loop in a cascade position control. The implementation of this cascade control scheme in real time has been done with a PC using WinTRAC (Adaptech 2004)[8].

The sampling period for the speed loop is *15 ms* (see Section 8.3). Since the dynamics of the position is significantly slower (presence of an integrator and of an important mechanical mass), a longer sampling time has to be used in the position control loop. A sampling time of *120 ms* has been chosen (it is a multiple of *15 ms*). In order to design the position controller it is necessary to identify the dynamic model between the reference of the speed loop and the axis position.

The PRBS for identification has been applied to the reference input of the controller. The characteristics of the PRBS used for identification are:

- Number of cells of the shift register: $N = 7$
- Clock frequency: $f_{PRBS} = f_s/2$ $(T_{PRBS} = 240 \ ms)$
- Magnitude: 2V (300 rpm)
- Length: 256 samples

The data acquired are shown in Figure 8.26 (after centering the data). The integrator behavior of the system is obvious.

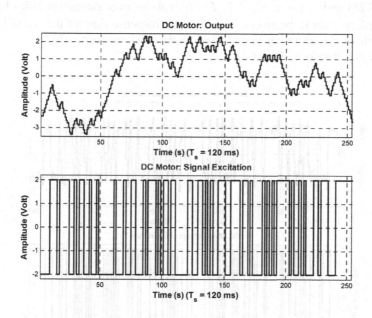

Figure 8.26. Input/output sequences for identification (input: speed reference; output: axis position)

Data Pre-Processing
Since it is known that the dynamic model contains an integrator (position is the integral of the speed) and taking into account the indications given in Chapter 7, Section 7.2.2, a model without integrator will be identified. In order to do that, a new input/output data set is considered: the input sequence is unchanged, whilst a new output sequence representing the variations of the output between the sampling instants $(y(t)-y(t-1))$ will be used. The corresponding I/O sequences (after elimination of the residual DC component) are represented in Figure 8.27.

One observes that the output level is about one fifth of the input level. Following the indications given in Section 7.2.3, the output should be multiplied by a factor of *5* in order to get a magnitude of the same order as the input. The resulting identified model will need to be scaled by dividing the coefficients of $B(q^{-1})$ with a factor of *5*.

Complexity Estimation
Using the order estimation procedure based on the use of the instrumental variable with delayed inputs (see Chapter 6, Section 6.5) and implemented in *estororderiv.sci (estororderiv.m)*, as well as in *WinPIM software* (Adaptech, 1996a), one gets for the model without integrator $n = max\ (n_A, n_B+d) = 2$ (see Figure 8.28) and $n_A = 1$, $n_B = 2$, $d = 0$, with an over estimation risk of 0.1%. These orders seem to be reasonable since the response time of the speed loop is about half of the sampling period used in the position loop, which leads to the presence of a fractional delay ($n_B = 2$).

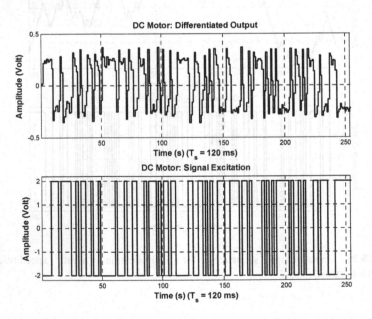

Figure 8.27. Input/output file for the system without integrator

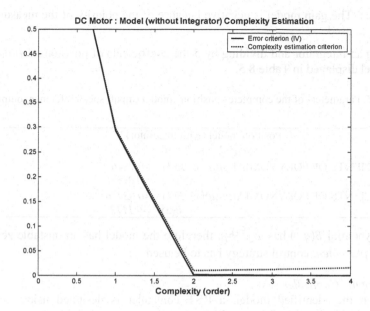

Figure 8.28. Order estimation from data for the model without integrator

Parameter Identification and Validation

The model (without integrator) identified with the *recursive least squares* (structure S1) does not pass the statistical validation tests. Selecting the structure S3 and the *output error with extended prediction model* with $n_A=1$, $n_B=2$, $d=0$ and $n_C=2$ one gets a model satisfying the whiteness validation test. The results are summarized in Table 8.4.

Table 8.4. Parameters of the model (without integrator)

Model without integrator and gain multiplied by 5 $n_A=1, n_B=2, n_C=2, d=0$
COEFFICIENTS OF POLYNOMIAL $A(q^{-1})$: $A(1) = 0.2154$
COEFFICIENTS OF POLYNOMIAL $B(q^{-1})$: $B(1) = -0.1737$ $B(2) = -0.5563$
COEFFICIENTS OF POLYNOMIAL $C(q^{-1})$: $C(1) = -0.3176$ $C(2) = -0.1878$
TEST DE VALIDATION: Whiteness of the residual error
Validation criterion: Theor. Value: $\lvert RN(i) \rvert \leq 0.136$, Pract. Value: $\lvert RN(i) \rvert \leq 0.15$
RN(1) = 0.0422   RN(2) = 0.0057
RN(3) = -0.136   RN(4) = 0.0497

Remarks:

1. The validation results can be improved using a variable forgetting factor, but the time responses of the two models are almost the same.

2.   The gain model is negative because of the polarity of the measurement used.

Inserting an integrator and dividing by 5 the coefficients of polynomial B one gets the model displayed in Table 8.5.

Table 8.5. Parameters of the complete "position" model (input: speed reference; output: axis position)

Position model (with integrator)
$n_A=2, n_B=2, d=0$
COEFFICIENTS OF POLYNOMIAL $A(q^{-1})$: $A(1) = -0.7846$
$A(2) = -0.2154$
COEFFICIENTS OF POLYNOMIAL $B(q^{-1})$: $B(1) = -0.03476$
$B(2) = -0.1112$

The polynomial $B(q^{-1})$ has $|b_2|>|b_1|$, therefore the model has an unstable zero and the *pole placement* control strategy has to be used.

Design of the Controller
Based on the identified model, a RST controller is designed using the pole placement with shaping of the sensitivity functions.
    The control specifications are:

- Dominant regulation dynamics: discretization of a second-order continuous time system with $\omega_0 = 3\ rad/s$ and $\zeta = 0.8\ (T_s = 120\ ms)$
- An integrator in the controller
- Opening of the loop at $0.5 f_s$
- Modulus margin: $\Delta M \geq 0.5$; delay margin: $\Delta\tau \geq T_s$
- Tracking dynamics: discretization of a second-order continuous time system with $\omega_0 = 4.5 rad/s$ and $\zeta = 0.9$

To get the integrator effect a fixed part $H_S(q^{-1}) = 1 - q^{-1}$ must be included. For opening the loop at $0.5 f_s$ a fixed part $H_R(q^{-1}) = 1 + q^{-1}$also has to be included.
    In the absence of the opening of the loop at $0.5 f_s$ (using a filter H_R), the modulus of the input sensitivity function would be very large at high frequencies, where the system has a very small gain (it is an integrator). This leads to an important stress on the actuator without having any effect on the output.
    Figures 8.29 and 8.30 give the modulus of the output and input sensitivity functions for the following three designs:

1.   without H_R and without auxiliary poles
2.   with H_R, but without auxiliary poles
3.   with H_R and 3 auxiliary poles at 0.1

The final controller corresponds to the third design. The auxiliary poles have been introduced in order to significantly reduce the stress on the actuator over $0.3 f_s$. One notes a small reduction of the modulus margin with respect to the design a and b, but which still has a value larger than 0.5 ($|S_{yp}(q^{-1})|_{max} < 6dB$). The coefficients of the controller corresponding to the third design are given in Table 8.6.

Table 8.6. RST controller for the position loop

Polynomial $R(q^{-1})$	Polynomial $S(q^{-1})$	Polynomial $T(q^{-1})$	Polynomial $B_m(q^{-1})$
$R(0)=-1.2139$ $R(1)=-0.7543$ $R(2)=0.9699$ $R(3)=0.5111$	$S(0)=1.0000$ $S(1)=-1.0222$ $S(2)=0.2836$ $S(3)=-0.2614$	$T(0)=-6.8479$ $T(1)=12.0843$ $T(2)=-7.0639$ $T(3)=1.4625$ $T(4)=-0.1255$ $T(5)=0.0038$	$B_m(0)=0.1057$ $B_m(1)=0.076381$ Polynomial $A_m(q^{-1})$ $A_m(0)=1.0000$ $A_m(1)=-1.1962$ $A_m(2)=0.3783$

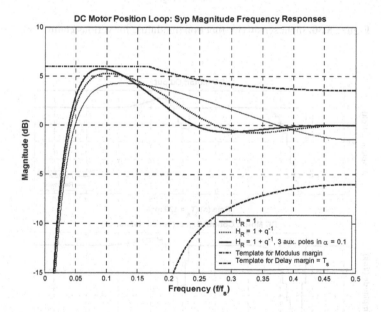

Figure 8.29. Output sensitivity function (modulus) for the various controllers

Figure 8.31 gives the simulation results (tracking and regulation) for the position loop. The rise time is slightly longer than $7T_s = 840\ ms$.

The experimental results with the two cascaded loops (speed and position) are illustrated in Figure 8.32. Figure 8.32a,b gives the evolution of the position for a step change on the reference for the position and the evolution of the output of

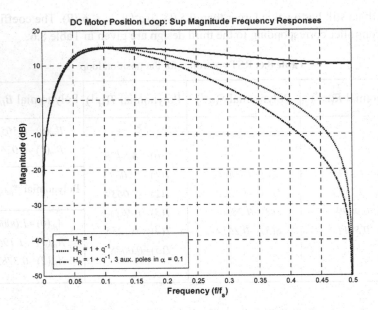

Figure 8.30. Input sensitivity function (modulus) for various controllers

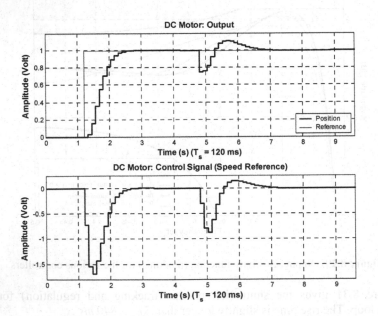

Figure 8.31. Tracking and regulation for the position loop (simulation)

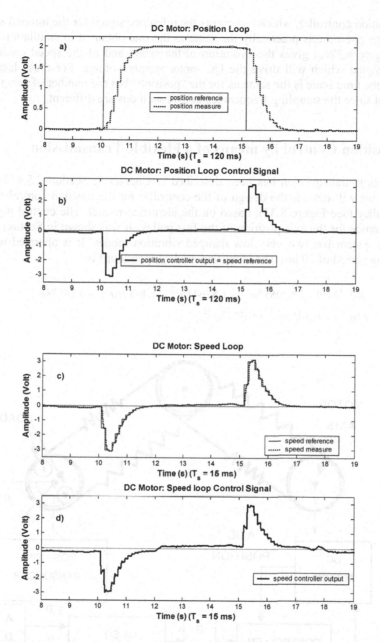

Figure 8.32a-d. Real time evolution of the variables in cascade position control of the DC motor: **a** position and reference for the position; **b** output of the position controller (speed reference); **c** speed and reference for speed; **d** output of the speed controller(command for DC motor supply voltage)

the position controller, which represents the reference signal for the internal speed loop. The rise time is practically the same as the one obtained in simulation (*840 ms*). Figure 8.32c,d gives the evolution of the speed and of the speed controller output signal which will drive the DC motor supply voltage. For these last two figures the time scale is the same as for the "position" but the number of samples is different since the sampling frequencies in the two loops are different.

8.5 Position Control by means of a Flexible Transmission

The flexible transmission has been described in Chapter 7, Section 7.5.4. In this Section we will discuss the design of the controller for the position control of the third pulley (see Figure 8.33), based on the identified model. The control input is the reference for the axis position of the first pulley. It was shown in Section 7.5.4 that the system has two very low damped vibration modes. It is operated with a sampling period of 50 *ms*. The identified model (open loop) is

$$A(q^{-1}) = 1 - 1.609555q^{-1} + 1.87644q^{-2} - 1.49879q^{-3} + 0.88574q^{-4}$$
$$B(q^{-1}) = 0.3053q^{-1} + 0.3943q^{-2}$$
$$d = 2$$

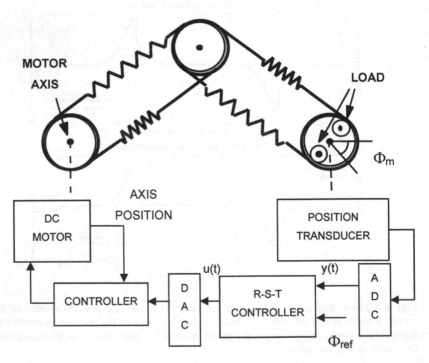

Figure 8.33. Digital control of a flexible transmission

The frequency characteristic of this model has been presented in Figure 7.24. The two vibration modes of the model are characterized by $\omega_0 = 11.949 \ rad/s$, $\zeta = 0.042$ and $\omega_0 = 31.462 \ rad/s$, $\zeta = 0.023$. The model has an unstable zero (because $|b_2|>|b_1|$).

Since the model has an unstable zero the *pole placement* strategy will be used for controller design. It is desired to get a regulation and a tracking behavior characterized by a pair of well damped dominant poles, corresponding to the first vibration mode of the flexible transmission. The controller should also assure certain robustness specifications in terms of modulus margin, delay margin and maximum value of the modulus of the input sensitivity function at high frequencies. The performance and robustness specifications are summarized next:

- Tracking dynamics: discretization of a second-order continuous time system with $\omega_0 = 11.94 \ rad/s$ and $\zeta = 0.9$ (Rise time $t_R \approx 0,285s$ which corresponds to $t_R \approx 6T_s$)
- Zero steady state error (controller should include an integrator)
- Dominant poles of the closed loop corresponding to the discretization of a second-order continuous time system with $\omega_0 = 11.94 \ rad/s$ and $\zeta = 0.8$;
- Modulus margin: $\Delta M \geq 0.5$; delay margin: $\Delta \tau \geq 0.05s$
- $|S_{up}(q^{-1})|_{max} \leq 10$ dB for $f \geq 0.35 f_s$

A first controller design, for which only the dominant poles have been specified, leads to the sensitivity functions S_{yp} and S_{up} shown in Figures 8.34 and 8.35 (curves A).

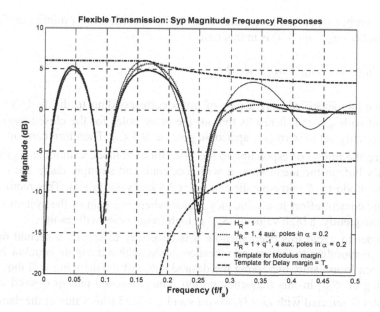

Figure 8.34. Output sensitivity functions for various controllers

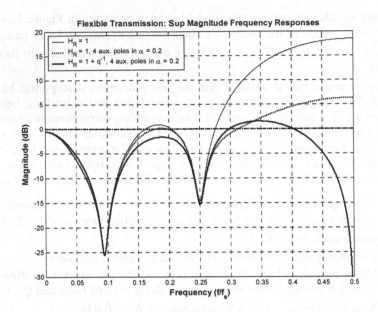

Figure 8.35. Input sensitivity functions for various controllers

The resulting modulus margin $\Delta M = 0.498$ is slightly less than the desired value. The delay margin $\Delta\tau = 0.043s$ is less than the desired value and the maximum of $|S_{up}(q^{-1})|$ at high frequencies is over the required value *10 dB* (see Table 8.7).

The number of closed loop poles that can be specified for a minimum size of the controller ($n_R = n_S = 4$) is in this case

$$n_P = n_A + n_B + n_{H_S} + d - 1 = 8$$

Since one has specified only a pair of poles corresponding to the first vibration mode but with $\zeta = 0.8$, it results that all the other poles for the closed loop have been implicitly set to zero (*i.e.* aperiodic poles at $0.5 f_s$). Therefore controller (A) will damp the poles corresponding to the first vibration mode without changing the frequency but, in the mean time, it will accelerate and strongly damp the second vibration mode ($z=0$ corresponding to $\omega_0=62.8$ *rad/s* and $\zeta=1$). This requires an important control effort in a frequency region where the gain of the system is low and, consequently, a high value of $|S_{up}|$ at high frequencies will result.

In order to avoid this phenomenon, it is necessary to specify a second pair of poles corresponding to the second vibration mode, with a damping equal or higher to the open loop value. Increasing the desired value of the damping will induce an increasing of $|S_{up}|$ in this frequency region. Thus a second pair of desired closed loop poles is selected with $\omega_0=31.46$ *rad/s* and $\zeta = 0.15$ (this value of the damping

Table 8.7. Specifications, and achieved modulus margin, delay margin and maximum $|S_{up}|$ for various controllers

	$H_S(q^{-1})$	$H_R(q^{-1})$	Closed loop poles		Modulus margin (dB)	Delay margin (s)	$\|S_{up}\|$ Max (dB)
			Dominant	Auxiliary			
A	$1-q^{-1}$		$\omega_0 = 11.94$ $\zeta = 0.8$		0.498 (-6.06)	0.043	18.43
B	$1-q^{-1}$		$\omega_0 = 11.94$ $\zeta = 0.8$	$\omega_0 = 31.46$ $\zeta = 0.15$ $(1-0.2q^{-1})^4$	0.522 (-5.65)	0.062	6.24
C	$1-q^{-1}$	$1+q^{-1}$	$\omega_0 = 11.94$ $\zeta = 0.8$	$\omega_0 = 31.46$ $\zeta = 0.15$ $(1-0.2q^{-1})^4$	0.544 (-5.29)	0.057	1.5

is not critical). For the remaining poles to be assigned, as indicated in Chapter 3, Section 3.6, it is wise to select aperiodic poles located on the real axis between *0.05* and *0.5*. These poles will have a beneficial effect upon the maximum value of $|S_{up}|$ at high frequencies. The remaining four poles have been set to 0.2 (since four poles have been already assigned). The frequency characteristics of S_{yp} and S_{up} for the new controller are represented in Figures 8.34 and 8.35 (curves B). The new controller satisfies the imposed specifications (see Table 8.7).

Figures 8.36 and 8.37 show the real time results obtained with controller B. Figure 8.36 corresponds to a step change on position reference. One observes that the rise time of the real system is close to the specified one ($6T_s$ plus the delay of $2T_s$). Figure 8.37 illustrates the behavior in regulation for a position disturbance followed by an instantaneous release (it is the return to the equilibrium position that is interesting for performance evaluation in regulation).

A detailed examination of the control signal (see Figures 8.36 and 8.37) shows the presence of a high frequency component without effect on the output (the position of the third pulley), since this frequency region is far beyond the band pass of the closed loop.

This phenomenon is even visible on the real system, where the position of the first pulley will follow to a large extent the control signal (the local position control of the first pulley has a high band pass). The input sensitivity function, in fact, will amplify by a few dB the high frequency measurement noise (which is interpreted as a disturbance). To counteract this effect it is sufficient to open the loop at high frequencies, by introducing in the controller a fixed part of the form $H_R = 1 + q^{-1}$ (the controller gain will be zero at $0.5f_s$), and to re-compute the controller with the same specifications.

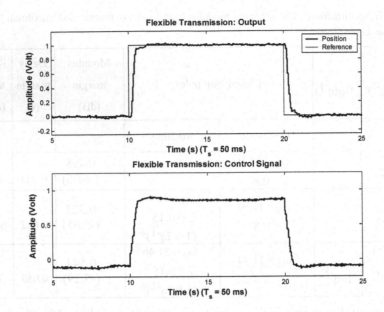

Figure 8.36. Position step response of the flexible transmission (controller B)

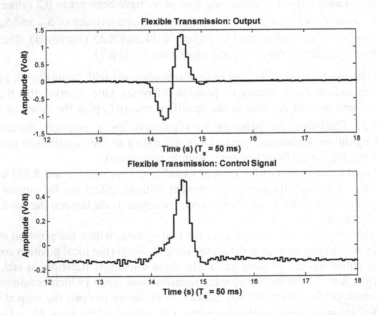

Figure 8.37. Flexible transmission – rejection of a position disturbance (controller B)

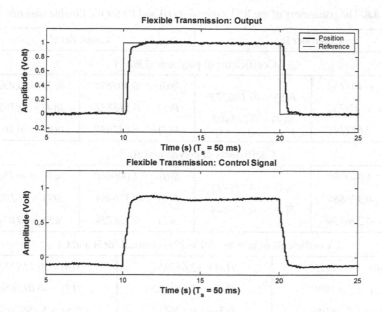

Figure 8.38. Position step response of the flexible transmission (controller C)

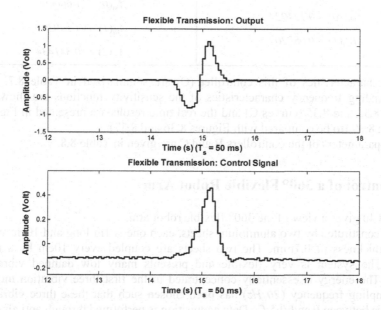

Figure 8.39. Flexible transmission – rejection of a position disturbance (controller C)

Table 8.8. The parameters of the RST controllers (B and C) for the flexible transmission

Controller B		Controller C	
Coefficients of polynomial $R(q^{-1})$			
$R(0) = 0.497161$	$R(3) = -0.186272$	$R(0) = 0.307895$	$R(3) = 0.452544$
$R(1) = -1.009731$		$R(1) = -0.515832$	$R(4) = -0.165905$
$R(2) = 0.622142$	$R(4) = 0.285405$	$R(2) = -0.037637$	$R(5) = 0.167641$
Coefficients of polynomial $S(q^{-1})$			
$S(0) = 1.000000$	$S(3) = -0.239733$	$S(0) = 1.000000$	$S(3) = -0.181950$
$S(1) = -0.376884$		$S(1) = -0.376884$	$S(4) = -0.109785$
$S(2) = -0.256739$	$S(4) = -0.126644$	$S(2) = -0.256739$	$S(5) = 0.074642$
Coefficients of polynomial $T(q^{-1})$ (identical for B and C)			
$T(0) = 1.429225$	$T(3) = -2.687952$		$T(6) = 0.118277$
$T(1) = -2.839061$	$T(4) = 1.594832$		$T(7) = -0.012656$
$T(2) = 3.181905$	$T(5) = -0.576414$		$T(8) = 5.488E-04$
Coefficients of polynomials $B_m(q^{-1})$ and $A_m(q^{-1})$ (identical for B and C)			
$B_m(0) = 0.124924$		$A_m(0) = 1.000000$	
$B_m(1) = 0.087209$		$A_m(1) = -1.129301$	
		$A_m(2) = 0.341434$	

The characteristics of this controller (C) are summarized in Table 8.7. The corresponding frequency characteristics of the sensitivity functions are shown in Figures 8.34 and 8.35 (curves C) and the real time results are presented in Figures 8.38 and 8.39 (to be compared with Figures 8.36 and 8.37).

The parameters of the controllers B and C are given in Table 8.8.

8.6 Control of a 360° Flexible Robot Arm

Figure 8.40 gives a view of the 360° flexible robot arm.

It is constituted by two aluminium sheets, each one is 1m long and 10cm wide, with a thickness of 0.7mm. The two sheets are coupled every 10cm by a rigid frame. The system is very flexible and presents many low damped vibration modes. The energy is essentially concentrated in the first three vibration modes. The sampling frequency (*20 Hz*) has been chosen such that these three vibration modes lie between *0* and *0.5 f_s* . Data acquisition is performed through anti aliasing filters. The block diagram of the control scheme is given in Figure 8.41. One of the extremities of the arm is directly coupled to the axis of a DC motor. The corresponding local position loop contains a cascade control of motor current, speed and position (measured by a potentiometer type transducer). The band pass

Figure 8.40. 360° Flexible robot arm (Laboratoire d'Automatique de Grenoble, INPG/CNRS/UJF)

of this loop is higher than the frequency of the first vibration mode.

The output of the system is the position of the free end of the arm. The measurement of the position of the free end is done by combining information upon the position of the motor axis (provided by an incremental transducer) and those provided by a carried on measurement device (including a light beam and a mirror), which gives the angular position with respect to the motor axis position (for details concerning the measurement device see Landau *et al.* 1996).

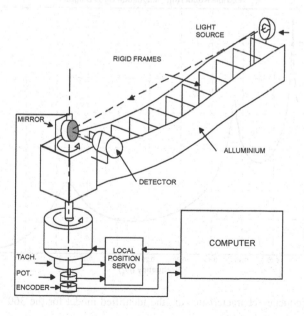

Figure 8.41. Position control scheme for the 360° flexible robot arm

The measurement system allows one to cover a rotation from 0 to 360°. The control signal provided by the computer is the reference position for the motor axis.

The identified and validated model for the case without load is (Langer and Landau 1999)

$$A(q^{-1}) = 1 - 2.1049\ q^{-1} + 1.04851\ q^{-2} + 0.33836\ q^{-3} + 0.46\ q^{-4}$$
$$- 1.5142\ q^{-5} + 0.7987\ q^{-6}$$
$$B(q^{-1}) = 0.0064\ q^{-1} + 0.0146\ q^{-2} - 0.0697\ q^{-3} + 0.044\ q^{-4}$$
$$+ 0.0382\ q^{-5} - 0.007\ q^{-6}$$
$$d = 0$$

The frequency characteristic of this model is shown in Figure 8.42.

This model is characterized by three very low damped vibration modes ($\omega_1 = 2.6173$, $\zeta_1 = 0.018$; $\omega_2 = 14.4027$, $\zeta_2 = 0.025$; $\omega_3 = 48.1169$, $\zeta_3 = 0.038$). The pole – zero map is shown in Figure 8.43. One notes the presence of unstable zeros. The unstable zeros with positive real part correspond to continuous time unstable zeros (non-minimum phase system). Since the model has unstable zeros the pole placement strategy will be used for the controller design.

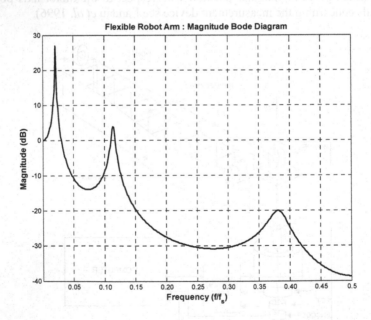

Figure 8.42. Frequency characteristics of the identified model for the 360° flexible robot arm

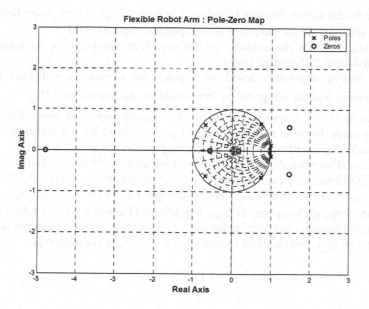

Figure 8.43. Pole – zero map of the identified model for the 360° flexible robot arm

The design of the controller will be done using the iterative pole placement design with shaping of the sensitivity functions, by simultaneous tuning of H_S (respectively H_R) and of auxiliary poles (see Chapter 3, Section 3.6 and (Prochazka and Landau 2003)). The effective computations have been carried on with *ppmaster* (MATLAB®)[9] (Prochazka and Landau 2001). The specifications are the same as in (Langer and Landau 1999):

- Tracking dynamics: discretization of a second-order continuous time system with $\omega_0 = 2.6173 rad/s$ and $\zeta = 0.9$
- Zero steady state error (controller should include an integrator)
- Dominant poles of the closed loop system corresponding to the discretization of a second-order continuous time system with $\omega_0 = 2.6173$ rad/s and $\zeta = 0.8$
- Modulus margin: $\Delta M \geq 0.5$ (-6 dB); delay margin: $\Delta \tau \geq 0.05s$ ($1T_s$)
- Constraints on the input sensitivity function $|S_{up} (q^{-1})|$: < 15 dB at low frequencies (< 4 Hz); <0 dB from 4.5 Hz to 6.5 Hz; < 15 dB from 6.5 Hz to 8 Hz; <10 dB from 8 to 10 Hz ($f_s = 20Hz$)

The low value imposed on $|S_{up}|$ between 4.5 Hz and 6.5 Hz (0.225 to 0.325 f/f_s) is a constraint resulting from the low value of the open loop gain and the uncertainties

---

[9] To be downloaded from *http//:landau-bookic.lag.ensieg.inpg.fr*

upon the model in this frequency region. The bound $|S_{up}|$ at high frequencies will limit the effect of the measurement noise upon the control signal.

The templates for the modulus of the sensitivity functions are represented in Figures 8.44 and 8.45 (dotted lines).

The desired dominant closed loop poles are chosen as indicated in the specifications, and an integrator is introduced in the controller ($H_{S_1} = 1 - q^{-1}$).

Note that it is not necessary to damp the second and third (high frequency) vibration mode because closed loop band pass (defined by the dominant closed loop poles), the disturbances and the tracking model dynamics are all are at low frequencies. Therefore these poles will be kept unchanged by specifying them as poles of the closed loop (see the partial internal model design, Section 3.5.5).

The result of this first design is a controller for which $|S_{yp}|$ and $|S_{up}|$ are far outside the imposed templates at high frequencies (Figures 8.44 and 8.45 – curves A). In such situations auxiliary poles have to be added. The total number of poles which can be specified (without increasing the size of the controller) is

$$n_P = n_A + n_B + d + n_{H_{S_1}} - 1 = 12$$

Six poles have been have been already assigned. It is therefore possible to add auxiliary poles of the form

$$P_F(z^{-1}) = (1 + p_1 z^{-1})^6 \quad ; \quad -0.5 \le p_1 \le -0.05$$

Taking $p_1 = -0.5$ one gets a controller leading to sensitivity functions which are slightly above the templates in two frequency regions (Figures 8.44 and 8.45 – curves B). $|S_{yp}|$ is above the template around *1 Hz* and $|S_{up}|$ is above the template between *4* and *6 Hz*. First, to improve the design, we will consider the introduction of a resonant pole–zero filter H_{S_2} / P_2 in S_{yp}. The continuous time filter, which will serve for the computation of the discrete time filter, is chosen with a resonance frequency $f_0 = 1Hz$ (*6.28 rad/s*). The damping for the denominator is chosen as $\zeta_{den} = 0.8$ (in order that the auxiliary poles which will be introduced be well damped). The desired attenuation is $M_t = -5.5 \ dB$, leading to $\zeta_{num} = 0.424$. The characteristics of the discrete time filters H_{S_2} and P_2 are given in Table 8.9[10]. The sensitivity functions obtained with the new controller are illustrated in Figures 8.44 and 8.45 – curves C. It remains to correct now the frequency characteristics of $|S_{up}|$

---

[10] Since the resonace frequency of the filter is below 0.17 f_S, the filter design can be done with an excellent precision directly in discrete time.

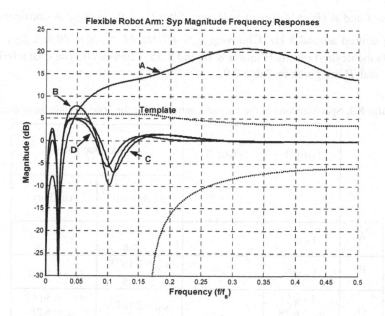

Figure 8.44. Flexible 360° arm. Output sensitivity function ($|S_{yp}|$) for various controllers

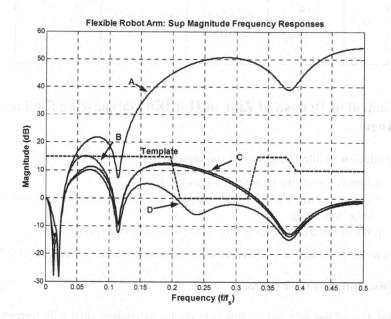

Figure 8.45. Flexible 360° arm. Input sensitivity function ($|S_{up}|$) for various controllers

between *4* and *6 Hz*. A resonant pole-zero filter H_{R_1} / P_3 in $|S_{up}|$ is considered. It will be centred around 5 Hz. Choosing $f_0 = 4.7$ *Hz*, $M_t = -16.4$ *dB* and $\zeta_{den} = 0.9$, one gets the desired results (Figures 8.44 and 8.45 – curves *D*). The characteristics of H_{R_1} and P_3 are given in Table 8.9.

Table 8.9. Specifications of the various controllers for the 360° flexible robot arm

| | $H_S(q^{-1})$ | $H_R(q^{-1})$ | Closed loop poles | |
			Dominant	Auxiliary
A	$1 - q^{-1}$	-	$\omega_0 = 2.1673$ $\zeta = 0.8$	-
B	$1 - q^{-1}$	-	$\omega_0 = 2.1673$ $\zeta = 0.8$	$(1 - 0.5q^{-1})^6$
C	$1 - q^{-1}$ $\omega_0 = 6.28$ $\zeta = 0.424$	-	$\omega_0 = 2.1673$ $\zeta = 0.8$	$(1 - 0.5q^{-1})^6$ $\omega_0 = 6.28$ $\zeta = 0.8$
D	$1 - q^{-1}$ $\omega_0 = 6.28$ $\zeta = 0.424$	$\omega_0 = 29.57$ $\zeta = 0.092$	$\omega_0 = 2.1673$ $\zeta = 0.8$	$(1 - 0.5q^{-1})^6$ $\omega_0 = 6.28$ $\zeta = 0.8$ $\omega_0 = 40.1$ $\zeta = 0.74$

8.7 Control of Deposited Zinc in Hot Dip Galvanizing (Sollac-Florange)

This application is interesting for several reasons:

- It clearly shows the benefit of a good control
- It points out the interest of using digital RST controllers for processes with a long time delay
- It illustrates the concept of the "open loop adaptive control"

A detailed presentation of this application can be found in Fenot *et al.* (1993a).

8.7.1 Description of the Process

The objective of the galvanizing line is to obtain galvanized steel with formability, surface quality and weldability equivalent to uncoated cold rolled steel. The variety of products is very large in terms of deposited zinc thickness and steel strip thickness. The deposited zinc may vary between 50 and 350 g/m^2 (each side) and the strip speed may vary from 30 to 180 m/mn.

The most important part of the process is the hot-dip galvanizing. The principle of the hot-dip galvanizing used at Sollac – Florange is illustrated in Figure 8.46 Preheated steel strip is passed through a bath of liquid zinc and then rises vertically out of the bath through the stripping "air knives" which remove the excess zinc before it solidifies (Figure 8.47). The remaining zinc on the strip surface solidifies before it reaches the rollers which guide the finished product. The effect of air knives depends on the air pressure, the distance between the air knives and the strip, and the speed of the strip. Nonlinear static models have been developed for computing the appropriate pressure, distance and speed for a given value of the desired deposited zinc.

The objective of the control is to assure a good uniformity of the deposited zinc whilst guaranteeing a minimum value of the deposited zinc per unit area. Tight control (*i.e.*, small variance of the controlled variable) will allow a more uniform coating and a reduction of the average quantity of deposited zinc per unit area. As a consequence, in addition to quality improvement, a tight control of the deposited zinc per unit area has an important economic impact since the average consumption for a modern galvanizing line is of the order of 40 tons per day.

The main difficulty for control results from the fact that measurement of the deposited zinc can be made only on the cooled finished strip. The transducers are located more than 100 m after the zinc bath, which results in an important delay between the action of the pressure at the level of air knives and the measurement of its effect on the finished product. The digital RST controller is well suited for the control of such processes with long delay for which PID control cannot be used.

In addition, the delay will depend upon the speed of the steel strip which may vary in a ratio 1 to 3. Furthermore, the dynamic behavior will also depend upon the position of the steel strip with respect to the air knives.

8.7.2 Process Model

The static model of the hot dip galvanizing process can be approached by

$$m = KD\sqrt{\frac{V}{P}} + \xi_m$$

where m is the deposited mass per unit area, K is a constant of proportionality, D is the distance between the air knives and the strip, P is the air pressure and V is the strip speed. ξ_m accounts for unpredictable effects and/or modelling errors. At SOLLAC Sainte Agathe, the control variable is the air pressure.

A linearized model around an operating point (P_0, D_0, V_0) can be obtained using a standard Taylor series expansion for variations of pressure (ΔP), speed (ΔV) and distance (ΔD). It has the form

$$m = KD_0 \sqrt{\frac{V_0}{P_0}} + \alpha\Delta D + \beta\Delta v - \mu\Delta P + \xi_m$$

where P_0, D_0, V_0 are the values of the pressure, distance and speed defining the operating point and $\Delta P, \Delta D, \Delta V$, are the variations of these variables. It can be seen that using the pressure as the control variable one can compensate for the disturbances created by variations of distance and speed as well as by the term ξ_m.

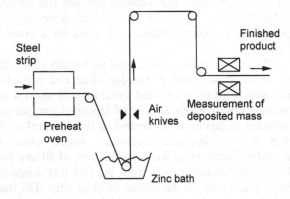

Figure 8.46. Scheme of the hot dip galvanizing process at Sollac - Florange

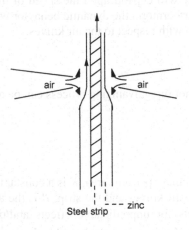

Figure 8.47. Details of the hot dip galvanizing process

The pressure in the air knives is regulated through a pressure loop, which can be approximated by a first-order system. The delay of the process will depend linearly on the steel strip speed. Therefore, for describing the relationship between the variations of pressure and the variations of the deposited mass, one can consider a continuous time dynamic model of the form

$$H(s) = \frac{Ge^{-s\tau}}{1+sT} \quad ; \quad \tau = \frac{L}{V}$$

where L is the distance between the air knives and the transducers and V is the strip speed.

When discretizing this model, the major difficulty comes from the variable time delay. In order to obtain a controller with a fixed number of parameters, the delay of the discrete-time model should remain constant. Therefore, the sampling period is tied to the strip speed (it is an "open loop adaptation") in order to get a discrete time model of constant complexity, using the formula

$$T_s = \frac{\left(\dfrac{L}{V} + \delta\right)}{d} \quad ; \quad d = \text{integer}$$

where d is the discrete-time delay (integer) and δ is an additional small time-delay due to implementation.

The corresponding linearized discrete-time model, which has been identified around various operating points, has the form

$$H(q^{-1}) = \frac{q^{-d}(b_1 q^{-1})}{1 + a_1 q^{-1}}$$

with $d = 7$. The fractional delay (which corresponds to the presence of an additional term $b_2 q^{-2}$) is negligible because of the way the sampling period T_S is selected and this was confirmed by the model identification procedure. However, the parameters of the model, given above, will depend on the distance D and on the speed V.

8.7.3 Model Identification

The plant is formed by the air pressure control loop and the coating process. The control input to the process is the reference of the air pressure control loop, the output of the process is the measured deposited mass per unit area (see Figure 8.48). The identification of the discrete time model of the plant has been done by superposing a PRBS to the reference for the pressure loop, as illustrated in Figure 8.48.

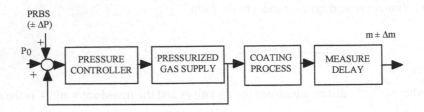

Figure 8.48. Block diagram of the identified plant (input: Δp, output: Δm)

The input used was a PRBS (Pseudo Random Binary Sequence) of a magnitude of $\pm 4\%$ with respect to the static pressure (Po). The PRBS was generated by a shift register with $N = 5$ cells and a clock frequency equal to half of the sampling frequency (length of the sequence: 64); 100 to 160 (average: 128) measurements have been used for the various identifications made in different regions of operation. The choice made for the PRBS allowed at least one full sequence to be sent for each experiment and yielded the largest pulse width (10 Ts) comparable with the rise time of the process (including the time delay). As both sides of the steel strip have to be galvanized, and because of the non symmetric position and physical realization of the two actuators, both "front" side and "back" side models have been identified. Data acquisition has been done using over sampling as indicated in Chapter 7, Section 7.1.3.

A comparative study has shown that for this application the *output error identification method* gives the best results in terms of model validation.

The identification of the process model around various operating points, defined by the strip speed and the distance between the air knives and steel strip, has shown a significant variability of the identified parameters. This has imposed the use of an "open loop adaptive control" in order to assure good performance in all operating points.

8.7.4 Controller Design

RST controllers (with integrator behavior) have been designed for the various operation regions using *tracking and regulation with independent objectives* (see Chapter 3, Section 3.4) since the model always has stable zeros (the sampling period is tied to the steel-strip speed). An important requirement is the achievement of a delay margin of $2\,T_s$.

The polynomial $P(q^{-1})$ defining the closed loop poles will have a maximum degree given by

$$\deg P(q^{-1}) \le \left(n_A + n_B + d + n_{H_S} - 1\right)$$

In this application one gets $\left(n_{H_S} = 1\right)$

$$n_A + n_B + d + n_{H_S} - 1 = 9$$

and therefore one can specify nine poles. The real poles of the model vary from 0.15 to 0.4 over the entire range of operating points. Two poles have been assigned to 0.2 and 0.3 (respectively 0.4 depending on the operating region and the values of the identified pole of the model) and the other seven auxiliary poles have been assigned at 0.1[11]. These auxiliary poles introduce an attenuation at high frequencies (outside the band pass of the system) and improve the robustness of the system (in particular, the delay margin is increased) without affecting the regulation performance at low frequencies.

Figure 8.49a gives the Nyquist plot for the case where only two poles are specified for $P(q^{-1})$ (0.2 and 0.3) and Figure 8.49b allows one to see the effect of the additional auxiliary poles (seven poles at 0.1). One observes an important increase of the delay margin from *10.9 s to 25.7 s* ($T_s = 12s$) since the Nyquist plot does no longer intersect the unit circle at high frequencies. The introduction of the auxiliary poles also produces a significant reduction of the magnitude of the input sensitivity function at high frequencies and, therefore, a reduction of the stress of the actuator in this frequency region. The other robustness margins have satisfactory values with or without the auxiliary poles ($\Delta M > -6dB$ $\Delta G > 2$ $\Delta\phi > 60 \, deg$).

Simulation tests have shown that the closed loop system tolerates delay variations of $\pm T_s$ (one sampling period corresponds to about 15% variations of the delay) and $\pm 50\%$ variations of other model parameters (b_1 and a_1).

Procedures for taking into account the quantization, the saturation and the bumpless transfer from open loop to closed loop (see Sections 8.1.3 through 8.1.5) have been implemented.

Since the speed of the steel strip is measured, it was also possible to make an "open loop compensation" of this disturbance. This has a beneficial effect upon the regulation transient; however, it is the integrator included in the controller which cancels the effect of this disturbance in steady state.

8.7.5 Open Loop Adaptation

To maintain the performance of the control system, the controller is adapted as a function of:

- Speed of the steel strip
- The distance between the air knives and the steel strip

In addition to the adaptation of the sampling frequency as a function of the speed of the steel strip, the operation range of the hot dip galvanization process has been divided into three speed regions (the speed of the strip may vary from 30 m/min to

---

[11] The choice made for the dominant and auxiliary poles is close to the choice made in the example of the internal model control of systems with time delay discussed in Chapter 3, Section 3.5.7.

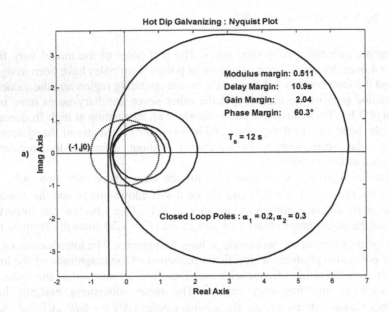

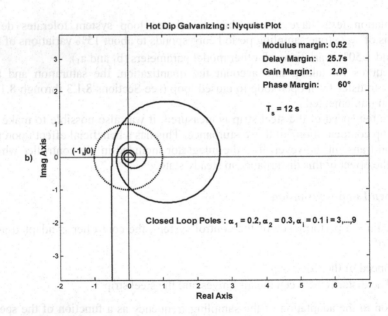

Figure 8.49a,b. Nyquist plot for hot dip galvanizing (a_1=-0.2, d=7): **a** Closed loop poles: 0.2; 0.3; **b** Closed loop poles. 0.2; 0.3; 7×0.1

180 m/min) and in three regions for the distance between the air knives and the steel strip which lead to a total of nine regions of operation, each one defined by a range of speeds and a range of distances. An identification of the plant model has been done for various operating points within each region of operation. A controller design on the basis of a model estimated for the central operating point within a region has been designed, and tested in simulation on the various identified plant models for the corresponding region. The controllers have been stored in a table. The controllers are switched when the system moves in a new region of operation. A bumpless transfer from one controller to another is implemented. In addition hysteresis is used to avoid unnecessary switching when one operates very close to the regions boundaries.

8.7.6 Results

Figure 8.50 shows one of the typical results obtained when one of the sides is under digital regulation and the other side is under computer aided manual control (the operator has on display a moving short time history of the deposited zinc and applied pressure).

The analysis of this curve points out two relevant facts when closed loop digital control is used:

- A smaller dispersion of the coating thickness (the standard deviation on the side where the feedback control was applied is about 30 % smaller than on the side under manual control). *This assures a better finished product quality* (extremely important for the use in the automotive industry, for example).
- The average quantity of deposited zinc is reduced by 3% still guaranteeing the specifications for minimum zinc deposit. *This corresponds to a significant reduction of the zinc consumption and to an important economic impact.*

Table 8.10 summarize the results obtained in different operation regions.

Table 8.10. Performances of the digital control of deposited zinc in the hot dip galvanizing (Sollac-Florange)

Distance (mm)	Speed (m/min)	Digital control			Computer aided manual control		
		Mean value %	Stan dev. %	Out of reference ± 10 %	Mean value %	Stan dev. %	Out of reference ± 10 %
10	60-90	100	4.7	4.3	102.5	6.7	10.1
10	95	100	3.3	0.7	103	4.5	5.9
10	85-100	100	2.9	0	99	5.2	5.9
10	110-116	100	4.3	2.5	105	4.3	3.9
10	100-117	100	5.1	6.4	100	6.3	8.4
15	70	100	1.5		102	2	

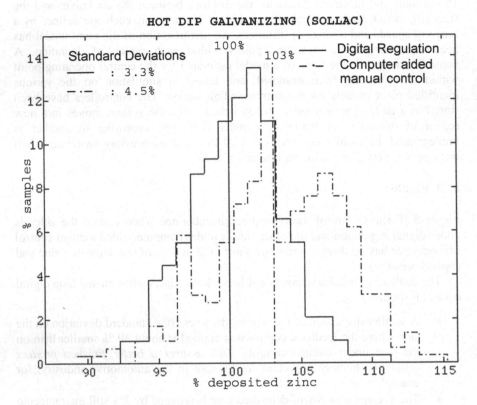

Figure 8.50. Typical performances for the digital control of deposited zinc in the hot dip galvanizing (SOLLAC-Florange)

8.8 Concluding Remarks

Digital controllers can be implemented on a large variety of hardware. This implementation, however, has to be done by paying attention to a number of issues listed here:

- Specification of the desired performances taking into account the actuator power and band pass, the time response of the plant in open loop and the desired robustness margins.
- Taking into account the computation delay.
- Taking into account the characteristics of the digital to analog converter.
- Introducing anti-saturation schemes.
- Implementation of a bumpless transfer procedure from open loop operation to closed loop operation.

For applications where significant variations of the dynamic model of the plant occur with the change of the operating points, it is recommended to proceed as follows:

- To identify the model of the plant in the various regions of operation.
- To design a controller for each region of operation assuring desired performance and robustness.
- To store the parameters of the controllers in a table.
- To change the parameters of the controllers when the system moves in a new region of operation by downloading the values of the controller parameters stored in the table.

This procedure is called "open loop adaptation".

In continuous production processes it is very useful to measure the quality of the control from the histogram of the regulated variables.

The application examples presented in this chapter (there are many others) have illustrated how the methodology for system identification and control design has to be used in order to implement a high performance control system.

8.9 Notes and References

For the implementation of digital controllers see also the C++ code (on the book website) as well as:

Åström K.J., Wittenmark B. (1997) Computer Controlled Systems - Theory and Design, 3rdedition, Prentice-Hall, Englewood Cliffs, N.J.
Franklin G.F., Powell J.D., Workman M.L. (1998): Digital Control of Dynamic Systems , 3rd edition, Addison Wesley, Reading, Mass.

For implementation of digital controllers on PLC see also:

Adaptech (2001a) Guidelines for RST controller implementation, St. Martin d'Hères, France.

For the implementation of digital controllers on a PC see:

Vieillard J.P. (1991) Machine automatique pour la fabrication de cables torsadés téléphoniques, La Lettre d'Adaptech, no. 2, pp. 1-2, Adaptech, St. Martin d'Hères, France.

For the digital control of DC motors see:

Landau I.D., Rolland F. (1993) Identification and digital control of electrical drives, Control Eng. Practice, vol. 1, no. 3, 539-546.
Adaptech (2004) Wintrack, Software for Data acquisition and real time RST digital control, St. Martin d'Hères, France.

For the control of deposited zinc in hot dip galvanizing see:

Fenot C., Rolland F., Vigneron G., Landau I.D. (1993a) Open loop adaptive feedback control of deposited zinc in hot dip galvanizing, Control Engineering Practice, vol. 1, no. 5.

Fenot C., Vigneron G., Rolland F., Landau I.D. (1993b) Régulation d'épaisseur de dépôt de zinc à Sollac, Florange, Revue Générale d'Electricité, no. 11, pp. 25-30, Dec.

For the control of the 360° flexible robot arm see:

Landau I.D., Langer J., Rey D., Barnier J. (1996) Robust control of a 360° flexible arm using the combined pole placement / sensitivity function shaping method, IEEE Trans. on Control Systems Technology, vol. 4, no. 4, July.

Langer J., Landau I.D. (1999) Combined pole placement / sensitivity function shaping method using convex optimization criteria, Automatica, vol. 55, no. 6, pp. 1111-1120.

Prochazka H., Landau I.D. (2003) Pole placement with sensitivity function shaping using second-order digital notch filters, Automatica, Vol. 59, n.6, pp. 1103-1107.

For the evaluation of the control performance see:

Rolland F., Landau I.D. (1991) Pour mieux réguler le PC va vous aider, Mesures, pp. 71-73, December,.

For an introduction to adaptive control see:

Landau I.D. (1986) La Commande Adaptative: Un Tour Guidé, Commande adaptative - Aspects pratiques et théoriques, (Landau, Dugard- Editors), Masson, Paris.

Landau I.D. (1979): Adaptive control – the model reference approach, Dekker, N.Y.

For a "state of the art" in adaptive control see:

Åström K.J., Wittenmark B. (1995) Adaptive Control, 2nd edition, Addison Wesley, Reaching, Mass.

Landau I.D. (1993): Evolution of Adaptive Control, A.S.M.E. Transactions, Journal D.S.M.C., vol. 115, no. 2, pp. 381-391, June.

Landau I.D., Lozano R., M'Saad M. (1997) Adaptive Control, Springer, London, UK.

For adaptive disturbance rejection techniques see:

Landau I.D., Constantinescu A., Rey D. (2005) Adaptive Narrow Band Disturbance Rejection Applied to an Active Suspension – An Internal Model Principle Approach, Automatica, vol.61, no.4, pp. 563-574.

9

Identification in Closed Loop

The techniques for model identification of plants operating in closed loop together with the corresponding validation techniques are presented in this chapter. The possibilities offered by the identification in closed loop for obtaining improved models for controller re-design leading to better performance are illustrated in the final part of the chapter.

9.1 Introduction

There exist situations in practice where the plant model should be identified in closed loop (i.e. in the presence of a controller):

- The first case is related to the plant dynamics characteristics. If the plant has an integrator or it is unstable, or an important drift of the operating point occurs, it is very difficult and, in some situations, very dangerous to operate it in open loop.
- The second case is related to systems where a controller is already operating and where it is neither possible nor recommended to open the loop in order to acquire data for the system identification.
- The third case corresponds to situation where the controller has been designed based on a plant model identified in open loop and where, for the purpose of improving the achieved performance, it is necessary to carry on model identification in closed loop for the re-design of the controller.

It is this last aspect that has gained importance in the last years since identification in closed loop on one hand provides in general better models for design than open loop identification and on the other hand allows tuning of a controller without opening the loop.

The objective of identification in closed loop is to obtain a plant model describing as precisely as possible the behavior of the real closed loop system for a given controller. It is also expected that this model identified in closed loop will allow redesigning of the controller in order to improve the performance of the real-time control system. In other words, (assuming that the real system is linear around an operating point), the objective of system identification in closed loop is to search for a plant model that in feedback with the controller operating on the true plant will lead to a closed loop transfer function (sensitivity function) that is as close as possible to that of the real closed loop system.

It has been shown in Landau and Karimi (1997b), Landau and Karimi (2001b), as well as in many other references, that identification in closed loop, provided that appropriate identification algorithms are used, leads in general to better models for controller design. It is effectively possible to improve the quality of the identified model in the frequency regions that are critical for controller design. These frequency regions are those where the Nyquist plot of the open loop transfer function is close to the critical point [-1, j0].

In order to understand the potential of the identification in closed loop as well as the difficulties which can be encountered, let us consider the case of plant model identification in closed loop where the external excitation is added to the controller output (see Figure 9.1a). Figure 9.1b shows an equivalent scheme that emphasizes the transfer function between the external excitation r_u and the plant input u, as well as the effect of the measurement noise upon the plant input. Assume that the external excitation is a PRBS that has almost constant frequency spectrum from 0 to $0.5f_s$.

One observes that the effective plant input corresponds to the external excitation filtered by the output sensitivity function S_{yp} whose magnitude has a maximum in the frequency regions close to the critical point [-1, j0] (see Chapter 2, Section 2.6.2). Therefore the frequency spectrum of the effective input applied to the plant will be enhanced in these frequency zones. As a consequence, the quality of the identified model in these critical regions for stability and performance will be improved. Unfortunately, in the meantime, the feedback introduces a correlation between the measurement noise and the plant input. This leads to an important *bias* on the estimated parameters if one would like to identify the plant model with open loop techniques based on uncorrelation (see Chapter 6). One may expect that the open loop identification techniques based on the whitening of the prediction error will still provide good results in closed loop operation. However, as a consequence of feedback, interdependence between the noise model and the plant model occurs and the parameter estimates will also be biased (Karimi and Landau 1998).

Therefore, for a good identification in closed loop operation one needs identification methods that take advantage of the "improved" characteristics of the effective excitation signal applied to the plant input but which are not affected by the noise in the context of feedback. An efficient solution for this problem is provided by the "closed loop output error" methods (CLOE) that will be presented in Section 9.2.

The schemes for data acquisition in closed loop operation have been already presented in Chapter 7 (Section 7.1).

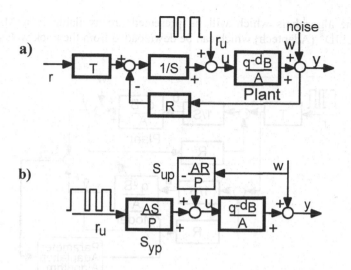

Figure 9.1a,b. Identification in closed loop: **a** excitation added to the control output; **b** equivalent representation

9.2 Closed Loop Output Error Identification Methods

9.2.1 The Principle

The principle of closed loop identification methods is illustrated in Figure 9.2. The upper part represents the true closed loop system and the lower part represents an adjustable predictor of the closed loop. This closed loop predictor uses a controller identical to the one used in the real time system.

The prediction error between the output of the real time closed loop system and the closed loop predictor (closed loop output error) is a measure of the difference between the true plant model and the estimated one. This error can be used to adapt the estimated plant model such that the closed loop prediction error is minimized (in the sense of a certain criterion). In other words the objective of the identification in closed loop is to find the best plant model which minimizes the prediction error between the measured output of the true closed loop system and the predicted closed loop output.

In Figure 9.2 the external excitation is superposed to the reference. However it can be superposed to the controller output as shown in Figure 9.3 (the reference is not represented). In both cases the same parameter adaptation algorithm is used, although the characteristics of the identified model will be slightly different since in the first case (Figure 9.2) one tries to approximate the sensitivity function S_{yr} while in the second case one tries to approximate the sensitivity function S_{yv}. Use of these methods requires knowledge of the controller.

All the algorithms which will be presented are available in a MATLAB®
toolbox CLID® (Adaptech) which can be downloaded from the book web site.

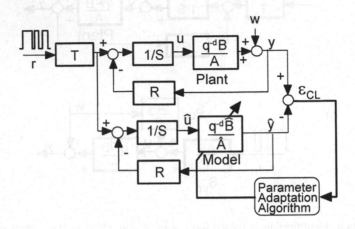

Figure 9.2. Closed loop output error identification method (excitation superposed to the reference)

9.2.2 The CLOE, F-CLOE and AF-CLOE Methods

The algorithms will be introduced using, as an example, a first order type plant model and then the general formulas will be given.

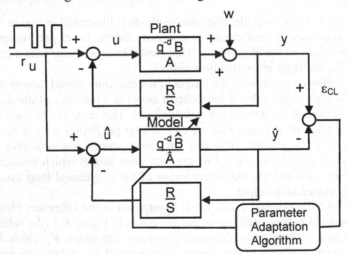

Figure 9.3. Closed loop output error identification method (excitation added to the controller output)

Closed Loop Output Error (CLOE)
The closed loop system is described by:

$$y(t+1) = -a_1 y(t) + b_1 u(t) = \theta^T \phi_0(t) \tag{9.2.1}$$

$$\theta^T = [a_1, b_1]; \quad \phi_0(t)^T = [-y(t), u(t)] \tag{9.2.2}$$

$$u(t) = -\frac{R(q^{-1})}{S(q^{-1})} y(t) + r_u(t) \tag{9.2.3}$$

The plant model is described by Equation 9.2.1 and the control u is given by Equation 9.2.3. $r_u(t)$ is the external excitation superposed to the controller output (see Figure 9.3).

In the case where the external excitation is superposed to the reference, the signal $r_u(t)$ is replaced by (see Figure 9.2)

$$r_u(t) = \frac{T(q^{-1})}{S(q^{-1})} r(t) \tag{9.2.4}$$

The adjustable closed loop predictor is described by

$$\hat{y}°(t+1) = -\hat{a}_1(t)\hat{y}(t) + \hat{b}_1(t)\hat{u}(t) = \hat{\theta}(t)^T \phi(t) \tag{9.2.5}$$

$$\hat{y}(t+1) = -\hat{a}_1(t+1)\hat{y}(t) + \hat{b}_1(t+1)\hat{u}(t) = \hat{\theta}(t+1)^T \phi(t) \tag{9.2.6}$$

$$\hat{\theta}(t)^T = [\hat{a}_1(t), \hat{b}_1(t)]; \quad \phi(t)^T = [-\hat{y}(t), \hat{u}(t)] \tag{9.2.7}$$

$$\hat{u}(t) = -\frac{R(q^{-1})}{S(q^{-1})} \hat{y}(t) + r_u(t) \tag{9.2.8}$$

where $\hat{y}°(t+1)$ and $\hat{y}(t+1)$ represent the *a priori* and the *a posteriori* outputs of the closed loop predictor. $\hat{u}(t)$ is the control signal delivered by the controller (which is identical to the one used on the true system) using the *a posteriori* output of the predictor and not the measured outputs (compare Equation 9.2.8 with Equation 9.2.3).

The closed loop prediction error is given by

$$\varepsilon_{CL}°(t+1) = y(t+1) - \hat{y}°(t+1) \quad a\ priori \tag{9.2.9}$$

$$\varepsilon_{CL}(t+1) = y(t+1) - \hat{y}(t+1) \qquad a\ posteriori \qquad (9.2.10)$$

The parameter adaption algorithm to be used is similar to those used for open loop identification:

$$\hat{\theta}(t+1) = \hat{\theta}(t) + F(t)\Phi(t)\varepsilon_{CL}(t+1) \qquad (9.2.11)$$

$$F(t+1)^{-1} = \lambda_1(t)F(t)^{-1} + \lambda_2(t)\Phi(t)\Phi(t)^T \ ;\ 0 < \lambda_1(t) \le 1; 0 \le \lambda_2(t) < 2 \quad (9.2.12)$$

$$\varepsilon_{CL}(t+1) = \frac{\varepsilon_{CL}^0(t+1)}{1 + \Phi(t)^T F(t)\Phi(t)} \qquad (9.2.13)$$

$$\Phi(t) = \phi(t) \qquad (9.2.14)$$

The fundamental differences with respect to the open loop output error are that the adjustable predictor and the regressor vector are different.

Note: If $S(q^{-1}) = 1$ and $R(q^{-1}) = 0$ one operates in open loop and the open loop output error algorithm is obtained as a particular case.

In the general case Equation 9.2.1 is replaced by

$$y(t+1) = -A^*(q^{-1})y(t) + B^*(q^{-1})u(t-d) + A(q^{-1})w(t+1). \qquad (9.2.15)$$

where w represents the noise effect. The noise w is supposed to be centered, of finite power and independent with respect to the external excitation r_u. Equations 9.2.5 and 9.2.6 keep the same form with

$$\hat{\theta}(t)^T = [\hat{a}_1(t), ..., \hat{a}_{n_A}(t), \hat{b}_1(t), ..., \hat{b}_{n_B}(t)] \qquad (9.2.16)$$

$$\phi(t)^T = [-\hat{y}(t), ..., -\hat{y}(t - n_A + 1), \hat{u}(t-d), ..., \hat{u}(t - n_B + 1 - d)] \qquad (9.2.17)$$

and the parameter adaptation algorithm is the one given by Equations 9.2.11 through 9.2.14.

The convergence of this algorithm (CLOE), in the absence of noise, is subject to a sufficient condition: the transfer function

$$\frac{S(z^{-1})}{P(z^{-1})} - \frac{\lambda_2}{2} \ ;\ 2 > \lambda_2 \ge \max \lambda_2(t) \qquad (9.2.18)$$

should be *strictly positive real*, where the polynomial

$$P(z^{-1}) = A(z^{-1})S(z^{-1}) + z^{-d}B(z^{-1})R(z^{-1}). \tag{9.2.19}$$

defines the poles of the closed loop.

It has been shown that, under the sufficient condition of Equation 9.2.18 , the closed loop output error (CLOE) gives asymptotically unbiased parameter estimates in the presence of noise (independent with respect to the external excitation) if the estimation model and the true model have the same structure.

Filtered Closed Loop Output Error (F-CLOE)

In order to relax the condition of Equation 9.2.18 one can filter the vector $\phi(t)$ through $S(q^{-1})/\hat{P}(q^{-1})$ where $\hat{P}(q^{-1})$ is an estimation of the polynomial defining the poles of the closed loop. If an estimated plant model $q^{-d}\hat{B}(q^{-1})/\hat{A}(q^{-1})$ is available, one can compute

$$\hat{P}(q^{-1}) = \hat{A}(q^{-1})S(q^{-1}) + q^{-d}\hat{B}(q^{-1})R(q^{-1}). \tag{9.2.20}$$

and in the parameter adaptation algorithm one uses

$$\Phi(t) = \frac{S(q^{-1})}{\hat{P}(q^{-1})}\phi(t). \tag{9.2.21}$$

We get in this way the F-CLOE method. In this case, the transfer function of Equation 9.2.18 is replaced by

$$\frac{\hat{P}(z^{-1})}{P(z^{-1})} - \frac{\lambda_2}{2} \quad ; \quad 2 > \lambda_2 \geq \max\lambda_2(t) \tag{9.2.22}$$

which should be *strictly positive real*. This condition is clearly easier to satisfy.

A first estimation of the plant model is necessary in order to compute the filter. This estimation can be provided by AF-CLOE or X-CLOE methods (described next) as well as by identification in closed loop, using an open loop type identification method or using the model identified in open loop (if it is available).

Adaptive Filtered Closed Loop Output Error (AF-CLOE)

One may consider filtering $\phi(t)$ through a time-varying filter $S(q^{-1})/\hat{P}(t,q^{-1})$ where $\hat{P}(t,q^{-1})$ is the estimation at time t of the polynomial defining the closed loop poles:

$$\hat{P}(t,q^{-1}) = \hat{A}(t,q^{-1})S(q^{-1}) + q^{-d}\hat{B}(t,q^{-1})R(q^{-1}) \tag{9.2.23}$$

computed from the estimations of $\hat{A}(t,q^{-1})$ and $\hat{B}(t,q^{-1})$ available at time t. This leads to the AF-CLOE algorithm.

In this case the vector $\Phi(t)$ given in Equation 9.2.21 is replaced by

$$\Phi(t) = \frac{S(q^{-1})}{\hat{P}(t,q^{-1})}\phi(t) \tag{9.2.24}$$

9.2.3 Extended Closed Loop Output Error (X-CLOE)

If it is assumed that the disturbance acting on the plant output can be represented by an ARMAX type model, Equations 9.2.1 and 9.2.3 become:

$$y(t+1) = -A^*(q^{-1})y(t) + B^*(q^{-1})u(t-d) + C^*(q^{-1})e(t) + e(t+1) \tag{9.2.25}$$

$$u(t) = -\frac{R(q^{-1})}{S(q^{-1})}y(t) + r_u(t) \tag{9.2.26}$$

where $e(t)$ is a Gaussian white noise and

$$C(q^{-1}) = 1 + q^{-1}C^*(q^{-1}) = 1 + c_1 q^{-1} + \dots + c_{n_C} q^{-n_C}$$

is an asymptotically stable polynomial.

X-CLOE is an identification method based on the whitening of the closed loop prediction error. An adjustable closed loop predictor of the following form is used:

$$\hat{y}^0(t+1) = -\hat{A}^*(t,q^{-1}) + \hat{B}^*(t,q^{-1})\hat{u}(t-d) + \hat{H}^*(t,q^{-1})\frac{\varepsilon_{CL}(t)}{S(q^{-1})} \tag{9.2.27}$$

$$= \hat{\theta}_e(t)^T \phi_e(t)$$

where

$$\hat{u}(t) = -\frac{R(q^{-1})}{S(q^{-1})}\hat{y}(t) + r_u(t) \tag{9.2.28}$$

$$\hat{\theta}_e(t)^T = [\hat{a}_1(t),\dots,\hat{a}_{n_A}(t),\hat{b}_1(t),\dots,\hat{b}_{n_B}(t),\hat{h}_1^*(t),\dots,\hat{h}_{n_H}^*(t)] \tag{9.2.29}$$

$$\phi_e(t)^T = [\phi(t)^T, \varepsilon_{CLf}(t),\dots,\varepsilon_{CLf}(t-n_H+1)] \tag{9.2.30}$$

$$\varepsilon_{CLf}(t) = \frac{\varepsilon_{CL}(t)}{S(q^{-1})} \qquad (9.2.31)$$

$$\varepsilon_{CL}^0(t+1) = y(t+1) - \hat{y}^0(t+1) \qquad (9.2.32)$$

$$\varepsilon_{CL}(t+1) = y(t+1) - \hat{\theta}_e(t+1)^T \phi_e(t) \qquad (9.2.33)$$

For

$$\hat{A}^*(t,q^{-1}) = A^*(q^{-1}), \hat{B}^*(t,q^{-1}) = B^*(q^{-1}) \text{ and } \hat{H}^*(t,q^{-1}) = H^*(q^{-1})$$

where[1]

$$H(q^{-1}) = 1 + q^{-1}H^*(q^{-1}) = 1 + C(q^{-1})S(q^{-1}) - P(q^{-1}) \qquad (9.2.34)$$

the closed loop prediction error is white noise (Landau and Karimi 2001b).

The parameter adaptation algorithm is given by Equations 9.2.11 through 9.2.14 where:

$$\hat{\theta}(t) = \hat{\theta}_e(t) \quad ; \quad \Phi(t) = \phi_e(t)$$

In the deterministic case (without noise) there is no any condition to be satisfied for stability. In the presence of an ARMA disturbance one obtains asymptotically unbiased parameter estimates under the sufficient condition that

$$\frac{1}{C(z^{-1})} - \frac{\lambda_2}{2} \quad ; \quad 2 > \lambda_2 \geq \max \lambda_2(t) \qquad (9.2.35)$$

is a *strictly positive real* transfer function (note that it is the same condition as in open loop identification using extended least squares or output error with extended prediction model).

9.2.4 Identification in Closed Loop of Systems Containing an Integrator

The identification methods presented previously allow a better estimation of a pure integrator than the open loop identification methods. However, if the existence of a pure integrator is known, it is better to take this into account. Two procedures can be used, and both require a modification of the controller used in the closed loop

---

[1] From Equation 9.2.34 it results that $n_H = \max(n_C + n_S, n_P)$.

adjustable predictor in order to preserve the input/output behaviour of the closed loop:

1. The plant input is replaced by its integral and the output remains unchanged. In this case the equations describing the adjustable predictor become

$$\hat{y}'^{0}(t+1) = -\hat{A}_1^*(t,q^{-1})\hat{y}'(t) + \hat{B}^*(t,q^{-1})\hat{u}'(t-d) \tag{9.2.36}$$

$$\hat{u}'(t) = \frac{1}{1-q^{-1}}\left[-\frac{R(q^{-1})}{S(q^{-1})}\hat{y}'(t) + r_u(t)\right] \tag{9.2.37}$$

with

$$A(q^{-1}) = (1-q^{-1})[1+q^{-1}A_1^*(q^{-1})] = (1-q^{-1})A_1(q^{-1}) \tag{9.2.38}$$

2. The plant output is replaced by its variations and the input remains unchaged. In this case the plant output in closed loop is

$$y'(t) = y(t) - y(t-1) = (1-q^{-1})y(t) \tag{9.2.39}$$

and the corresponding equations of the adjustable predictor are

$$\hat{y}'^{0}(t+1) = -\hat{A}_1^*(t,q^{-1})\hat{y}'(t) + \hat{B}^*(t,q^{-1})\hat{u}'(t-d) \tag{9.2.40}$$

$$\hat{u}'(t-d) = -\frac{R(q^{-1})}{S(q^{-1})(1-q^{-1})}y'(t) + r_u(t) \tag{9.2.41}$$

The closed loop prediction error is in this case

$$\varepsilon_{CL}^{0}(t+1) = y'(t) - \hat{y}'^{0}(t+1) \tag{9.2.42}$$

Figure 9.4a,b illustrates the modifications for the case when the external excitation is applied to the controller output.

9.2.5 Model Validation in Closed Loop

As in open loop identification, it is the model validation that will tell us on one hand if the identified model is acceptable and on the other hand it will allow us to select the best model among the models provided by various identification methods.

The objective of the model validation in closed loop is to find what plant model combined with the current controller provides the best prediction of the behavior of the closed loop system. The model validation in closed loop will depend upon the controller which will be used.

Four validation procedures can be defined:

1. Statistical validation tests on the closed loop output error (uncorrelation test between $\varepsilon_{CL}(t+1)$ and $\phi(t)$)
2. Closeness of the computed and identified poles of the closed loop system
3. Closeness of the computed and identified sensitivity functions of the closed loop system
4. Time response validation (comparison of time response of the real closed loop system and of the closed loop predictor)

Statistical Validation

The statistical validation follows the same principles as for open loop identification. However in this case one considers the residual prediction error between the output of the plant operating in closed loop and the output of the closed loop predictor. An uncorrelation test will be used.

Using the schemes shown in Figure 9.2 (or Figure 9.3) where the predictor is given by Equations 9.2.5 through 9.2.8, one computes with the identified values of the parameters:

- The correlations between the residual closed loop output error $\varepsilon_{CL}(t+1)$ and the components of the predictor regressor vector $\phi(t)$ ($\hat{y}(t)$, $\hat{u}(t-d)$ and their delayed values)
- The covariance of the residual closed loop output error

This type of test is motivated on one hand by the fact that uncorrelation between the observations (the components of $\phi(t)$ filtered or not) and the closed loop prediction error leads to unbiased parameter estimates and on the other hand this uncorrelation implies the uncorrelation between the closed loop output error and the external excitation. This means that the residual prediction error does not contain any information which depends upon the external excitation and therefore all the correlations between the external excitation and the output of the closed loop system are captured by the closed loop predictor.

One defines

$$R(i) = \frac{1}{N}\sum_{t=1}^{N}\varepsilon_{CL}(t)\hat{y}(t-i) \qquad (9.2.43)$$

and one computes

$$RN(i) = \frac{R(i)}{\left[\left(\frac{1}{N}\sum_{t=1}^{N}\hat{y}^2(t)\right)\left(\frac{1}{N}\sum_{t=1}^{N}\varepsilon_{CL}^2(t)\right)\right]^{1/2}}; \quad i=0,1,2,\ldots,\max(n_A,n_B+d) \qquad (9.2.44)$$

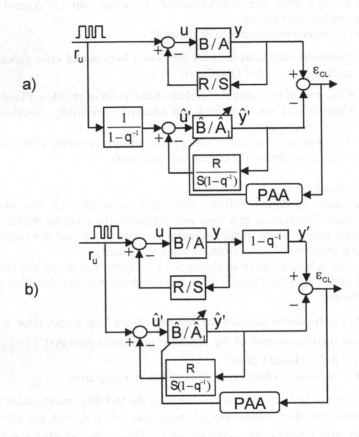

Figure 9.4a,b. Identification in closed loop of systems containing a pure integrator ($A(q^{-1}) = (1-q^{-1})A_1(q^{-1})$): **a** plant input replaced by its integral; **b** plant output replaced by its variations

As a validation test one uses the criterion (Landau *et al.* 1997)

$$|RN(i)| \le \frac{\alpha}{\sqrt{N}} \; ; i = 1,2,...,i_{max}$$

where α is the confidence interval (a typical value is 2.17 which corresponds to 97% level of confidence) and N is the number of data (see Chapter 6).

In many practical situations, from one set of input/output data several models can be identified using various methods (even a model identified in open loop may be available). A comparative validation is necessary in order to select the best model. The comparison indicators are the variance of the residual closed loop prediction error and $\max|RN(i)|$ for each model.

Pole Closeness Validation
If the model identified in closed loop in feedback with the controller used during identification allows one to construct a good predictor for the real system, this implies that the poles of the closed loop system and of the closed loop predictor are close (assuming that a persistent excitation has been applied for identification). As a consequence, the closeness of the closed loop predictor poles (which can be computed) and those of the real closed loop system (which can be identified by an open loop type identification between the external excitation and the output) will give an indication of the quality of the identified model.

The closeness of the two sets of poles can be judged by a visual examination of the poles chart. It is however possible to quantify this closeness by evaluating the distance between the closed loop transfer function of the real system and of the predictor (see next).

Sensitivity Functions Closeness Validation
From the same arguments as above it results that if the identified model is good, the sensitivity functions of the closed loop predictor (which can be computed) are close to the sensitivity functions of the real system (which can be identified by an open loop type identification between the external excitation and the output).

To some extent the closeness of the sensitivity functions can be assessed by visual inspection. Moreover it is possible to quantify rigorously the distance between two transfer functions by computing the Vinnicombe distance (see Appendix D).

Extensive simulations and a large number of experimental results have shown that the statistical tests and the poles or sensitivity functions closeness give coherent results and allow a clear comparison between several models (Landau and Karimi 1997b).

Time Domain Validation
For the validation in the time domain, one compares the time responses of the closed loop system and of the closed loop predictor. Unfortunately in practice it is in general not easy to compare accurately several models using this technique. In fact a good validation by poles or sensitivity functions closeness will imply a good superposition of the time domain responses while the reciprocal is not always true.

9.3 Other Methods for Identification in Closed Loop

1. Using Open Loop Identification Techniques on Filtered Data
The procedures try to approximate the methods presented in Section 9.2. Identification algorithms for open loop identification will process the plant input/output (u,y) data generated through feedback, but in general these data will be filtered. The filters are estimations of various sensitivity functions. These methods require the knowledge of the controller (Landau and Karimi 1997b, Landau and Karimi 2001b).

2. Direct identification
One tries to identify the plant model from plant input/output data ignoring the effect of the feedeback (the external excitation being added either to the reference or to the controller output). Open loop identification algorithms are used and in particular those assuming the structure S3 for the noise (ARMAX), like extended least squares, output error with extended prediction model, *etc*. This approach can be viewed as a particular case of the previous one.

The quality of the results is very variable. Despite the fact that the knowledge of the controller is not explicitly required, the results will strongly depend on the characteristics of the controller and the level of noise. As a general rule a "soft" controller which essentially tries to stabilize the operating point has to be used.

3. Multi steps identification
These methods have as objective on one hand to eliminate the need for the knowledge of the controller and on the other hand to eliminate the effect of the measurement noise (which is critical when using open loop type identification methods) (Van den Hof and Shrama 1993).

To understand this approach it is necessary to refer to Figure 9.1. The implementation of this approach is done in several steps:

1. The sensitivity function $S_{yp} = AS/P$ between the external excitation $r_u(t)$ and and the plant input $u(t)$ is identified.

2. One filters the external excitation $r_u(t)$ through the estimated sensitivity function $\hat{S}_{yp}$ and one gets the instrumental variable $\hat{u}(t)$ which is not correlated with the noise

$$\hat{u}(t) = \hat{S}_{yp}(q^{-1})r_u(t)$$

3. One uses the open loop output error identification method on $\hat{u}(t), y(t)$.

9.4 Identification in Closed Loop: A Simulated Example

The model used for simulation is

$$A(q^{-1})y(t) = B(q^{-1})u(t) + C(q^{-1})e(t)$$

with

$$A(q^{-1}) = 1 - 1.5q^{-1} + 0.7q^{-2}; \quad B(q^{-1}) = q^{-1} + 0.5q^{-2}; \quad d = 0$$
$$C(q^{-1}) = 1 + 1.6q^{-1} + 0.9q^{-2}$$

where $e(t)$ is a gaussian white noise sequence (zero mean, finite variance).

The controller used (stored in the text file SIMUBF_RST.reg[2]) is defined by

$$R(q^{-1}) = 0.8659 - 1.2763q^{-1} + 0.5204q^{-2}$$
$$S(q^{-1}) = 1 - 0.6283q^{-1} - 0.3717q^{-2}$$
$$T(q^{-1}) = 0.11$$

The excitation signal superposed to the reference is a PRBS generated by a shift register with N=7 and a frequency divider equal to 2 ($f_{PRBS}=0.5f_s$). A total of 1024 samples are used. A high noise level has been voluntarily chosen (approximately 20% noise to signal ratio).

With 1024 data it is not possible to reach the asymptotic results; however the relative performance of the various methods are clearly emphasized. The external excitation, the plant input and the plant output are stored in the file SIMUBF4.ACQ[2] and represented in Figure 9.5.

Table 9.1 gives the identified parameters and the statistical validation results for the models obtained with various methods. The F-CLOE algorithm has provided the best model in terms of validation tests (the model identified with AF-CLOE has been used for computing the data filter).

The last row of Table 9.1 gives the results with the best model (in terms of validation in closed loop) obtained by a direct open loop type identification using the plant input and output (and ignoring the feedback). The extended least squares method provided the best results. One can observe that this model does not pass the closed loop validation tests and the maximum value of the normalized cross-correlations is much larger than the values obtained with the methods dedicated to the identification in closed loop.

Table 9.1. Models identified in closed loop

Method	a_1	a_2	b_1	b_2	Closed loop error variance R(0)	Normalized cross-correlations valid. threshhold: 0.068 |RN(max)|
Nominal model	-1.5	0.7	1	0.5		
AF-CLOE	-1.47	0.6708	0.954	0.5218	2.63e-4	0.0097
CLOE	-1.4761	0.667	0.959	0.4858	0.00187	0.0284
F-CLOE	-1.4691	0.6703	0.9593	0.5152	4.51e-5	0.0085
X-CLOE	-1.4695	0.6524	0.942	0.3849	2.42e-4	0.083
OL type identification (ELS)	-1.423	0.6216	0.9408	0.4898	0.0025	0.0749

---

[2] Available from the web site: *http//:landau-bookic.lag.ensieg.inpg.fr*

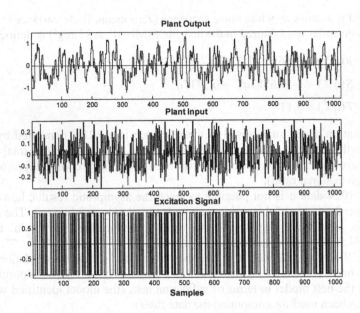

Figure 9.5. The external excitation, the plant input and output (file SIMUBF4.ACQ) for identification in closed loop (simulated example)

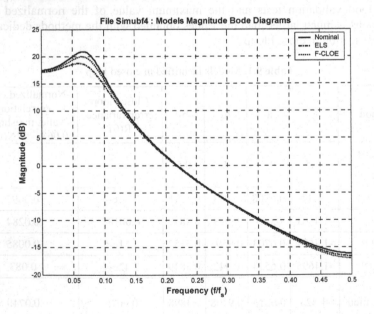

Figure 9.6. Frequency characteristics of various identified models

Figure 9.6 illustrates the frequency characteristics of the nominal model, the model identified with F-CLOE and the model identified with the extended least squares (ELS).

It is the model identified by F-CLOE which is the closest to the nominal model (it is also the one which gives the best results for statistical validation in terms of uncorrelation).

Augmenting the number of data, the models obtained with F-CLOE (or CLOE, AF-CLOE) will approach more and more the nominal model. See Landau and Karimi (1997b) and Landau *et al.* (1997).

9.5 Identification in Closed Loop and Controller Re-Design (the Flexible Transmission)[3]

We will illustrate next the advantage of identification in closed loop for the improvement of the achieved control performances for the case of a flexible transmission (see also Chapters 7 and 8). The system has been described in Chapter 7, Section 7.5.2 and in Chapter 8, Section 8.5. The block diagram of the system is recalled in Figure 9.7 where in addition the point of application of the external excitation in closed loop is indicated.

The controller used is the R-S-T controller (B) designed in Chapter 8, Section 8.5 (see Tables 8.7 and 8.8) and is stored in the text file flex_rst.reg. The external excitation (same PRBS used for open loop identification, the data are stored in the text file ibfs_sor.c) has been added to the output of the controller (see Figure 9.7).

The model identified in closed loop with F-CLOE (the model identified with AF-CLOE has been used for computing the data filter) has provided the best validation results. The validation results for various models are given in Table 9.2 and the parameters of the model considered for further use (F-CLOE) are

Table 9.2. Statistical validation of the various identified models for the flexible transmission

Method	Closed loop error variance R(0)	Normalized cross-correlations valid. threshhold: 0.136 |RN(max)|
AF-CLOE	0.01164	0.0498
CLOE	0.01310	0.0998
F-CLOE	0.01117	0.0147
OL identified model	0.01455	0.3821

---

[3] This example has been worked out with the WinPIM® (Adaptech) identification software.

$$d = 2$$
$$B(q^{-1}) = 0.42415q^{-1} + 0.27288q^{-2}$$
$$A(q^{-1}) = 1 - 1.5782q^{-1} + 1.8181q^{-2} - 1.41982q^{-3} + 0.84288q^{-4}$$

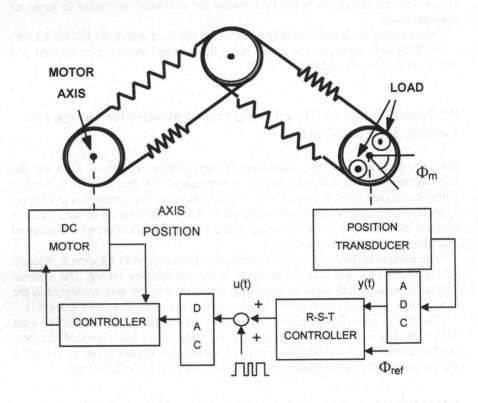

Figure 9.7. Closed loop identification scheme for the flexible transmission

The statistical validation in closed loop (see Table 9.2) clearly indicates that the model identified in closed loop is better than the model identified in open loop. The frequency characteristics of the model identified in open loop (see Chapter 7, Section 7.5.2) and the model identified in closed loop (F-CLOE) are shown in Figure 9.8. One observes a difference in the values of the damping factors for the two vibration modes (the model identified in open loop is less damped). The conclusions drawn from statistical validation are confirmed also by the pole closeness validation between the identified and computed poles of the closed loop. In Figure 9.9 are represented the estimated "true" poles of the closed loop and the poles of the closed loop computed by using the model identified in open loop and the controller (B) designed in Section 8.5. The "true" poles are obtained by open loop type identification between the external excitation and the output of the closed loop system (using the same data as for identification of the plant model in closed loop). One observes first a qualitative difference since the computed poles correspond to a pair of poles with the damping $\zeta = 0.8$ and $\omega_0 = 11.94 \, \text{rad/s}$ while the "true" dominant poles are an aperiodic pole at 0.763 and a pair of poles with the damping $\zeta = 0.523$ and $\omega_0 = 11.6 \, \text{rad/s}$.

In Figure 9.10 the identified "true" poles of the closed loop and the closed loop poles computed using the model identified in closed loop (F-CLOE) and the same controller as before are shown. One observes an almost superposition of the dominant poles and of one pairs of auxiliary poles. Comparing Figures 9.9 and 9.10 one can conclude that the model identified in closed loop gives a better description of the behavior of the closed loop.

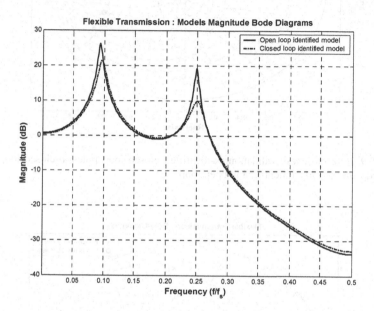

Figure 9.8. Frequency characterisitics of the models of the flexible transmission identified in open loop (OL) and in closed loop(CL)

The identified model of the "closed loop" whose poles are used in Figures 9.9 and 9.10 has been obtained with the recursive maximum likelihood method (structure S3) using decreasing adaptation gain and an initialization horizon of 120 samples. The size of the model results from the complexity of the model of the flexible transmission and of the controller which has been used ($n_A = 8, n_B = 8, n_C = 8, d = 0$).

The better quality of the plant model identified in closed loop is also confirmed by the time domain validation. Figures 9.11 and 9.12 provide comparisons of the simulated and achieved time response to a step on the reference. One observes a better coherence between the simulation and the real time response in Figure 9.12 where the simulation uses the model identified in closed loop (F-CLOE).

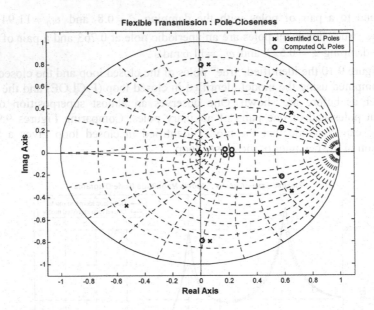

Figure 9.9. Pole closeness validation: x-identified closed loop poles; o-closed loop poles computed with the model identified in open loop

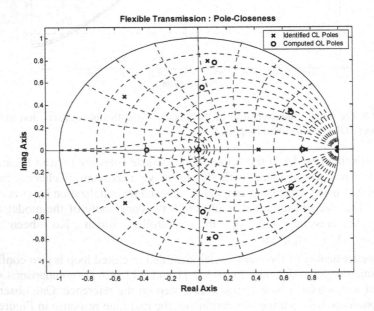

Figure 9.10. Pole closeness validation: x-identified closed loop poles; o-closed loop poles computed with the model identified in closed loop

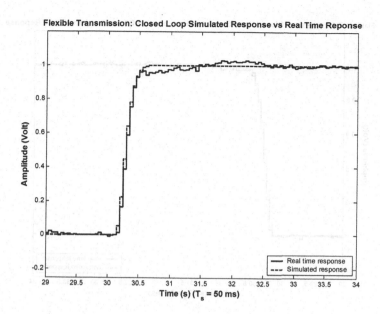

Figure 9.11. Time domain validation: (--) simulation using the model identified in open loop; (-) real time response

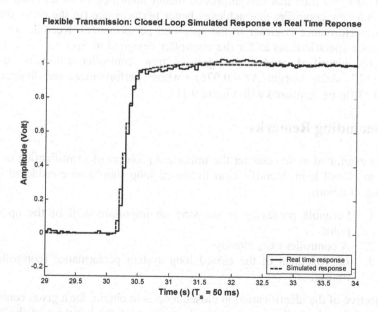

Figure 9.12. Time domain validation: (--) simulation using the model identified in closed loop; (-) real time response

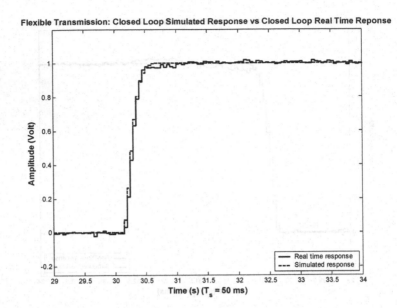

Figure 9.13. Improvement of the performance with the controller designed using the model identified in closed loop: (--) simulation; (-) real time

We will now illustrate that this improved model identified in closed loop allows designing a new controller that achieves better performance (in the sense that the achieved performance is closer to the specified performance). Keeping the same performance specifications as for the controller designed in Section 8.5 using the open loop identified model one gets a new controller (modulus margin $\Delta M = 0.537$, delay margin $\Delta \tau = 0.075s$) whose performances are illustrated in Figure 9.13 (to be compared with Figure 9.11).

9.6 Concluding Remarks

We have examined in this chapter the important problem of identification of plant models in closed loop. Identification in closed loop should be considered in the following situations:

1. Unstable processes or showing an important drift of the operating point
2. A controller exits already
3. Improvement of the closed loop system performance (controller re-tuning)

The objective of the identification in closed loop is to obtain, for a given controller, a model of the plant allowing the best description of the behavior of the closed loop system.

Identification in closed loop, provided that appropriate algorithms are used, allows to obtain better models for the controller design.

The closed loop output error identification methods (CLOE) are well suited for the identification of the plant models in closed loop operation. These methods require the knowledge of the controller.

As for open loop identification, the model identified in closed loop should be validated. Several techniques for model validation in closed loop have been defined:

- Statistical validation (uncorrelation test)
- Pole closeness validation (evaluating the closeness of the true closed loop poles and those of the closed loop predictor)
- Sensitivity functions closeness (evaluating the closeness of the true sensitivity functions and those of the closed loop predictor)
- Time domain validation (comparison of real time response and of the simulated time response).

The choice of the best model for re-design of the controller is done by comparing the validation results for all available models. It is useful to take also in account the model identified in open loop for this comparative validation (if it is available).

9.7 Notes and References

More details on closed loop identification methods can be found in:

Landau I.D., Karimi A. (1997b) Recursive algorithms for identification in closed-loop – a unified approach and evaluation, Automatica, vol. 33, n. 8, pp. 1499-1523.

Landau I.D., Lozano R., M'Saad M. (1997) Adaptive Control, Chap. 9, Springer, London, UK.

Karimi A., Landau I.D. (1998) Comparison of the closed loop identification methods in terms of the bias distribution, Systems and Control Letters, vol. 34, pp. 159-167.

Landau I.D., Karimi A. (2001) Identification des modèles de procédé en boucle fermée, in Identification de systèmes, (I.D. Landau, A. Besançon-Voda, Eds), pp. 213-244, Hermès, Paris.

Adaptech (1997b) WinPIM-BF – Software for closed loop identification, St. Martin d'Hères, France.

Adaptech (1999b) CLID® – Plant model identification in closed loop (toolbox for MATLAB®), 4 rue de la Tour de l'Eau, St. Martin d'Hères, France.

Ljung L. (1999) System Identification – Theory for the user, 2nd edition, Prentice Hall, London.

Forsell U., Ljung L. (1999) Closed loop identification revisited, Automatica, vol. 35, n. 7, pp. 1215-1241.

Soderström T, Stoica P. (1989) System Identification, Prentice Hall, London.

The original reference for the closed loop output error method is:

Landau I.D., Karimi A. (1997a): An output error recursive algorithm for unbiased identification in closed loop, Automatica, vol. 33, no. 8, pp. 933-938.

The original reference for the multi steps identification is:

Van den Hof P., Shrama R. (1993): An indirect method for transfer function estimation from closed loop data, Automatica, vol. 29, no. 6, pp. 1523-1528.

The importance of identification in closed loop for the re-design of a better performing system is discussed in:

Gevers M. (1993): Towards a joint design of identification and control. In Essays in Control (H.L. Trentelman, J.C. Willems, Eds), Birkhäuser, Boston, USA, pp. 111-152.

Landau I.D. (2001a): Identification in closed loop: a powerful design tool (better models, simple controllers), Control Engineering Practice, vol. 9, pp. 51-65.

Langer J., Landau I.D. (1996): Improvement of robust digital control by identification in closed loop. Application to a 360° flexible arm, Control Engineering Practice, vol. 8, no. 4, pp. 1079-1088.

and (Landau et al 2001b).

10

Reduction of Controller Complexity

Controller complexity reduction is an issue in many applications. Techniques for controller complexity (order) reduction based on the estimation in closed loop of a reduced order controller will be presented. Methods for the validation of the estimated reduced order controllers are also presented. The use of these techniques is illustrated by their application to the complexity reduction of a controller for a flexible transmission.

10.1 Introduction

The complexity (order of the polynomials R and S) of the controllers designed on the basis of identified models depends upon:

- The complexity of the identified model
- The performance specifications
- The robustness constraints

The controller will have a minimum complexity equal to that of the plant model but as a consequence of performance specifications and robustness constraints this complexity increases (often up to the double of the size of the model, in terms of number of parameters, and in certain cases even more).

In many applications the reduction of the controller complexity results from constraints on the computational resources (reduction of the number of additions and multiplications). These computational resources constraints come from various reasons:

- Price in mass production (ex: cars)
- Miniaturization and reduction of the energy consumption in certain embedded systems
- High sampling frequency

One should also add that often in practice one may be interested in tuning an existing digital PID even if the order of the plant model to be controlled is higher than 2.

Therefore one should ask the question: can we obtain a simpler controller with almost the same performance and robustness properties as the nominal one (designed on the basis of the plant model)?

Let us recall first that the polynomial defining the closed loop poles results from the equation

$$P(q^{-1}) = A(q^{-1})H_S(q^{-1})S'(q^{-1}) + q^{-d}B(q^{-1})H_R(q^{-1})R'(q^{-1}) \quad (10.1.1)$$

where A, B, d correspond to the model of the plant to be controlled and H_R and H_S correspond to the fixed parts of the controller (introduced for performance and robustness reasons). The controller polynomials $R(q^{-1})$ and $S(q^{-1})$ are given by

$$R(q^{-1}) = H_R(q^{-1})R'(q^{-1}) \qquad\qquad (10.1.2)$$

$$S(q^{-1}) = H_S(q^{-1})S'(q^{-1}) \qquad\qquad (10.1.3)$$

and the corresponding minimal orders are

$$n_R = n_A + n_{H_S} + n_{H_R} - 1 \qquad\qquad (10.1.4)$$

$$n_S = n_B + d + n_{H_S} + n_{H_R} - 1 \qquad\qquad (10.1.5)$$

The concrete objective will be to reduce the order of n_R and n_S.

The basic rule for developing procedures for controller complexity reduction is to search for controllers of reduced order which preserve as much as possible the properties of the closed loop. A direct simplification of the controller transfer function by traditional techniques (cancellation of poles and zeros which are close, approximations in the frequency domain, balanced reduction, *etc.*) without taking into account the properties of the closed loop leads in general to unsatisfactory results.

Two approaches can be considered for the controller complexity reduction:

1. Indirect Approach
This approach is implemented in two steps:

1. Reduction of the complexity of the model used for design, trying to preserve the essential characteristics of the model in the critical frequency regions for design.
2. Design of the controller on the basis of the reduced model.

2. Direct Approach
Search for a reduced order approximation of the nominal controller which preserves the properties of the closed loop.

The indirect approach has a number of drawbacks:

- Does not guarantee the complexity of the resulting controller (since the robustness specifications will be more severe when using reduced models).
- The errors resulting from model reduction will propagate in the design of the controller.

The direct approach seems the most appropriate for the reduction of the controllers complexity since the approximation is done in the last stage of the design and the resulting performance can be easily evaluated. It is this approach which will be developed in this chapter.

Two criteria can be considered for direct reduction of the controller complexity:

- *Closed loop input matching (CLIM).* In this case one would like that the control generated in closed loop by the reduced order controller be as close as possible to the control generated in closed loop by the nominal controller.
- *Closed loop output matching (CLOM).* In this case one would like that the closed loop output obtained with the reduced order controller be as close as possible to the closed loop output obtained with the nominal controller.

These two criteria are illustrated in Figure 10.1a,b where the nominal controller is denoted by K:

$$K = \frac{R(q^{-1})}{S(q^{-1})} \tag{10.1.6}$$

with

$$R(q^{-1}) = r_0 + r_1 q^{-1} + \ldots + r_{n_R} q^{-n_R} \; ; \; S(q^{-1}) = 1 + s_1 q^{-1} + \ldots + r_{n_S} q^{-n_S} \tag{10.1.7}$$

the reduced controller by $\hat{K}$:

$$\hat{K} = \frac{\hat{R}(q^{-1})}{\hat{S}(q^{-1})} \tag{10.1.8}$$

with

$$\hat{R}(q^{-1}) = \hat{r}_0 + \hat{r}_1 q^{-1} + \ldots + \hat{r}_{n_{\hat{R}}} q^{-n_{\hat{R}}} \; ; \; \hat{S}(q^{-1}) = 1 + \hat{s}_1 q^{-1} + \ldots + \hat{r}_{n_{\hat{S}}} q^{-n_{\hat{S}}} \tag{10.1.9}$$

and the plant model by $\hat{G}$:

$$\hat{G} = \frac{q^{-d}\hat{B}(q^{-1})}{\hat{A}(q^{-1})} \tag{10.1.10}$$

with

$$\hat{A}(q^{-1}) = 1 + \hat{a}_1 q^{-1} + \ldots + \hat{a}_{n_A} q^{-n_A} \quad ; \quad \hat{B}(q^{-1}) = \hat{b}_1 q^{-1} + \ldots + \hat{b}_{n_B} q^{-n_B} \tag{10.1.11}$$

For closed loop input matching (Figure 10.1a) one tries to find a reduced controller $\hat{K}$ which will minimize the difference between the input sensitivity function of the nominal simulated system

$$\hat{S}_{up} = \frac{K}{1 + K\hat{G}} \tag{10.1.12}$$

computed with K and $\hat{G}$ and the input sensitivity function of the simulated system using the reduced controller

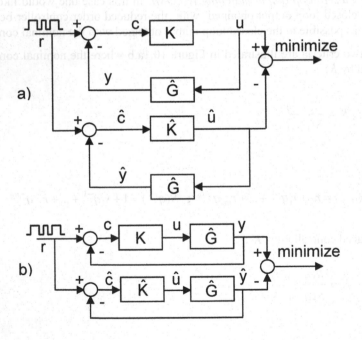

Figure 10.1a,b. Criteria for controller complexity reduction: **a** tracking of the nominal control; **b** tracking of the nominal output

$$\hat{S}_{up} = \frac{\hat{K}}{1 + \hat{K}\hat{G}}$$ (10.1.13)

computed with $\hat{K}$ and $\hat{G}$. This is equivalent to the search of a reduced controller $\hat{K}$ which minimizes the error between the two loops (in the sense of a certain criterion) for a white noise type excitation (like PRBS).

For the tracking of the nominal output (Figure 10.1b) the principle remains the same except that in this case one tries to minimize the difference between[1]

$$\hat{S}_{yr} = \frac{K\hat{G}}{1 + K\hat{G}}$$ (10.1.14)

computed with K and $\hat{G}$ and the sensitivity function of the simulated system using the reduced controller

$$\hat{S}_{yr} = \frac{\hat{K}\hat{G}}{1 + \hat{K}\hat{G}}$$ (10.1.15)

computed with $\hat{K}$ and $\hat{G}$.

One can see immediately that in both cases the problem of finding a reduced order controller can be formulated as an identification in closed loop (see Chapter 9) where the plant model is replaced by the reduced order controller to be estimated and the controller is replaced by the available estimated model of the plant (dual problem).

The reduction procedures and the validation techniques for reduced order controllers to be presented next are available in the MATLAB® toolbox REDUC® (Adaptech 1999a) which can be downloaded from the book web site[2].

Design of a reduced order RST controller and of a digital PID for the flexible transmission presented in Section 7.5.4, using the reduction of a nominal RST controller (designed by pole placement), will illustrate the use of the techniques presented in this chapter. For the use of these techniques for controller reduction applied to an active suspension see (Landau et al. 2001).

---

[1] Note that $\hat{S}_{yr} - \hat{S}_{yr} = \hat{S}_{yp} - \hat{S}_{yp}$ and therefore the tracking of the nominal output is equivalent to the minimization of the difference between the nominal and reduced output sensitivity functions.

[2] http://landau-bookic.lag.ensieg.inpg.fr

10.2 Estimation of Reduced Order Controllers by Identification in Closed Loop

10.2.1 Closed Loop Input Matching (CLIM)

The principle is illustrated in Figure 10.2.

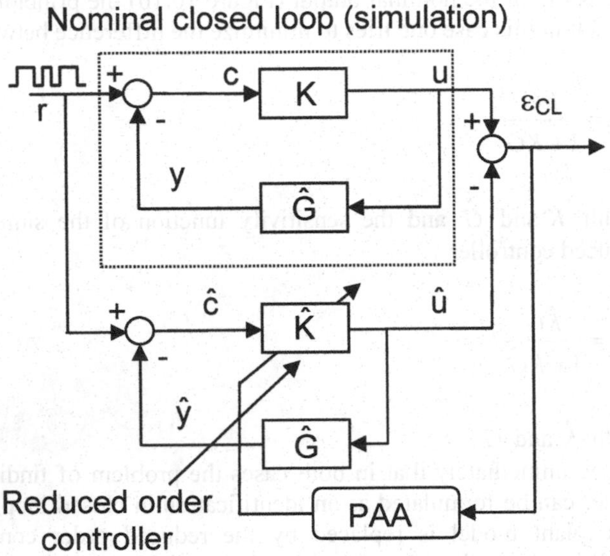

Figure 10.2. Estimation of reduced order controllers by the method of closed loop input matching (CLIM). Use of simulated data

The upper part represents the simulated nominal closed loop system. It is made up of the nominal controller (K) and the best identified plant model ($\hat{G}$). This model should assure the best closeness behavior of the true closed loop system and the nominal simulated one. Identification of this plant model in closed loop can be considered if the nominal controller can be implemented.

The lower part is made up of the estimated reduced order controller ($\hat{K}$) in feedback connection with the plant model ($\hat{G}$) used in the nominal simulated system. The parameter adaptation algorithm (PAA) will try to find the best reduced order controller which will minimize the closed loop input error. The closed loop input error is the difference between the plant input generated by the nominal simulated closed loop system and the plant input generated by the simulated closed loop using the reduced order controller.

The output of the nominal controller (plant input) is given by

$$u(t+1) = -S^*(q^{-1})u(t) + R(q^{-1})c(t+1) \tag{10.2.1}$$

with

$$c(t+1) = r(t+1) - y(t+1) \tag{10.2.2}$$

The *a priori* predicted output of the reduced order controller is given by

$$\hat{u}^0(t+1) = -S^*(t,q^{-1})\hat{u}(t) + \hat{R}(t,q^{-1})\hat{c}(t+1) = \hat{\theta}(t)^T \phi(t) \tag{10.2.3}$$

and the *a posteriori* predicted output is given by

$$\hat{u}(t+1) = \hat{\theta}(t+1)^T \phi(t) \tag{10.2.4}$$

where

$$\hat{\theta}(t)^T = \left[\hat{s}_1(t),...,\hat{s}_{n_{\hat{S}}}(t),\hat{r}_0(t),...,\hat{r}_{n_{\hat{R}}}(t) \right] \tag{10.2.5}$$

$$\phi(t)^T = \left[-\hat{u}(t),...,-\hat{u}(t-n_{\hat{S}}+1),\hat{c}(t+1),...,\hat{c}(t-n_{\hat{R}}+1) \right] \tag{10.2.6}$$

$$\hat{c}(t+1) = r(t+1) - \hat{y}(t+1) = r(t+1) + A^*(q^{-1})\hat{y}(t) - B^*(q^{-1})u(t-d) \tag{10.2.7}$$

The closed loop input error is given by

$$\varepsilon_{CL}^0(t+1) = u(t+1) - \hat{u}^0(t+1) \qquad \textit{(a priori)} \tag{10.2.8}$$

$$\varepsilon_{CL}(t+1) = u(t+1) - \hat{u}(t+1) \qquad \textit{(a posteriori)} \tag{10.2.9}$$

and the parameter adaptation algorithm is expressed as

$$\hat{\theta}(t+1) = \hat{\theta}(t) + F(t)\Phi(t)\varepsilon_{CL}(t+1) \tag{10.2.10}$$

$$F^{-1}(t+1) = \lambda_1(t)F^{-1}(t) + \lambda_2(t)\Phi(t)\Phi(t)^T$$
$$0 < \lambda_1(t) \le 1 \; ; \; 0 \le \lambda_2(t) < 2 \; ; \; F(0) > 0 \tag{10.2.11}$$

$$\varepsilon_{CL}(t+1) = \frac{\varepsilon_{CL}^0(t+1)}{1 + \Phi(t)^T F(t)\Phi(t)} \tag{10.2.12}$$

Specific algorithms are obtained for different choices of the observation vector:

- CLIM algorithm: $\Phi(t) = \phi(t)$

- F-CLIM algorithm: $\Phi(t) = \dfrac{\hat{A}(q^{-1})}{\hat{P}(q^{-1})}\phi(t)$

where

$$\hat{P}(q^{-1}) = \hat{A}(q^{-1})S(q^{-1}) + q^{-d}\hat{B}(q^{-1})R(q^{-1})\qquad(10.2.13)$$

The introduction of the filtering of $\phi(t)$ is motivated by the elimination of a sufficient condition for convergence which, in the case of the CLIM algorithm, depends on $\hat{A}(z^{-1})/\hat{P}(z^{-1})$. A detailed analysis of the properties of these algorithms can be found in Landau *et al.* (2001).

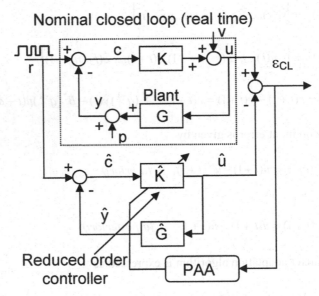

Figure 10.3. Estimation of reduced order controllers using real data (CLIM)

The estimation of reduced order controllers is also possible by using real time data (if the prototype of the nominal controller can be implemented on the real system). It is shown in Landau *et al.* (2001) that the use of real data allows one to take in account the error between the model and the true plant model in certain frequencies regions.

10.2.2 Closed Loop Output Matching (CLOM)

The principle of this method is illustrated in Figure 10.4.

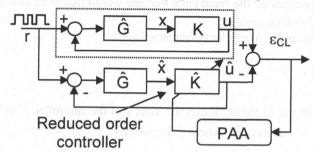

Figure 10.4. Estimation of reduced order controllers by the method of closed loop output matching (CLOM). Use of simulated data

Despite that the point where the external excitation is applied and the output variable is different with respect to Figure 10.1b, the transfer function between $r(t)$ and $u(t)$ is still the one given by Equation 10.1.14. This means that in the absence of disturbances (it is the case in simulation) $u(t)$ generated by the upper part of the scheme given in Figure 10.4 is equal to $y(t)$ generated in Figure 10.1b. This allows one to use for closed loop output matching the CLIM (or F-CLIM) algorithm with the only difference that $c(t)$ in Equation 10.2.1 is replaced by:

$$x(t) = \hat{G}(q^{-1})(r(t) - u(t))$$
(10.2.14)

and accordingly $\hat{c}(t)$ in 10.2.3 and 10.2.6 is replaced by $\hat{x}(t)$:

$$\hat{x}(t) = \hat{G}(q^{-1})(r(t) - \hat{u}(t))$$
(10.2.15)

One should note that the order of the blocks in the upper part of Figure 10.4 can be interchanged (like the upper part of Figure 10.1b) without affecting the operation of the algorithm. This observation is of interest when real time data are used.

10.2.3 Taking into Account the Fixed Parts of the Nominal Controller

It is often required that the reduced order controller contains some of the fixed filters incorporated in the nominal controller (for example: integrator, opening of the loop at $0.5f_s$ or at other frequency). In order to do this one first factorizes the nominal controller under the form[3]

$$K = K_F K'$$

(10.2.16)

where K_F represents all the fixed parts that one would like to be also incorporated in the reduced order controller.

The reduced order controller is factorized as

$$\hat{K} = K_F \hat{K}'$$

(10.2.17)

One replaces in the CLIM algorithm the input $\hat{c}$ of the controller $\hat{K}$ by the input to the controller $\hat{K}'$, denoted $\hat{c}'$, where $\hat{c}'$ is given by

$$\hat{c}'(t) = K_F(q^{-1})\hat{c}(t)$$

(10.2.18)

and in $\phi(t)$, $\hat{c}$ is replaced by $\hat{c}'$. In the CLOM algorithm one replaces $\hat{x}(t)$ by $\hat{x}'(t)$ given by

$$\hat{x}'(t) = K_F(q^{-1})\hat{G}(q^{-1})[r(t) - y(t)]$$

(10.2.19)

10.2.4 Re-Design of Polynomial $T(q^{-1})$

Once a controller of reduced order is obtained and validated, one should re-compute the polynomial $T(q^{-1})$ using the new polynomial defining the closed loop poles

$$\hat{P}(q^{-1}) = \hat{A}(q^{-1})\hat{S}(q^{-1}) + q^{-d}\hat{B}(q^{-1})\hat{R}(q^{-1})$$

(10.2.20)

10.3 Validation of Reduced Order Controllers

Once a reduced order controller has been estimated, it should be validated before considering its implementation on the real system.

---

[3] If the nominal controller has unstable poles (which is certainly not desirable), they should be maintained as they are and therefore they should be included in K_F.

10.3.1 The Case of Simulated Data

It is assumed that the nominal controller stabilizes the nominal plant model (used for controller reduction). One implicitly assumes that the model uncertainties have been taken into account in the design of the nominal controller.

The reduced order controller should satisfy the following conditions:

- It stabilizes the nominal plant model.
- The *reduced* sensitivity functions (computed with the reduced order controller) are close to the *nominal* sensitivity functions in the critical frequency regions for performance and robustness. In particular output and input sensitivity functions should be examined.
- The *generalized stability margin* (see Appendix D) of the system using the reduced order controller should be close to the generalized stability margin of the nominal closed loop. This condition is expressed as

$$\left| b(K,\hat{G}) - b(\hat{K},\hat{G}) \right| < \varepsilon \quad ; \quad \varepsilon > 0$$

where $b(K,\hat{G})$ and $b(\hat{K},\hat{G})$ are the generalized stability margins corresponding to the nominal controller and to the reduced order controller and ε is a small positive number. The closeness of the two stability margins allows maintaining the robustness properties of the initial design.

The proximity or the nominal and reduced sensitivity functions can be judged by visual examination of their frequency characteristics. There exists however the possibility to make a numerical evaluation of this proximity by computing the *Vinnicombe distance* (*ν gap*) between these transfer functions (see Appendix D). The *Vinnicombe distance* allows one with one number (between 0 and 1), to make a first evaluation of the proximity of the reduced and nominal sensitivity functions which is useful for the comparative evaluation of the various reduced order controllers.

10.3.2 The Case of Real Data

The use of real data (in the case where the prototype of the nominal controller can be implemented) allows completion of the validation done with simulated data.

From Figure 10.3 it results that the objective is to test to what extent the simulated closed loop using the reduced order controller is close to the real closed loop system using the nominal controller.

Initial information is provided by the variance of the residual closed loop input error. Subsequent information can be obtained by identifying the transfer function of the real closed loop system (between r_u and u) and then comparing the frequency characteristics of this transfer function with that of the simulated closed loop system using the reduced order controller. In this case, using the Vinnicombe distance one can also evaluate rapidly the quality of various reduced order controllers.

10.4 Practical Aspects

The quality of the resulting reduced order controller will depend upon the ability of the plant model in feedback with the nominal controller to reproduce the behavior of the real feedback system.

If one has access to the real system and the nominal controller can be implemented, it is wise to identify a new plant model in closed loop and to validate comparatively in closed loop the various available models in order to select the best one.

Note also that the same input/output data used for identification in closed loop can be in general used for controller reduction. For more details on these aspects see Landau and Karimi (2002).

Once the best plant model is selected, the procedure for controller reduction is implemented (with simulated or real data).

Using the methods described in this chapter, all the controllers with order of the polynomials R and S lower than the nominal one, can be estimated with only one set of data. Then, using the validation techniques, one select those reduced order controllers which assure a good approximation of the behavior of the nominal closed loop system.

10.5 Control of a Flexible Transmission – Reduction of Controller Complexity

The algorithms for reducing the controller complexity described in the previous sections will be applied to the flexible transmission already presented in Chapter 7, Section 7.5.4. On the basis of the fourth-order model identified for the flexible transmission (following an open loop identification procedure, see Section 7.5.4), a controller that guarantees the specified performances (standard robustness margins and regulation/tracking closed loop dynamics) has been computed by pole placement with the sensitivity functions shaping method in Chapter 8, Section 8.5.

In this section we consider the *RST* controller obtained by requiring the same specifications (*i.e.* same assigned closed loop poles, controller fixed parts and robustness constraints) considered in Chapter 8, Section 8.5 (design C) but with the model identified in closed loop in Chapter 9, Section 9.5. It is characterized by the orders $n_R = n_S = 5$ for the polynomials R and S. We will refer to this controller as the *nominal controller*.

The first objective of the controller reduction will be to find a good controller with orders for the polynomials R and S lower than the minimum orders which can be obtained in model based control design when only the integrator is imposed ($n_R = n_S = 4$).

Consider now the following problem: is it possible to design a digital PID controller for the flexible transmission which will provide closed loop performance and robustness close to those provided by the *nominal controller*? The answer to this control design problem is not easy if one tries to tune directly the PID parameters. An effective solution can be obtained by reducing the complexity of the full controller to $n_R = n_S = 2$ using the techniques presented in this chapter[4]. Therefore the second objective of controller reduction will be to tune a digital PID.

The procedure for finding a controller of reduced complexity involves the following steps:

1. Load the *RST full controller*, the flexible transmission model identified in closed loop and the excitation used to estimate the reduced order controller.
2. Specify the fixed parts of the reduced controller together with the desired polynomials orders (n_R and n_S).
3. Run an appropriate algorithm for the reduction of the controller complexity in closed loop (CLIM or CLOM).
4. Validate the results in terms of comparison of the nominal and reduced: sensitivity functions and generalized stability margin.

In this specific application, we will look for reduced order controllers with

- $n_R = 3, n_S = 3$
- $n_R = 2, n_S = 2$ (digital PID)

and with the fixed parts:

- $H_S = 1 - q^{-1}$ (integrator), $H_R = 1$

In Table 10.1 three reduced order controllers are considered. The reduced order RST controller K_3 has a lower complexity than the minimum complexity which can be obtained in a model based control design and has been obtained using the CLOM reduction algorithm (simulated data).

The orders of the polynomials R and S have been reduced from 5 to 3. Controllers K_1 and K_2 correspond to digital PID controllers and they have been obtained using CLOM (K_1) and CLIM (K_2) algorithms respectively. These reduced controllers are compared with the nominal controller (K_n).

The values found for the Vinnicombe normalized distance[5] (δ_v, rows 1 and 2) give a first indication of the proximity between the sensitivity functions computed with K_n and those computed with K_1, K_2 and K_3. As the first objective is to maintain the robustness properties and performance with respect to output external disturbances, the proximity of the reduced output sensitivity function S_{yp}^i with respect to the nominal one is the main criterion for selecting the reduced order controller. Row 3 gives the generalized stability margin for each controller. The last two rows of the table give the frequency and the amplitude of the peak of the output sensitivity functions.

Clearly controller K_3 provides very close performances with respect to the nominal one in terms of the gap on S_{yp}, generalized stability margin and maximum of the output sensitivity function.

---

[4] More details about this example and the files to run the simulations can be found in Appendix G, Section G.6.

[5] For definitions of the Vinnicombe normalized distance (v-gap) and the generalized stability margin see Appendix D.4.

Table 10.1. Comparison of the reduced order (PID and RST) controllers with the nominal controller (validation)

Controller		K_n $n_R = 5$ $n_S = 5$	K_1 $n_R = 2$ $n_S = 2$ (CLOM)	K_2 $n_R = 2$ $n_S = 2$ (CLIM)	K_3 $n_R = 3$ $n_S = 3$ (CLOM)
1	$\delta_v(S_{up}^n, S_{up}^i)$	-	0.6331	0.5280	0.1416
2	$\delta_v(S_{yp}^n, S_{yp}^i)$	-	0.1443	0.2396	0.0111
3	$b(k)$	0.2875	0.2490	0.2446	0.2854
4	$Max(S_{yp})$ (dB)	5.13	6.31	6.35	5.25
5	f_{max} (Hz)	0.93	0.99	2.85	0.93

The PID obtained with CLOM algorithm (K_1) provides a smaller gap for the S_{yp} than the one obtained using CLIM algorithm (K_2) (this result was expected taking in account the criterion which is minimized – see Section 10.1). This is confirmed by a visual inspection of the output sensitivity frequency characteristics (see Figure 10.5). In terms of closeness of nominal and reduced S_{up}, controller K_1 is slightly less good than K_2 (see Figure 10.6).

Note that since in the reduced order controllers the opening of the loop at $0.5\,f_s$ has not been imposed (in order to get a low order controller), there are important differences at $0.5\,f_s$. However the level of obtained S_{up} is acceptable.

The generalized stability margins $(b(K))$ for the reduced PID controllers are still close to that of K_n, and then the robustness properties will be only slightly changed. The magnitude of the peak of S_{yp} is almost the same for both PID controllers but for the K_1 controller the frequency is very close to that of K_n. The peak for K_1 is only 0.31 dB over the standard 6 dB level required for robustness margin.

In Figure 10.7 time domain simulations are given for comparing both tracking and regulation behavior of the reduced controllers considered in Table 10.1 with respect to the full RST controller (for a fair comparison with the PID, the full controller uses $T = R(1)$ which corresponds to having same performance specifications in tracking and regulation). The K_3 controller also uses $T = R(1)$. A PID2 structure (see Chapter 3, Section 3.2.4) has been used (with $T = R(1)$). In the case of the controller K3, the number of parameters for R and S has been reduced from 11 to 7 and in the case of the PID the reduction is from 11 to 5.

The simulations show that controller K3 gives almost the same performances as the full controller and that K1 controller provides a better performance than K2 controller. The performance of controller K1 is acceptable both in tracking and regulation (filtering the reference can further reduce the overshoot while providing the same time response as the full controller).

Real time results are shown in Figure 10.8 and they confirm the results obtained in simulation.

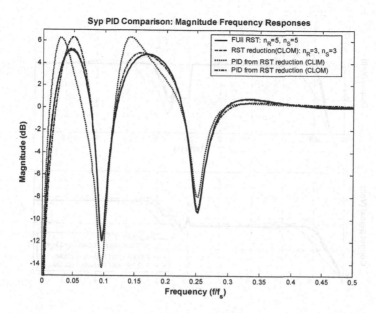

Figure 10.5. Syp comparison between the reduced order (PID and RST) controllers and the full RST controller

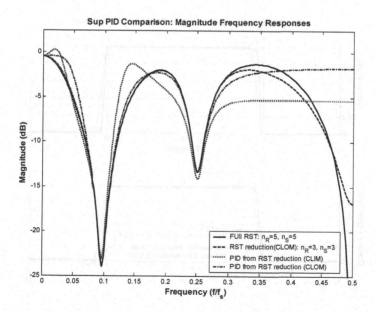

Figure 10.6. Sup comparison between the reduced order (PID and RST) controllers and the full RST controller

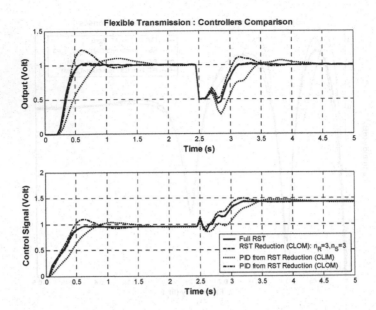

Figure 10.7. Time domain comparison (both tracking and regulation) between the reduced order (PID and RST) controllers and the nominal RST controller (simulation)

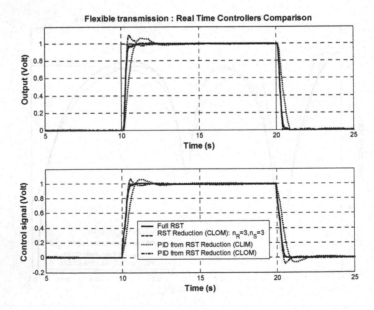

Figure 10.8. Real time domain comparison between the reduced order (PID and RST) controllers and the nominal RST controller

10.6 Concluding Remarks

In this chapter the bases of controller reduction have been presented together with a set of methods allowing one to effectively achieve controller order reduction. These methods are based on the use of closed loop identification techniques for estimating a reduced order controller. The objective is to find a reduced order controller such that the characteristics of the closed loop is as close as possible to the closed loop characteristics obtained with the nominal controller.

Two specific objectives for controller reduction have been considered:

*   Closed loop input matching (CLIM)
*   Closed loop output matching (CLOM)

The reduced order controller should be validated before its effective use. Dedicated validation techniques have been presented.

In addition to the practical considerations which require reduction of the controller complexity, the use of the reduction techniques allows one to provide indirect answers to the following two important questions:

*   What is the optimal choice of the closed loop poles for a design by pole placement?
*   How to compute a reduced order controller (and in particular a digital PID) for a plant characterized by a high order dynamical model?

10.7 Notes and References

For basic references in controller reduction see:

Anderson B.D.O., Liu Y. (1989) Controller reduction: concepts and approaches, IEEE Trans. on Automatic Control, vol. 34, no. 8, pp. 802-812.

Anderson B.D.O. (1993) Controller reduction: moving from theory to practice, IEEE Control Magazine, vol. 13, pp. 16-25.

The main references for the techniques presented in this chapter are:

Landau I.D., Karimi A., Constantinescu A. (2001) Direct controller reduction by identification in closed loop, Automatica, vol. 37, no. 11, pp. 1689-1702.

Adaptech (1999a) REDUC® – Controller order reduction by closed-loop identification (Toolbox for MATLAB®), Adaptech, 4 rue du Tour de l'Eau, St. Martin d'Hères, France

Experimental results for the controller reduction applied to an active suspension can be found in the above references.

Interaction between controller complexity reduction and closed loop identification is discussed in:

Landau I.D., Karimi A. (2002) A unified approach to closed-loop plant identification and direct controller reduction, European J. of Control, vol. 8, no.6, pp 561-572.

A

A Brief Review of Some Results from Theory of Signals and Probability

A.1 Some Fundamental Signals

For the analysis of dynamic systems in the time domain the knowledge of some typical signals is necessary.

Dirac Pulse
It is a *fundamental* signal, since all the other signals are obtained by passing the *Dirac pulse* through an appropriate filter. The *Dirac pulse* is defined as the limit when $\Delta \rightarrow 0$, of a pulse of unit surface having duration Δ and magnitude $1/\Delta$ (see Figure A.1). For $\Delta \rightarrow 0$, the magnitude goes towards infinity.

The *Dirac pulse* is written as

$$u(t) \begin{cases} = \lim_{\Delta \to 0} \dfrac{1}{\Delta}, & 0 \leq t \leq \Delta \\[2mm] = 0, & t < 0, t > \Delta \end{cases}$$

Despite the fact that it is an abstract mathematical object, in many situations, the duration of a pulse applied to a system is very short compared to the time constants of the system, and it can be approximated by a Dirac pulse. The response of a system to a Dirac pulse is known as the *impulse response*.

The discrete time equivalent of the Dirac pulse (also known as *delta (δ)* function or Kronecker's function) is

$$u(t) \begin{cases} = 1 & t = 0 \\ = 0 & t \neq 0 \end{cases}$$

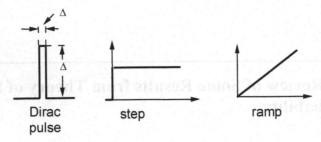

Figure A.1. Dirac pulse, step and ramp signals

Step

The *unit step* is defined as follows:

$$u(t)\begin{cases} =0 & t<0 \\ =1 & t\geq 0 \end{cases}$$

The step can be interpreted as the integral of the Dirac pulse.

The *discrete time step* is defined in a similar way:

$$u(t)\begin{cases} =0 & t<0 \\ =1 & t=0,1,2,3,\ldots \end{cases}$$

The discrete time step can be obtained by the numerical integration of the discrete time Dirac pulse (δ function).

The response of a system to a *step* is known as *step response*.

Ramp

The *ramp* is defined as follows:

$$u(t)\begin{cases} =0 & t<0 \\ =t & t\geq 0 \end{cases}$$

The *ramp* corresponds to the integral of the unit step or to the double integration of the Dirac pulse. In discrete time, the ramp is defined as

$$u(t)\begin{cases} =0 & t<0 \\ =t & t=0,1,2,3\ldots \end{cases}$$

A.2 The z - transform

Let $f(k)$ a discrete time sequence defined for all $k \geq 0$, $(k = 0,1,2,...)$.

The z - transform of the sequence $f(k)$ is defined by

$$F(z) = \sum_{k=0}^{\infty} f(k)z^{-k}$$

If $U(z)$ is the z-transform of the discrete time input sequence $u(k)$ applied to a linear system and $Y(z)$ is the z- transform of the discrete time output sequence $y(k)$, one can define the pulse transfer function of a discrete time system as

$$H(z) = Y(Z)/U(z)$$

(for an equivalent definition see Chapter 2, Section 2.3.2).

The z-transform of the discrete time Dirac pulse is 1. Therefore the pulsed transfer function $H(z)$ of a linear discrete time system is equal to the z-transform of the output sequence when the input is a discrete time Dirac pulse.

A.3 The Gauss Bell

It is extremely important for a random variable (or for an ergodic stochastic process) to be able to specify what is the probability that this variable takes a value within a certain interval, or to specify what is the percentage of points, over a set of measurements of a random variable, which will be within a certain interval. If the random variable is gaussian (normal) or if the ergodic stochastic process is gaussian, this information is obtained from the measure of the area under the "Gauss bell" between the coordinates (values) x_1 and x_2 (see Figure A.2).

The expression for the "Gauss bell" is:

$$f_X(x) = \frac{1}{\sigma\sqrt{2\pi}} e^{-\frac{(x-MV)^2}{2\sigma^2}} \qquad -\infty \leq x \leq +\infty \tag{A.1}$$

where, for the case of gaussian stochastic processes, MV is the expectation (or the *mean value*)

$$MV = E\{X(t)\} = \lim_{N \to \infty} \frac{1}{N} \sum_{t=1}^{N} X(t) \tag{A.2}$$

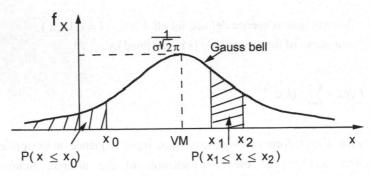

Figure A.2. Gauss bell

and σ is the *standard deviation* related to the variance of the stochastic process by the relationship

$$Var. = E\left\{(X(t)-MV)^2\right\} = \lim_{N \to \infty} \frac{1}{N} \sum_{t=1}^{N} (X(t)-MV)^2 = \sigma^2 \qquad (A.3)$$

From Equation A.1, it results that the "Gauss bell" shrinks around the mean value and its maximum will grow as the standard deviation σ decreases. This is illustrated in Figure A.3.

The probability that X takes a value between $-\infty$ and $+\infty$ being equal to 1, it results that the total area under the "Gauss bell" is equal to 1 (or 100 %).

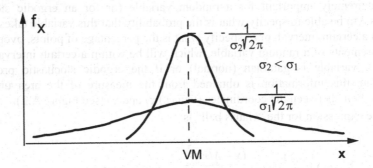

Figure A.3. Gauss bells for two different values of the standard deviation

The probability that X takes a value $\le x_0$ is equal to the area under the Gauss bell for $-\infty < x \le x_0$. This probability (denoted P) is a function which depends upon x. It is called *probability distribution function*.

$$F_X(x) = P\{X \le x\} = P\{-\infty < X \le x\} \qquad (A.4)$$

From Figure A.2, it results that

$$P\{x_1 < X \leq x_2\} = P\{X \leq x_2\} - P\{X \leq x_1\} = F_X(x_2) - F_X(x_1) \qquad (A.5)$$

The derivative of the probability distribution function (if it exists), is called the *probability density function* of the random variable.

$$f_X(x) = \frac{dF_X(x)}{dx} \qquad (A.6)$$

where

$$dF_X(x) = P\{x \leq X \leq x + dx\}$$

$F(x)$ being represented by the area under the "Gauss bell", one concludes that the Gauss bell gives the probability density for gaussian random variables (or for ergodic guassian stochastic processes).

From Equation A.6, it results that

$$dF_X(x) = P\{x \leq X \leq x + dx\} = f_X(x)dx \qquad (A.7)$$

which means that the probability that a random gaussian variable takes a value between x and $x + dx$ is equal to $f_X(x)\, dx$.

Let us return now to the expression of the density of probability given by Equation A.1. If we will center the random variable (or the realization of the gaussian stochastic process) by subtracting the mean value, the Gauss bell will be centered around 0. In Figure A.4, a Gauss bell centered on 0 has been represented. The value of the area between various values of x has been indicated. This will give the percentage of points which will be found within these values (or the probability that a gaussian random variable takes a value within this interval).

Some significant values concerning the Gauss bell are summarized in Table A.1.

Table A.1. Percentage of points verifying $|X| \leq \alpha, \alpha \geq 0$ for an ergodic gaussian stochastic process

α	σ	$1{,}808\sigma$	$1{,}96\sigma$	2σ	$2{,}17\sigma$	$2{,}58\sigma$	3σ
$|X| \leq \alpha$ (%)	63	93	95	95,5	97	99	99,7

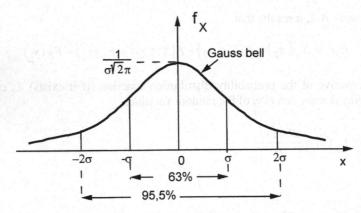

Figure A.4. Centered Gauss bell. Percentage of points located within two values of x

B

Design of RST Digital Controllers in the Time Domain

B.1 Introduction

In Chapter 3, the design of digital controllers in a deterministic environment has been discussed using exclusively the plant and controller representation by transfer functions (operators) in order to assign the poles of the closed loop.

The same results can be obtained by designing the controllers in the time domain. The time domain approach emphasizes the *predictive* character of digital control by pole placement (or by other related control strategies presented in Chapter 3). This approach allows one on one hand to get a deeper understanding of the control of systems with delay since it clearly shows the presence of a *predictor* inside the controller and, on the other hand, it allows one to understand better the connection with the synthesis of digital controllers in a stochastic environment.

An essential aspect in the design of digital controllers in the time domain is the synthesis of an (explicit or implicit) predictor for discrete time systems. Another important issue is that the resulting control $u(t)$ should depend only upon the information available up to and including time t.

If one takes the point of view of the controller design in the time domain, there are a number of specific control strategies known under the generic name *model predictive control*. Comparing the time domain control objectives for various model predictive control strategies, one can classify these strategies into two categories:

1. One step ahead model predictive control
In these strategies one computes a prediction of the plant output at $t+d+1$ (d-integer delay of the plant) namely $\hat{y}(t + d +1) = f(u(t), u(t-1), ... y(t), y(t-1),..)$ and one computes $u(t)$ such that the control objectives expressed in terms of $\hat{y}(t + d +1)$ be satisfied.

2. Long range model predictive control
In these strategies the control objective is expressed in terms of future values of the output over a certain horizon and of a sequence of future control values. In order to solve this problem one need to compute the following predictions:

$$\hat{y}(t+d+1) = f_1(y(t), y(t-1), ..u(t), u(t-1), ..)$$

$$\vdots$$

$$\hat{y}(t+d+j) = f_j(y(t), y(t-1), ..u(t), u(t-1), ..) + g_j(u(t+1), ...u(t+j-1)); \; g_1 = 0$$

Note that predictions of the output beyond *d+1* will depend on the future values of the control. The sequence of present and future values of the control *i.e.* *u(t)*, *u(t+1)*, ...*u(t+j-1)* is computed in order to satisfy the control objective. However only *u(t)* is applied to the plant and the procedure is restarted at t+1. This is known as the *receding horizon* procedure.

All these strategies will lead to a linear controller as long as constraints on the values of the admissible control applied to the plant will not be considered. In all cases the resulting design can be interpreted as a pole placement. The closed loop poles will be defined by the design criterion used in time domain. The control strategies presented in Chapters 3 and 4 belong to the category of one step ahead predictive control (*i.e.*: there is a time domain control objective associated to each of these strategies).

This appendix is organized as follows. Predictors for linear discrete time systems will be presented in Section B.1. The design of RST controllers in the time domain using one step ahead model predictive control strategies will be illustrated in Section B.2 for the tracking and regulation with independent objectives and pole placement. In Section B.3 it will be shown that the RST controller for a system with delay *d* can be decomposed in a d steps ahead predictor for the plant output and an RST controller for the plant without delay, using the predictor output instead of the measured output. Section B.4 will present briefly the long range model predictive control strategies which will be illustrated by the *generalized predictive control* strategy.

B.2 Predictors for Discrete Time Systems

The discrete-time model of a sampled-data system is described in the general case by

$$y(t) = -\sum_{i=1}^{n_A} a_i \, y(t-i) + \sum_{i=1}^{n_B} b_i \, u(t-d-i) \tag{B.1}$$

The linear discrete time model given in Equation B.1 is in fact a *one step ahead predictor*. Moving from *t* to *t+1* its expression becomes

$$y(t+1) = -\sum_{i=1}^{n_A} a_i y(t+1-i) + \sum_{i=1}^{n_B} b_i u(t+1-d-i) = A^*(q^{-1})y(t) + q^{-d}B^*(q^{-1})u(t)$$

$$= f[y(t), y(t-1),..., u(t), u(t-1),...] \tag{B.2}$$

where $A^*(q^{-1})$ and $B^*(q^{-1})$ are given by Equations 2.3.22 and 2.3.24. One can observe that $y(t+1)$ depends only on the values of y and u up to the instant t.

An important problem is the prediction of the output y for $t+1, t+2,...$ using the information (measures) available up to the instant t. This type of problem occurs for the control of systems with a discrete time delay d. In this case, we would like either to predict the value of the output with $d+1$ steps in advance, using only the information available at instant t, or to compute the value of $u(t)$ allowing us to obtain a specified value of the output y at instant $t+d+1$. In fact we are looking for an expression of the form

$$\hat{y}(t+j) = \hat{y}(t+j/t) = f[y(t), y(t-1),..., u(t), u(t-1),...]; \quad j \leq d+1 \tag{B.3}$$

where $\hat{y}(t+j/t)$, denoted in a simplified notation by $\hat{y}(t+j)$, represents the prediction of the output at the instant $t+j$ based on the knowledge of y and u up to and including t.

In order to illustrate the concept and the synthesis of a predictor, first let consider an example.

Example: Let

$$y(t+1) = -a_1 y(t) + b_1 u(t-1) \tag{B.4}$$

which corresponds to a system with a discrete time delay $d = 1$.

We are interested in predicting the output value $y(t+2)$ at instant t on the basis of information available at instant t. In order to compute this predicted value, one expresses first the output $y(t+2)$ as a function of the available information at instant t. From Equation B.4 one gets

$$y(t+2) = -a_1 y(t+1) + b_1 u(t) \tag{B.5}$$

which can also be written

$$A(q^{-1})y(t+2) = B^*(q^{-1})u(t); \quad A(q^{-1}) = 1 + a_1 q^{-1}; \quad B^*(q^{-1}) = b_1 \tag{B.6}$$

Note that $y(t+2)$, given by Equation B.5, depends upon $y(t+1)$ which is unknown at instant t. Therefore this expression does not allow a two steps ahead prediction. But $y(t+1)$ can be replaced in Equation B.5 by its expression given by Equation B.4. One then gets

$$y(t+2) = -a_1[-a_1 y(t) + b_1 u(t-1)] + b_1 u(t) = f_0 y(t) + (1 + e_1 q^{-1}) b_1 u(t)$$
$$= F(q^{-1}) y(t) + E(q^{-1}) B^*(q^{-1}) u(t) \tag{B.7}$$

with

$$F(q^{-1}) = f_0 = a_1^2; \ E(q^{-1}) = 1 + e_1 q^{-1} = 1 - a_1 q^{-1} \tag{B.8}$$

One observes that the right hand member of Equation B.7 depends only on the information available at instant t and therefore the expression of the two steps ahead predictor will be given by

$$\hat{y}(t+2) = F(q^{-1}) y(t) + E(q^{-1}) B^*(q^{-1}) u(t) \tag{B.9}$$

where $F(q^{-1})$ and $E(q^{-1})$ are given by Equation B.8.

This technique of successive substitution of the one step ahead prediction can be generalized for any d, A and B. However, it is possible to directly find the polynomials $E(q^{-1})$ and $F(q^{-1})$. Using Equation B.6 in Equation B.7 for replacing the term $B^*(q^{-1}) u(t)$ one gets

$$y(t+2) = F(q^{-1}) y(t) + E(q^{-1}) A(q^{-1}) y(t+2)$$
$$= [A(q^{-1}) E(q^{-1}) + q^{-2} F(q^{-1})] y(t+2) \tag{B.10}$$

In order that the two sides of Equation B.10 be equal, $E(q^{-1})$ and $F(q^{-1})$ should verify the polynomial equation

$$1 = A(q^{-1}) E(q^{-1}) + q^{-2} F(q^{-1}) \tag{B.11}$$

In other terms, this means that the coefficients of the polynomials $E(q^{-1})$ and $F(q^{-1})$ required for the computation of the predicted value $\hat{y}(t+2)$ at instant t, are the solutions of the polynomial Equation B.11. This approach can be generalized for any A, B and d.

In general, we look for the expression of a *filtered prediction*

$$P(q^{-1}) \hat{y}(t+d+1) = f_P[y(t), y(t-1),..., u(t), u(t-1),...] \tag{B.12}$$

where $P(q^{-1})$ is an asymptotically stable monic polynomial (the coefficient of power q^0 is 1). The discrete time system B.2 can be re-written in the form:

$$A(q^{-1})y(t+d+1) = B^*(q^{-1})u(t) \tag{B.13}$$

We will search for an expression of the output having the form

$$P(q^{-1})y(t+d+1) = F(q^{-1})y(t) + E(q^{-1})B^*(q^{-1})u(t) \tag{B.14}$$

where

$$F(q^{-1}) = f_0 + f_1 q^{-1} + \ldots + f_{n_F} q^{-n_F} \tag{B.15}$$

$$E(q^{-1}) = 1 + e_1 q^{-1} + \ldots + e_{n_E} q^{-n_E} \tag{B.16}$$

Using Equation B.13, Equation B.14 becomes:

$$P(q^{-1})y(t+d+1) = [A(q^{-1})E(q^{-1}) + q^{-d-1}F(q^{-1})]y(t+d+1) \tag{B.17}$$

and therefore $E(q^{-1})$ and $F(q^{-1})$ are solutions of the polynomial equation

$$P(q^{-1}) = A(q^{-1})E(q^{-1}) + q^{-d-1}F(q^{-1}) \tag{B.18}$$

with

$$n_F = \max(n_A - 1, n_P - d - 1) \; ; \; n_E = d \tag{B.19}$$

in order to obtain minimum unique solution.

Equation B.18 can also be expressed in matrix form

$$Mx = p \tag{B.20}$$

where M is a lower triangular matrix of dimension $(n_A + d + 1) \times (n_A + d + 1)$, the vector x contains the coefficients of $E(q^{-1})$ and $F(q^{-1})$ and the vector p contains the coefficients of the polynomial $P(q^{-1})$.

$$
\begin{bmatrix}
\overbrace{}^{n_A+d+1} \\
\begin{array}{ccccccccccc}
1 & 0 & . & . & . & . & . & . & . & . & 0 \\
a_1 & 1 & 0 & . & . & . & . & . & . & . & . \\
a_2 & a_1 & 1 & 0 & . & . & . & . & . & . & . \\
. & . & . & . & . & . & . & . & . & . & . \\
. & . & . & . & . & . & . & . & . & . & . \\
. & . & . & . & . & . & . & . & . & . & . \\
a_d & a_{d-1} & . & . & . & a_1 & 1 & . & . & . & . \\
a_{d+1} & a_d & . & . & . & . & a_1 & 1 & . & . & . \\
a_{d+2} & a_{d+1} & . & . & . & . & a_2 & 0 & . & . & . \\
. & . & . & . & . & . & . & . & . & . & 0 \\
0 & 0 & . & . & . & 0 & a_{n_A} & 0 & 0 & 1 &
\end{array}
\end{bmatrix}
\Big\} n_A + d + 1
$$

$$
\underbrace{}_{d+1} \qquad \underbrace{}_{n_A}
$$

$$
x^T = [1, e_1 \ldots e_d, f_0, f_1 \ldots f_{n_F}] \tag{B.21}
$$

$$
p^T = [1, p_1, p_2 \ldots p_{N_A}, p_{N_A+1} \ldots p_{N_A+d}] \tag{B.22}
$$

In general $n_p \le n_A + d$. If $n_p < n_A + d$ some of the components p_i in B.22 will be null.

The computation of $E(q^{-1})$ and $F(q^{-1})$ requires uniquely the knowledge of d and $A(q^{-1})$. For effective computations the functions: *predisol.sci* (Scilab) et *predisol.m* (MATLAB®)[1] can be used (see also Chapter 3, Section 3.4).

The d+1 steps ahead predictor will be given by

$$
P(q^{-1})\hat{y}(t+d+1) = F(q^{-1})y(t) + E(q^{-1})B^*(q^{-1})u(t) \tag{B.23}
$$

and the *prediction error* defined by

$$
\varepsilon(t+1) = y(t+1) - \hat{y}(t+1) \tag{B.24}
$$

will satisfy the equation

$$
P(q^{-1})\varepsilon(t+d+1) = 0 \tag{B.25}
$$

---

[1] To be downloaded from the book web site.

which is obtained by subtracting Equation B.23 from Equation B.14. In other words, the prediction error will tend towards zero for any initial prediction error provided that $P(q^{-1})$ is asymptotically stable. For $P(q^{-1}) = 1$ the prediction error becomes zero at $t + d + 1$.

B.3 One Step Ahead Model Predictive Control

Tracking and Regulation with Independent Objectives
Since an expression of the predicted output with $d + 1$ steps ahead is available, it is now possible to compute the control $u(t)$ allowing us to obtain at $t + d + 1$ an output $y(t + d + 1) = y^*(t + d + 1)$ where $y^*(t + d + 1)$ is the desired output at $t + d + 1$.

Let us considers the example of Equation B.4. The condition $y(t + 2) = y^*(t + 2)$ can be expressed using Equations B.7 and B.8 as

$$y^*(t + 2) = a_1^2 y(t) - a_1 b_1 u(t - 1) + b_1 u(t)$$

from which one gets

$$u(t) = \frac{y^*(t + 2) - a_1^2 y(t) + a_1 b_1 u(t - 1)}{b_1}$$

In the general case one would like to obtain

$$P(q^{-1})y(t + d + 1) = P(q^{-1})y^*(t + d + 1) \tag{B.26}$$

and one gets from Equation B.14:

$$P(q^{-1})y^*(t + d + 1) = F(q^{-1})y(t) + E(q^{-1})B^*(q^{-1})u(t) \tag{B.27}$$

and accordingly

$$u(t) = \frac{P(q^{-1})}{E(q^{-1})B^*(q^{-1})} y^*(t + d + 1) - \frac{F(q^{-1})}{E(q^{-1})B^*(q^{-1})} y(t) \tag{B.28}$$

Introducing the notations

$$R(q^{-1}) = F(q^{-1}) \tag{B.29}$$

$$S(q^{-1}) = b_1 + q^{-1}S^*(q^{-1}) = E(q^{-1})B^*(q^{-1}) \tag{B.30}$$

$$T(q^{-1}) = P(q^{-1}) \tag{B.31}$$

Equation B.26 becomes

$$S(q^{-1})u(t) = T(q^{-1})y^*(t+d+1) - R(q^{-1})y(t) \tag{B.32}$$

that corresponds to an RST digital controller given in Chapter 3, Section 3.4.

Since in the general case $u(t)$ has been computed such that Equation B.27 be satisfied at each sampling instant, it results from Equations B.24 and B.27 that

$$P(q^{-1})\hat{y}(t+1) = P(q^{-1})y^*(t+1) \tag{B.33}$$

and taking in account the definition of $\varepsilon(t+1)$ given in Equation B.24 as well as Equation B.25, one gets

$$P(q^{-1})[y(t+d+1) - y^*(t+d+1)] = 0 \quad , \forall t > 0 \tag{B.34}$$

which can be considered as the time domain objective of the tracking and regulation with independent objectives.

To summarize: the problem of the design of the control law is to compute

$$u(t) = f_u(y(t), y(t-1),..., u(t-1), u(t-2),...) \tag{B.35}$$

such that Equation B.34 be satisfied (with $P(q^{-1})$, an asymptotically stable polynomial).

The solution is obtained in two stages:

Stage I: A $d+1$ step ahead predictor of the output is designed such that the prediction error satisfies

$$P(q^{-1})\varepsilon(t+d+1) = 0$$

Stage II: Compute $u(t)$ such that:

$$P(q^{-1})\hat{y}(t+d+1) = P(q^{-1})y^*(t+d+1) \quad , \forall t > 0$$

Note that the same result is obtained if the criterion of Equation B.35 is replaced by the one step ahead quadratic criterion

$$\min_{u(t)} J(t+d+1) = \left\{ P(q^{-1}) \left[y(t+d+1) - y^*(t+d+1) \right] \right\}^2$$

Pole Placement
To get the pole placement, the time domain objective B.35 is replaced by

$$P(q^{-1})[y(t+d+1) - \beta B^*(q^{-1})y^*(t+d+1)] = 0 \quad \forall t > 0 \qquad (B.36)$$

with $\beta = \dfrac{1}{B^*(1)}$ and imposing the restriction that $B^*(q^{-1})$ should not appear in the denominator of the controller. This restriction requires a different structure for the output predictor to be considered:

$$P(q^{-1})y(t+d+1) = B^*(q^{-1})R(q^{-1})y(t) + B^*(q^{-1})S(q^{-1})u(t) \qquad (B.37)$$

Using Equation B.13, Equation B.37 becomes

$$P(q^{-1})y(t+d+1) = [AS + q^{-d-1}B^* R]y(t+d+1) \qquad (B.38)$$

and therefore polynomials S and R are solutions of the polynomial equation

$$P(q^{-1}) = A(q^{-1})S(q^{-1}) + q^{-d-1}B^*(q^{-1})R(q^{-1}) \qquad (B.39)$$

which is exactly the pole placement equation discussed in Chapter 3, Section 3.3. Introducing Equation B.38 into Equation B.36 and taking into account Equation B.13 one gets the RST controller of Equation B.32 where S and R are solutions of Equation B.39 and $T(q^{-1}) = \beta P(q^{-1})$.

B.4 An Interpretation of the Control of Systems with Delay

We will show next that the RST controller for a system with a delay d can be decomposed in a d step ahead predictor for the plant output and an RST controller for the system without delay which uses instead of the measured output, the output of the predictor.

Consider first a system without delay $(d = 0)$ characterized by the discrete time model

$$A(q^{-1}) y(t+1) = B^*(q^{-1}) u(t) \qquad (B.40)$$

The desired closed loop poles are defined by a polynomial $P(q^{-1})$. Using the *pole placement*, the controller equation is

$$u(t) = \frac{T(q^{-1})y*(t+1) - R_0(q^{-1})y(t)}{S_0(q^{-1})} \quad \text{(B.41)}$$

where $S_0(q^{-1})$ and $R_0(q^{-1})$ are the solution of the polynomial equation

$$A(q^{-1})S_0(q^{-1}) + q^{-1}B*(q^{-1})R_0(q^{-1}) = P(q^{-1}) \quad \text{(B.42)}$$

and $T(q^{-1}) = \beta P(q^{-1})$.

The transfer operator between the reference trajectory and the output has the expression

$$H_0(q^{-1}) = \frac{q^{-1}T(q^{-1})B*(q^{-1})}{A(q^{-1})S_0(q^{-1}) + q^{-1}B*(q^{-1})R_0(q^{-1})} = \frac{q^{-1}T(q^{-1})B*(q^{-1})}{P(q^{-1})} \quad \text{(B.43)}$$

Consider now a system with an integer delay d characterized by

$$A(q^{-1})y(t+d+1) = B^*(q^{-1})u(t) \quad \text{(B.44)}$$

The desired closed loop poles are defined by the same polynomial $P(q^{-1})$ as above. Using the *pole placement*, the controller equation will be

$$u(t) = \frac{T(q^{-1})y*(t+d+1) - R_d(q^{-1})y(t)}{S_d(q^{-1})} \quad \text{(B.45)}$$

where $R_d(q^{-1})$ et $S_d(q^{-1})$ are solutions of the polynomial equation

$$A(q^{-1})S_d(q^{-1}) + q^{-d-1}B*(q^{-1})R_d(q^{-1}) = P(q^{-1}) \quad \text{(B.46)}$$

The transfer operator between the reference trajectory and the output has the expression

$$H_d(q^{-1}) = \frac{q^{-d-1}T(q^{-1})B*(q^{-1})}{A(q^{-1})S_d(q^{-1}) + q^{-d-1}B*(q^{-1})R_d(q^{-1})} = \frac{q^{d-1}TB*}{P} \quad \text{(B.47)}$$

We will show next that this controller can be equivalently represented as shown in Figure B.1a. The controller of Equation B.47 is replaced by the controller for the system without delay given by Equation B.41 which uses instead of the measured output, the d step ahead predicted output given by

$$\hat{y}(t+d/t) = F(q^{-1})y(t) + q^{-1}E(q^{-1})B^*(q^{-1})u(t) \quad \text{(B.48)}$$

where $E(q^{-1})$ and $F(q^{-1})$ are solutions of the polynomial equation (for prediction)[2]

$$A(q^{-1}) E(q^{-1}) + q^{-d} F(q^{-1}) = 1 \qquad (B.49)$$

Using the predictor Equation B.48, the scheme given in Figure B.1a takes the equivalent form shown in Figure B.1b.

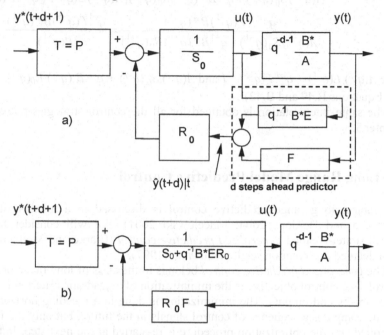

a)

b)

Figure B.1a,b. Equivalent representation of RST digital controllers for systems with delay

The control signal is given by Equation B.41 in which $y(t)$ is replaced by the predicted output $\hat{y}(t + d / t)$, which leads to

$$S_0(q^{-1})u(t) = T(q^{-1})y^*(t+1) - R_0(q^{-1})\hat{y}(t+d/d) \qquad (B.50)$$

Introducing in Equation B.50 the expression of $\hat{y}(t + d / t)$ given by Equation B.48 one gets

$$[S_0(q^{-1}) + q^{-1}E(q^{-1})B^*(q^{-1})R_0(q^{-1})]u(t) = T(q^{-1})y^*(t+d+1) - R_0(q^{-1})F(q^{-1})y(t)$$

---

[2] If additional auxiliary poles are considered, they can be interpreted as poles of the predictor. In this case in Equation B.49, the right hand side is replaced by a polynomial defining the auxiliary poles. See Landau *et al.* (1997).

which corresponds to the controller shown in Figure B.3b. If one computes now from Figure B.1b, the transfer operator between the reference trajectory and the output one gets

$$H_d = \frac{q^{-d-1}T(q^{-1})B*(q^{-1})}{A(q^{-1})S_0(q^{-1})+q^{-1}B*(q^{-1})R_0(q^{-1})[A(q^{-1})E(q^{-1})+q^{-d}F(q^{-1})]}$$

$$= \frac{q^{-d-1}T(q^{-1})B*(q^{-1})}{A(q^{-1})S_0(q^{-1})+q^{-1}B*(q^{-1})R_0(q^{-1})} = \frac{q^{-1}T(q^{-1})B*(q^{-1})}{P(q^{-1})}$$

since $A(q^{-1}) E(q^{-1}) + q^{-d} F(q^{-1}) = 1$ and $A(q^{-1}) S_0(q^{-1}) + q^{-1} B^*(q^{-1}) R_0(q^{-1}) = P(q^{-1})$ (see Equations B.49 and B.42).

The same interpretation is obtained for all the control strategies presented in Chapter 3.

B.5 Long Range Model Predictive Control

The subject long range predictive control is discussed in a number of books (Camacho and Bordons 2004; Macejowski 2001). We will consider here the control strategy called *generalized predictive control* (Clarke and Mothadi 1987). For a detailed presentation see Landau *et al.* (1997).

The *generalized predictive control* belongs to the class of long range predictive control. The control objective is the minimization of a quadratic criterion involving future inputs and outputs. The minimization is done in a receding horizon sense, *i.e.* one computes a sequence of control signals in the future, but only the first one is applied and the optimization procedure is re-started at the next step. Indeed the resulting control law is stationary and the controller takes the form of a linear digital RST controller. With an appropriate formulation of the performance criterion, generalized predictive control can be interpreted as a generalization of the various *one step ahead* control strategies. However the generalized predictive control can be also viewed as a way of assigning the desired closed loop poles starting from a time domain performance criterion.

In order to make a direct connection with pole placement it is convenient to make the following change of variables:

$$e_y(t + j) = y(t + j) - \beta B*(q^{-1})y*(t + j) \tag{B.51}$$

$$e_u(t + j) = u(t + j) - \beta A(q^{-1})y*(t + j + d + 1) \tag{B.52}$$

where e_y and e_u define a new output and a new input respectively. They are related by

$$A(q^{-1})e_y(t + d + 1) = B*(q^{-1})e_u(t) \tag{B.53}$$

One considers as control objective the minimization of the quadratic criterion

$$J(t,h_p,h_c,h_i) = \sum_{j=h_i}^{h_p} \left\{ \left[e_y(t+j)\right]^2 + \lambda \left[e_u(t+j-h_i)\right]^2 \right\}; \quad \lambda > 0 \tag{B.54}$$

where

- h_p – prediction horizon
- h_c – control horizon
- h_i – initial horizon ($\geq d + 1$)

The criterion of Equation B.54 should be minimized in a receding horizon sense with respect to the vector of control signals

$$E_u^T(t+h_c-1) = [e_u(t), e_u(t+1),....e_u(t+h_c-1)] \tag{B.55}$$

subject to the constraint:

$$e_u(t+i) = 0 \quad h_c \leq i \leq h_p$$

The initial horizon h_i is taken equal or larger than the delay $d+1$. The prediction horizon h_p is usually taken close to the time response of the system. The control horizon h_c will define the complexity of the optimization procedure and the objective is to take it as small as possible. However instabilities may occur for too short control horizons. If $h_p, h_c \to \infty$ one gets the infinite horizon quadratic criterion and this is discussed in Appendix C.

It is important to note that for $h_p = h_i = d + 1$, $h_c = 1$ one step ahead control strategies will be obtained. If in addition $\lambda = 0$ one obtains the pole placement. For a more detailed discussion see Landau *et al.* (1997).

The minimization of the criterion of EquationB.54 involves two steps:

1. Computation of the future values of $e_y(t+j)$ for $j \in [h_i, h_p]$ as

$$e_y(t+j) = f[E_u(t+h_c-1] + \hat{e}_y^0(t+j/t) \tag{B.56}$$

where $\hat{e}_y^0(t+j/t)$ is the prediction based on the available information up to the instant t:

$$\hat{e}_y^0(t+j/t) = f[e_y(t), e_y(t-1),...e_u(t-1), e_u(t-2),....]$$

2. Minimization of the criterion of Equation B.54 with respect to $E_u(t + h_c - 1)$.

We will give now the details of these two steps for the computation of the control law.

Step 1: Computation of the Future Values of $e_y(t + j)$

Using the polynomial equation

$$P_D = AE_j + q^{-j}F_j \tag{B.57}$$

with $\deg E_j = j - 1$ and $\deg F_j = \max(n_{P_D} - j, n_A - 1)$, one can express the filtered values of $e_y(t + j)$, which are used in the criterion of Equation B.54, as follows (taking also in account Equation B.53):

$$P_D e_y(t + j) = E_j B^* e_u(t + j - d - 1) + F_j e_y(t) \tag{B.58}$$

The effect of using a filtered prediction equation is that the resulting closed loop characteristic polynomial will contain a factor equal to the P_D polynomial. In other words one can assign by this choice a number of poles of the closed loop.

To get the desired form of the prediction indicated in Equation B.56 one should use a second polynomial equation

$$B^* E_j = P_D G_{j-d} + q^{-j+d} H_{j-d} \tag{B.59}$$

where

$$\deg G_{j-d} = j - d - 1$$
$$\deg B^* E_j = n_{B^*} - j - 1$$
$$\deg H_{j-d} = n_{B^*} - d - 1$$

The polynomials G_{j-d} and H_{j-d} will have the form

$$G_{j-d} = g_0 + g_1 q^{-1} + ... + g_{j-d-1} q^{-j-d-1} \tag{B.60}$$

$$H_{j-d} = h_0^{j-d} + h_1^{j-d} q^{-1} + ... + h_{n_{B^*}-d-1}^{j-d} q^{-n_{B^*}-d-1} \tag{B.61}$$

Using Equation B.59, the filtered prediction $e_y(t + j)$ given in Equation B.58 can be written

$$P_D e_y(t+j) = P_D G_{j-d} e_u(t+j-d-1) + H_{j-d} e_u(t-1) + F_j e_y(t) \qquad \text{(B.62)}$$

Passing this expression through the transfer operator $1/P_D$ and denoting

$$P_D \hat{e}_y^0(t+j/t) = H_{j-d} e_u(t-1) + F_j e_y(t) \qquad \text{(B.63)}$$

one gets

$$e_y(t+j) = G_{j-d} e_u(t+j-d-1) + \hat{e}_y^0(t+j/t) \qquad \text{(B.64)}$$

with the important remark that the term $G_{j-d} e_u(t+j-d-1)$ will contain $e_u(t), e_u(t+1), ... e_u(t+j-d-1)$, i.e. only future values of e_u since $j \geq h_i \geq d+1$ [3].

Step 2 Minimization of the Criterion
Introducing the notation

$$E_y^T = [e_y(t+h_i), ... e_u(t+h_p)] \qquad \text{(B.65)}$$
$$E_u^T = [e_u(t), ... e_u(t+h_c-1)] \qquad \text{(B.66)}$$

the criterion of Equation B.54 takes the form:

$$J = E_y^T E_y + \lambda E_u^T E_u \qquad \text{(B.67)}$$

Taking in account the expression of $e_y(t+j)$ given in Equation B.64 and introducing the notation

$$E_y^{0^T} = [\hat{e}_y^0(t+h_i/t), ... \hat{e}_y^0(t+h_p/t)] \qquad \text{(B.68)}$$

one can re-write Equation B.65 using Equations B.64 and B.68 as

$$E_y = G E_u + E_y^0 \qquad \text{(B.69)}$$

where G is a matrix formed with the coefficients of the polynomials G_{j-d} :

---

[3] At instant t, $e_y(t)$ is available before the computation of $e_u(t)$.

$$G = \begin{bmatrix} g_{h_i-d-1} & \cdots & g_0 & 0 & \cdots & 0 \\ g_{h_i-d} & \cdots & g_1 & g_0 & 0 & 0 \\ \vdots & & & & \ddots & \\ g_{h_c-d} & \cdots & \cdots & \cdots & \cdots & g_0 \\ g_{h_d-d-1} & \cdots & \cdots & \cdots & \cdots & g_{h_p-h_c-1} \end{bmatrix} \qquad (B.70)$$

The criterion of Equation B.67 will take the form

$$J(t, h_p, h_c, h_i) = [GE_u + E_y^0]^T [GE_u + E_y^0] + \lambda E_u^T E_u \qquad (B.71)$$

Minimization of this criterion is obtained by searching for E_u assuring $\dfrac{\delta J}{\delta E_u} = 0$, which yields

$$E_{u\,opt} = -[G^T G + \lambda I_{h_c}]^{-1} G^T E_y^0 \qquad (B.72)$$

This formula clearly indicates that the problem always has a solution provided that $\lambda > 0$ (however the stability of the resulting closed loop has to be checked).

For an effective implementation of the control one only needs the first component of E_u, namely $e_u(t)$, and, as a consequence, only the first row of the matrix $[G^T G + \lambda I_{h_c}]^{-1} G^T$ is of interest. Therefore one obtains

$$e_u(t) = -[\gamma_{h_i}, \ldots, \gamma_{hp}] E_y^0 \qquad (B.73)$$

where γ_i are the coefficients of the first row of $[G^T G + \lambda I_{h_c}]^{-1} G^T$.

Using Equations B.63, Equation B.73 can be further rewritten as:

$$P_D e_u(t) = -P_D [\gamma_{h_i}, \ldots, \gamma_{hp}] E_y^0$$

$$= -\left(\sum_{j=h_i}^{j=h_p} \gamma_j H_{j-d} \right) e_u(t-1) - \left(\sum_{j=h_i}^{j=h_p} \gamma_j F_j \right) e_y(t)$$

or as

$$\left[P_D + q^{-1} \sum_{j=h_i}^{j=h_p} \gamma_j H_{j-d} \right] e_u(t) + \left(\sum_{j=h_i}^{j=h_p} \gamma_j F_j \right) e_y(t) = \qquad (B.74)$$

$$= S(q^{-1}) e_u(t) + R(q^{-1}) e_y(t) \qquad\qquad = 0$$

where the polynomials S and R correspond to

$$S(q^{-1}) = P_D(q^{-1}) - q^{-1} \sum_{j=h_i}^{j=h_p} \gamma_j H_{j-d}(q^{-1}) \qquad (B.75)$$

$$R(q^{-1}) = \sum_{j=h_i}^{j=h_p} \gamma_j F_j(q^{-1}) \qquad (B.76)$$

Using now the expressions for $e_y(t)$ and $e_u(t)$ given by Equations B.51 and B.52, one gets the expression of the RST control law:

$$S(q^{-1})u(t) = T(q^{-1})y^*(t+d+1) - R(q^{-1})y(t) \qquad (B.77)$$

where

$$T(q^{-1}) = \beta P = \beta[AS + q^{-d-1}B^*R] \qquad (B.78)$$

and P defines the resulting closed loop poles corresponding to the minimization of the criterion of Equation B.54. The controller structure is the same as for the pole placement (see Chapter 3, Section 3.3). It can be shown that the resulting closed loop poles are (Landau $et\ al.$ 1997)

$$P(q^{-1}) = P_D(q^{-1})P_{GPC}(q^{-1})$$

where:

$$P_{GPC}(q^{-1}) = A(q^{-1}) + q^{-d-1} \sum_{j=h_i}^{h_p} \gamma_j q^j [B^*(q^{-1}) - A(q^{-1})G_{j-d}(q^{-1})] \quad .$$

and P_D is the polynomial defining the poles of the predictor.

If P_D is chosen such that it defines the dominant poles of the closed loop, the poles introduced by the minimization of the criterion of Equation B.54 can be interpreted as auxiliary poles. However if $P_D=1$, then the poles of the closed loop will be exclusively defined by the minimization of the criterion of Equation B.54. The values of the closed loop poles will depend of how λ, h_i, h_c and h_p will be selected.

B.6 Notes and References

For a time domain approach to internal model control of systems with delay see:

Landau I.D. (1995) Robust digital control of systems with time delay (the Smith predictor revisited), Int. J. of Control, vol. 62, no. 22, pp. 325-347.

For more details on long range model predictive control strategies see:

Landau I.D., Lozano R., M'Saad M. (1997) Adaptive control, (Chapter 7), Springer, London, UK.

Clarke D., Mothadi C. (1987) Generalized predictive control, Automatica, vol 23, pp. 137-160.
Clarke D., Mothadi C. (1989) Properties of generalized predictive control, Automatica, vol 25, pp. 859-876.
Camacho E.F., Bordons C. (2004) Model predictive control, 2nd edition, Springer, London, UK.
Macejowski J.M.(2001) Predictive control with constraints, Prentice Hall, N.J.

C

State-Space Approach for the Design of RST Controllers

C.1 State-Space Design

In Chapter 3 we discussed the pole placement design using a two degrees of freedom polynomial RST controller. The assignment of the closed loop poles is a problem that can be solved also taking a state-space formulation of the problem.

The same type of controller is obtained by designing the controller using a state-space approach. The state-space approach will emphasize the interpretation of the RST controller as a state estimator followed by a state feedback using the estimated state. This approach also provides an interesting interpretation for the dominant and auxiliary closed loop poles. Furthermore the relation for passing from a controller designed in state-space for SISO systems to a polynomial controller will be provided. This result can be used for transforming state-space controllers designed by different approaches in a polynomial form.

A detailed discussion of regulation by state feedback is beyond the purposes of this book. The reader can find exhaustive descriptions of this technique in (Phillips and Nagel 1995; Franklin *et al.* 2000). Here we just recall some fundamental facts. We will consider in this appendix the state space solutions for *pole placement* and *linear quadratic control.*

Let consider the controllable canonical space state representation of a single input single output discrete time model characterized by the transfer operator $B(q^{-1})/A(q^{-1})$ [1] (it is assumed that $A(q^{-1})$ and $B(q^{-1})$ do not have common factors, *i.e.* the system is both controllable and observable)

---

[1] The delay is included in the B polynomial (i.e. the first coefficients b_i of B will be null up to the value d of the integer delay).

$$x_c(t+1) = \begin{bmatrix} -a_1 & \cdots & \cdots & -a_n \\ 1 & & & \\ & \ddots & & \\ & & 1 & 0 \end{bmatrix} x_c(t) + \begin{bmatrix} 1 \\ 0 \\ \vdots \\ 0 \end{bmatrix} u(t)$$

(C.1)

$$y(t) = \begin{bmatrix} b_1 & \cdots & \cdots & b_n \end{bmatrix} x_c(t)$$

Suppose that the desired closed loop characteristic polynomial is specified by

$$P_c(z^{-1}) = 1 + p_1 z^{-1} + \ldots + p_n z^{-n} \tag{C.2}$$

If the full state is available, the control law that allows to obtain the specified closed loop poles is

$$u(t) = -kx_c(t) + y^*(t+d) \tag{C.3}$$

where

$$k = [k_1 \quad k_2 \quad \ldots \quad k_n] \tag{C.4}$$

is a state-space feedback gain matrix (which has to be computed) and $y^*(t)$ is the external reference.

The closed loop system in the state-space form is found by introducing Equation C.3 in Equation C.1:

$$x_c(t+1) = \begin{bmatrix} (-a_1 - k_1) & \cdots & \cdots & (-a_n - k_n) \\ 1 & & & \\ & \ddots & & \\ & & 1 & 0 \end{bmatrix} x_c(t) + \begin{bmatrix} 1 \\ 0 \\ \vdots \\ 0 \end{bmatrix} y^*(t+d)$$

$$\overset{\Delta}{=} \bar{A} x_c(t) + \bar{b} y^*(t+d)$$

(C.5)

The characteristic polynomial of the closed loop system thus is

$$P_{cl}(z^{-1}) = 1 + (a_1 + k_1)z^{-1} + \ldots + (a_n + k_n)z^{-n} \tag{C.6}$$

and it can be easily observed that, comparing Equations C.2 and C.6, the closed loop poles can be assigned with an appropriate choice of vector k ($k_i = p_i - a_i$).

If the full state $x_c(t)$ is not directly available a state observer will be implemented. We consider the canonical observable form of the plant model

$$x_o(t+1) = \begin{bmatrix} -a_1 & 1 & & \\ \vdots & & \ddots & \\ & & & 1 \\ -a_n & & & 0 \end{bmatrix} x_o(t) + \begin{bmatrix} b_1 \\ \vdots \\ \vdots \\ b_n \end{bmatrix} u(t)$$

(C.7)

$$y(t) = \begin{bmatrix} 1 & 0 & \dots & 0 \end{bmatrix} x_o(t)$$

An estimate of the state $\hat{x}_o(t)$ can be provided by the so-called *current estimator* (see Phillips and Nagel 1995; Franklin *et al.* 2000), where a description of the standard *prediction estimator* is also provided), that is described by the following expressions:

$$\bar{x}_o(t+1) = \begin{bmatrix} -a_1 & 1 & & \\ \vdots & & \ddots & \\ & & & 1 \\ -a_n & & & 0 \end{bmatrix} \hat{x}_o(t) + \begin{bmatrix} b_1 \\ \vdots \\ \vdots \\ b_n \end{bmatrix} u(t)$$

(C.8)

$$\hat{x}_o(t+1) = \bar{x}_o(t+1) + \begin{bmatrix} l_1 \\ \vdots \\ \vdots \\ l_n \end{bmatrix} [y(t+1) - \begin{bmatrix} 1 & \dots & \dots & 0 \end{bmatrix} \bar{x}_o(t+1)]$$

where $\bar{x}_o(t+1)$ is the *a priori* state estimation at time $t+1$, on the basis of the signals acquired at time t, and $\hat{x}_o(t+1)$ is the estimation depending on the current measurement at time $t+1$. The coefficients l_i are the observer gains acting on the error between the measured output and the a priori predicted output.

Consider now the state estimation error $e(t) = x_o(t) - \hat{x}_o(t)$.
Then

$$
\begin{aligned}
e(t+1) &= & x_o(t+1) & & - & & \hat{x}_o(t+1) \\
&= & Ax_o(t) & & + & & bu(t) - [A - lcA]\hat{x}_o(t) \\
& & - [b - lcb]u(t) & & - & & lc[Ax_o(t) + bu(t)] \\
&= & [A - lcA][x(t) - \hat{x}_o(t)] & & = & & [A - lcA]e(t)
\end{aligned}
$$

where

$$A = \begin{bmatrix} -a_1 & 1 & & \\ \vdots & & \ddots & \\ & & & 1 \\ -a_n & & & 0 \end{bmatrix} \quad ; \quad b = \begin{bmatrix} b_1 \\ \vdots \\ \vdots \\ b_n \end{bmatrix}$$

$$c = \begin{bmatrix} 1 & \dots & \dots & 0 \end{bmatrix} \quad ; \quad l = \begin{bmatrix} l_1 \\ \vdots \\ \vdots \\ l_n \end{bmatrix}$$

The state estimation error will converge asymptotically to zero if the matrix $[A - lcA]$ is asymptotically stable. This property will be defined by the roots of the characteristic equation

$$\left| z^{-1}I - A + lcA \right| = 0 \Rightarrow |z| < 1$$

and the estimation error can be made to converge to zero with a desired dynamics. Once the desired dynamics is imposed the parameters of l can be computed accordingly.

Assuming that the desired estimator dynamics is defined by the polynomial

$$P_E(z^{-1}) = 1 + p_{E1}z^{-1} + \dots + p_{En}z^{-n} \tag{C.9}$$

then the coefficients l_i are given by

$$l = \begin{bmatrix} l_1 \\ l_2 \\ \vdots \\ l_n \end{bmatrix} = P_E(A) \begin{bmatrix} cA \\ cA^2 \\ \vdots \\ cA^n \end{bmatrix}^{-1} \begin{bmatrix} 0 \\ 0 \\ \vdots \\ 1 \end{bmatrix}$$

where $P_E(A)$ is a matrix polynomial whose coefficients are those of $P_E(z)$:

$$P_E(A) = A^n + p_{E_1}A^{n-1} + \dots + p_{E_n}I$$

Note that, if the new state representation

$$x_{CL} = \begin{pmatrix} x(t) \\ e(t) \end{pmatrix}$$

is considered, the closed loop system matrix is expressed by

$$\begin{pmatrix} x(t+1) \\ e(t+1) \end{pmatrix} = \begin{pmatrix} A-bk & +bk \\ 0 & A-lcA \end{pmatrix} \begin{pmatrix} x(t) \\ e(t) \end{pmatrix}$$

from which it results that the closed loop poles are a combination of the estimator poles and the control poles (specified by Equations C.2 and C.9 respectively).

The characteristic equation of the closed loop system is thus defined by

$$\left| z^{-1}I - A + lcA \right| \left\| z^{-1}I - A + bk \right| = P_C(z^{-1})P_E(z^{-1}) = 0 \qquad \text{(C.10)}$$

We remark the rapprochement between the RST pole-placement and the state-space approach in terms of the assigned closed loop poles. In the RST pole placement we have specified the closed loop poles as a combination of dominant poles (slow poles) and auxiliary poles (fast poles). In the state-space pole placement the control poles (defined by Equation C.2) correspond to the dominant poles while the estimator poles (defined by Equation C.8) play the role of the auxiliary poles. Note however that one could choose the dominant poles as the estimator poles and the auxiliary poles as control poles (this is the case in the predictive control approach).

We recall that in Chapter 3 several techniques for selecting auxiliary poles have been proposed in order to obtain a robust control design. One notes that the choice of the estimator poles could be not immediate if robustness considerations are not taken into account.

Thus, the full controller will be given by the current estimator and the state feedback as shown in Figure C.1.

The corresponding pulse transfer function of the digital controller resulting from the state-space approach is given by

$$K(z^{-1}) = -\frac{U(z^{-1})}{Y(z^{-1})} = z^{-1}k\left[z^{-1}I - A + lcA + bk - lcbk\right]^{-1}l \qquad \text{(C.11)}$$

It is now straightforward to show the equivalence between the digital controller $K(q^{-1})$ resulting from the state-space approach and the digital controller $R(q^{-1})/S(q^{-1})$ obtained by pole placement design in polynomial form, when imposing the same closed loop dynamics.

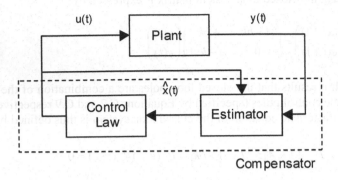

Figure C.1. State-space control scheme

Consider the transfer function of the *RS* controller

$$\frac{R(q^{-1})}{S(q^{-1})} = \frac{r_{\partial} + r_1 z^{-1} + \ldots r_{n_R} z^{-n_R}}{s_{\partial} + s_1 z^{-1} + \ldots s_{n_S} z^{-n_S}} \tag{C.12}$$

The numerator and the denominator of the *RS* controller transfer function are

$$R(z^{-1}) = z^{-1} K \, adj \left[z^{-1} I - A + lcA + bk - lcbk \right] L \tag{C.13}$$
$$S(z^{-1}) = \left| z^{-1} I - A + lcA + bk - lcbk \right|$$

Introduction of the Internal Model

Note that with the state-space design no special compensation of specific classes of disturbances (for example constant or sinusoidal disturbances) is done unless the pole placement problem in state-space is reformulated.

The basic idea in order to introduce the internal model of the disturbance in the controller is to augment the state-space representation of the model by adding the internal model and then apply the procedure presented above. If, for example, a compensation for constant disturbances is desired, then an integral action is required. This can be accomplished by augmenting the model of the plant with an integrator and recalculating the feedback control gains for the augmented state. Of course in the implementation the internal model will appear in the controller.

In the same way, for a generic class of disturbances modeled by a system matrix A_d, the augmented system matrix should contain the model of the disturbances that is

$$A_{aug} = \begin{bmatrix} A & C_d \\ 0 & A_d \end{bmatrix}$$

where C_d is the matrix that describes how the disturbance is injected into the system.

Taking into Account Tracking Performance Specifications
So far we have discussed the regulation problem. A common control problem has specifications that also involve tracking properties. A controller that takes into account both regulation and tracking requirements generally has a two degrees of freedom structure (feedback and feedforward compensator).

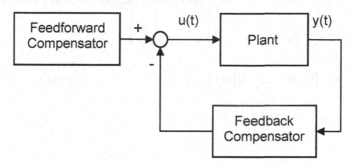

Figure C.2. Two degrees of freedom control scheme

The control signal generated by the state feedback is designed in order to have a closed loop system with desired regulation and robustness properties. The feedforward compensator is designed in order to match the desired tracking performances.

In order to obtain the standard RST control law (Equation 2.5.9, which takes into account both tracking and regulation performances) starting from a state space representation, one considers the plant model described by

$$A(q^{-1})y(t+d+1) = B*(q^{-1})u(t) \tag{C.14}$$

and the following variable transformation:

$$e_y(t) = y(t) - \frac{B(q^{-1})}{B(1)} y*(t)$$

$$e_u(t) = u(t) - \frac{A(q^{-1})}{B(1)} y*(t+d+1) \tag{C.15}$$

Left multiplying the first row of Equation C.15 with $A(q^{-1})$ and considering an instant $t+d+1$ we obtain the following relation (by taking in account Equation C.14):

$$A(q^{-1})e_y(t+d+1) = B*(q^{-1})e_u(t) \tag{C.16}$$

Note that e_y and e_u are related to each other by the same transfer operator as the plant model.

The state-space representation in observable form of the system described by Equation C.16 is:

$$x(t+1) = \begin{bmatrix} -a_1 & 1 & & \\ \vdots & & \ddots & \\ & & & 1 \\ -a_n & & & 0 \end{bmatrix} x(t) + \begin{bmatrix} \bar{b}_1 \\ \vdots \\ \vdots \\ \bar{b}_n \end{bmatrix} e_u(t) \quad = Ax(t) + \bar{b}' e_u(t)$$

$$e_y(t) = \begin{bmatrix} 1 & 0 & \cdots & 0 \end{bmatrix} x(t) \qquad\qquad = c\, x(t)$$

$$\text{(C.17)}$$

where

$$\bar{b}_i = 0 ; 0 \le i \le d ; \quad \bar{b}_{i+d} = b_i ; 1 \le i \le n_B$$

If we apply to the system of Equation C.14 both state-feedback and observer Equations C.3 and C.8 an RS controller of the form of Equation C.12 is found.

The control law for the system of Equation C.14 becomes:

$$S(q^{-1}) e_u(t) + R(q^{-1}) e_y(t) = 0 \qquad\qquad \text{(C.18)}$$

We now rewrite this equation in the initial variables $u(t)$, $y(t)$ and $y*(t)$. The control law is thus

$$S(q^{-1})u(t) = -R(q^{-1})y(t) + \left[\frac{A(q^{-1})S(q^{-1})}{B(1)} y*(t+d+1) + \frac{B(q^{-1})R(q^{-1})}{B(1)} y*(t) \right]$$

$$= -R(q^{-1})y(t) + T(q^{-1})y*(t+d+1)$$

$$\text{(C.19)}$$

where

$$T(q^{-1}) = \frac{\left[A(q^{-1})S(q^{-1}) + B(q^{-1})R(q^{-1}) \right]}{B(1)} = \frac{P(q^{-1})}{B(1)} \qquad\qquad \text{(C.20)}$$

Equation C.19 represents the two degrees of freedom RST control law, as obtained in Chapter 3, Section 3.3.

C.2 Linear Quadratic Control

Consider again the system of Equation C.14. The objective is to find an admissible control $u(t)$ which minimizes the following quadratic criterion:

$$J(t,T) = \lim_{T \to \infty} \frac{1}{T} \sum_{j=1}^{T} [y(t+j) - y*(t+j)]^2$$
$$+ \lambda [u(t+j-1+d)]^2 \qquad (C.21)$$

However in order to obtain directly a two degrees of freedom controller it is useful to use again the change of variables given in Equation C.15 and to consider the criterion:

$$J(t,T) = \lim_{T \to \infty} \frac{1}{T} \sum_{j=1}^{T} e_y(t+j)^2$$
$$+ \lambda [e_u(t+j-1+d)]^2 ; \qquad \lambda > 0 \qquad (C.22)$$

The criterion of Equation C.22, with the state-space representation of Equation C.17, takes the form:

$$J(t,T) = \lim_{T \to \infty} \frac{1}{T} \sum_{j=1}^{T} x^T(t+j)CC^T x(t+j) + \lambda e_u^2(t+j) \qquad (C.23)$$

Assuming that the state $x(t)$ is measurable, the optimal control law $e_u(t)$ is given by

$$e_u(t) = -\frac{b^T \Gamma A}{b^T \Gamma A + \lambda} x(t) \qquad (C.24)$$

where Γ is the positive definite solution of the Algebraic Riccati Equation (Åström and Wittenmark 1997):

$$A^T \Gamma A - \Gamma - A^T \Gamma b (b^T \Gamma b + \lambda)^{-1} b^T \Gamma A + c^T c = 0 \qquad (C.25)$$

The existence and unicity of the solution for Equation C.25 is guaranteed by the following conditions:

The pair (A,B) is stabilizable
The pair (A,C) is detectable

Since the state is not measurable, the certainty equivalence principle will be used and the non-accessible states will be replaced by the estimated states given by Equation C.8. A two degrees of freedom RST controller will be obtained just by replacing the controller gains in Equation C.3 by those resulting from Equation C.24.

For further details on this control strategy see Landau *et al.* (1997); Åström and Wittenmark (1997).

C.3 Notes and References

State space approach for the design of digital controllers is presented in many books. We mention:

Åström K.J., Wittenmark B. (1997) Computer-Controlled Systems, Theory and Design, 3rd edition, Prentice Hall, NJ.

Phillips C.L., Troy Nagel H. (1997) Digital Control System Analysis and Design, 3rd edition, Prentice Hall, NJ.

Goodwin G.C., Sin K.S. (1984)   Adaptive Filtering Prediction and Control, Prentice Hall, NJ.

Franklin G.F., Powell J.D., Workman M. (2000) Digital Control of Dynamic Systems, 3rd edition, Addison Wesley, CA.

Landau I.D., Lozano R., M'Saad M. (1997) Adaptive Control, Springer, London, UK.

D

Generalized Stability Margin and Normalized Distance Between Two Transfer Functions

D.1 Generalized Stability Margin

In Chapter 2, Section 2.6.2 we presented the *modulus margin*, which gives the minimum distance between the Nyquist plot of the open loop transfer function and the critical point [*-1, j0*] and is a good indicator of the robustness of the closed loop system. The modulus margin has the expression

$$\Delta M = \left(\left| S_{yp}(e^{-j\omega}) \right|_{\max_{\omega}} \right)^{-1} = \left\| S_{yp}(e^{-j\omega}) \right\|_{\infty}^{-1} \quad, \text{ for } \omega = 0 \text{ to } \pi \, f_s \qquad \text{(D.1)}$$

Effectively the maximum of the modulus of a transfer function corresponds to "H infinity norm" and is denoted H_{∞}. In other words the modulus margin is the inverse of the H_{∞} norm of the output sensitivity function.

In Section 2.6 it has also been mentioned that stability of the closed loop system requires that all the sensitivity functions be asymptotically stable. Furthermore it was shown in Section 2.6 that the uncertainties tolerated on the plant model depend upon the sensitivity functions. More specifically, the admissible uncertainties will be smaller as the maximum of the modulus of the various sensitivity functions grows.

One can ask if it is not possible to give a global characterization of the stability margin of a closed loop system and its robustness, taking simultaneously into account all the four sensitivity functions. This problem can be viewed as the generalization of the *modulus margin*.

Denoting the controller by

$$K = \frac{R(z^{-1})}{S(z^{-1})} \tag{D.2}$$

and the transfer function of the plant model by

$$G = \frac{z^{-d}B(z^{-1})}{A(z^{-1})} \tag{D.3}$$

one defines for the closed loop system (K,G) the matrix of sensitivity functions $(z = e^{j\omega})$

$$\mathbf{T}(j\omega) = \begin{vmatrix} S_{yr}(j\omega) & S_{yv}(j\omega) \\ -S_{up}(j\omega) & S_{yp}(j\omega) \end{vmatrix} \tag{D.4}$$

where $S_{yr}, S_{yv}, S_{up}, S_{yp}$ have been defined in Chapter 2, Section 2.5 (Equations 2.5.14 through 2.5.18).

In order to be able to give an interpretation similar to the modulus of a transfer function, one should factorize the matrix $\mathbf{T}(j\omega)$ in the form (singular value decomposition)

$$\mathbf{T}(j\omega) = \mathbf{U}\mathbf{S}(j\omega)\mathbf{V}^* \tag{D.5}$$

where $\mathbf{U}$ and $\mathbf{V}$ are orthonormal (unit) matrices with the property that $\mathbf{U}\mathbf{U}^* = \mathbf{U}^*\mathbf{U} = \mathbf{I}$, $\mathbf{V}\mathbf{V}^* = \mathbf{V}^*\mathbf{V} = \mathbf{I}$ (* indicates the complex conjugate) and

$$\mathbf{S}(j\omega) = \begin{bmatrix} \sigma_1(j\omega) & 0 & 0 & 0 \\ 0 & \sigma_2(j\omega) & 0 & 0 \\ 0 & 0 & \sigma_3(j\omega) & 0 \\ 0 & 0 & 0 & \sigma_4(j\omega) \end{bmatrix} \tag{D.6}$$

is a diagonal matrix (of transfer functions), with the property

$$|\sigma_1(j\omega)| \ge |\sigma_2(j\omega)| \ge |\sigma_3(j\omega)| \ge |\sigma_4(j\omega)| \quad \text{from } \omega = 0 \text{ to } \pi f_s \tag{D.7}$$

$\sigma_1(j\omega)$ is called "the largest singular value of $\mathbf{T}(j\omega)$" and it is denoted by

$$\bar{\sigma}(j\omega) = \sigma_1(j\omega) \tag{D.8}$$

One defines the *modulus* of $\mathbf{T}(j\omega)$ as follows:

$$|T(j\omega)| = |\bar{\sigma}(j\omega)| \qquad (D.9)$$

and the *maximum of the modulus* of $\mathbf{T}(j\omega)$ as

$$|T(j\omega)|_{\underset{\omega}{max}} = |\bar{\sigma}(j\omega)|_{\underset{\omega}{max}} = \|T(j\omega)\|_{\infty} \quad , \text{for } \omega = 0 \text{ to } \pi f_s \qquad (D.10)$$

Similarly to the modulus margin, one defines the *generalized stability margin* as

$$b(K,G) = \begin{cases} \left(|T(j\omega)|_{\underset{\omega}{max}}\right)^{-1} = \|T(j\omega)\|_{\infty}^{-1} & \text{if } (K,G) \text{ is stable} \\ 0 & \text{if } (K,G) \text{ is unstable} \end{cases} \qquad (D.11)$$

The generalized stability margin can be computed with the function *smarg.m* from the toolbox REDUC® (Adaptech 1999a)[1].

As the value of $b(K,G)$ decreases, the closed loop system will be close to instability and it will be less robust with respect to the variations (or uncertainties) of the plant nominal transfer function.

In what follows it will be shown that the generalized stability margin allows one to characterize the tolerated uncertainties on the plant transfer function for which the closed loop remains stable. To do this, first it is necessary to introduce the concept of normalized distance between two transfer functions.

D.2 Normalized Distance Between Two Transfer Functions

Consider a transfer function G. Let denotes the number of unstable zeros by n_{z_i} and the number of unstable poles by n_{p_i}. The number of encirclements of the origin of the complex plane is given by

$$wno\,(G) = n_{z_i}(G) - n_{p_i}(G) \qquad (D.12)$$

(positive value = counter clockwise encirclements; negative value = clockwise encirclements.). It is possible to compare two transfer functions G_1, G_2 only if they satisfy the following property:

$$wno\,(1 + G_2^* G_1) + n_{p_i}(G_1) - n_{p_i}(G_2) - n_{P_1}(G_2) = 0 \qquad (D.13)$$

---

[1] To be downloaded from the book website.

where G_2^* is the complex conjugate of G_2 and $n_{P_1}(G_2)$ is the number of poles G_2 located on the unit circle[2].

The *normalized distance* between two transfer functions satisfying the property of Equation D.13 is called the *Vinnicombe distance* or $v-gap$ (Vinnicombe 1993).

Let defines the normalized difference between two transfer functions $G_1(j\omega)$ and $G_2(j\omega)$ as

$$\Psi[G_1(j\omega),G_2(j\omega)] = \frac{G_1(j\omega)-G_2(j\omega)}{\left(1+|G_1(j\omega)|^2\right)^{1/2}\left(1+|G_2(j\omega)|^2\right)^{1/2}} \qquad (D.14)$$

The normalized distance (Vinnicombe distance) is defined by

$$\delta_v(G_1,G_2) = \left| \Psi[G_1(j\omega),G_2(j\omega)] \right|_{max}^{\omega} = \left\| \Psi[G_1(j\omega),G_2(j\omega)] \right\|_{\infty} \qquad (D.15)$$
$$\text{for } \omega = 0 \text{ to } \pi f_s$$

One observes immediately from the structure of Ψ that

$$0 \le \delta_v(G_1,G_2) < 1 \qquad (D.16)$$

If the condition of Equation D.13 is not satisfied, by definition

$$\delta_v(G_1,G_2) = 1$$

The Vinnicombe distance can be computed with the function *smarg.m* from the toolbox REDUC® (Adaptech 1999a)[3].

D.3 Robust Stability Condition

Using the *generalized stability margin* and the *Vinnicombe distance* between two transfer functions, one can express a robust stability condition (sufficient condition) for a controller K designed on the basis of the nominal model G_1 as follows. *Controller K which stabilizes model G_1 will also stabilize model G_2 if*

$$\delta_v(G_1,G_2) \le b(K,G_1) \qquad (D.17)$$

---

[2] The condition of Equation D.13 is less restrictive than the condition used in Section 2.6 where two transfer functions with the same number of unstable poles and with the same number of encirclements of the origin have been considered.

[3] To be downloaded from the book website.

This condition can be replaced by a less restrictive condition, but which should be verified at all frequencies[4]:

$$\left| \Psi[G_1(j\omega), G_2(j\omega)] \right| \leq |T(j\omega)|^{-1} \quad for\ \omega = 0\ to\ \pi\ f_s \tag{D.18}$$

D.4 Notes and References

The concepts of *Vinnicombe distance* (v–gap) and *generalized stability margin* are extremely useful in practice since with one number it is possible to characterize either the distance between two transfer functions or the robustness of a controller. The original reference is:

Vinnicombe G. (1993) Frequency domain uncertainty and the graph topology, IEEE Trans. on Automatic Control, vol. 38, no. 9, pp. 1371-1383.

For a good pedagogical presentation, but with extensive use of the H_∞ norm, see :

Zhu K.,(1998) Essentials of robust control, Prentice Hall, N.J., U.S.A.

These concepts have been very useful for various applications. In particular they have been used for the validation of controllers resulting from controller reduction techniques (see Chapter 10) and for the validation of models identified in closed loop (see Chapter 9). Details and examples can be found in :

Landau I.D., Karimi A., Constantinescu A. (2001) Direct controller reduction by identification in closed loop, Automatica, vol. 37, no. 11, pp. 1689-1702.
Adaptech (1999a) REDUC® – Controller order reduction by closed-loop identification (Toolbox for MATLAB®), Adaptech, 4 rue de la Tour de l'Eau, St. Martin d'Hères, France.

---

[4] This condition has to be compared with the conditions given in Section 2.6.2 (Equations 2.6.7, 2.6.9 and 2.6.10). Equation D.18 can be interpreted as a generalization of these conditions.

E

The Youla–Kučera Controller Parametrization

E.1 Controller Parametrization

The Youla-Kučera parametrization allows characterizing the set of all stabilizing controllers for a given plant model. This result has interesting implications in the design of RST controllers.

Let consider a plant model characterized by the transfer function

$$G(z^{-1}) = \frac{z^{-d}B(z^{-1})}{A(z^{-1})} \tag{E.1}$$

a stabilizing controller $R_0(z^{-1})/S_0(z^{-1})$ also termed the "central controller" and an asymptotically stable rational proper function

$$Q(z^{-1}) = \frac{\beta(z^{-1})}{\alpha(z^{-1})} \tag{E.2}$$

The poles of the closed loop when using the central controller are defined by the polynomial

$$P = AS_0 + z^{-d}BR_0 \tag{E.3}$$

Youla – Kucera Parametrization

All rational stabilizing controllers for the plant model $G(z^{-1})$ are given by :

$$\frac{R(z^{-1})}{S(z^{-1})} = \frac{R_0 + AQ}{S_0 - z^{-d}BQ} \tag{E.4}$$

This controller will lead to the following expression of the polynomial defining the closed loop poles:

$$A(\alpha S_0 - \beta z^{-d} B) + z^{-d} B(\alpha R_0 + A\beta) = \alpha(AS_0 + z^{-d} BR_0) \tag{E.5}$$

Comparing Equation E.5 with Equation E.3 and taking in account the hypotheses made upon $Q(z^{-1})$, it results that the closed loop poles will be asymptotically stable. One can also see that the denominator of $Q(z^{-1})$ appears in the characteristic polynomial defining the closed loop poles (*i.e.* it can serve to introduce additional poles at the desired location).

The structure of the Youla–Kučera parametrized controller is shown in Figure E.1.

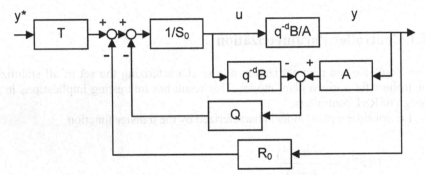

Figure A.5.1. The Youla – Kucera RST parametrized controller

For a review of Youla-Kučera parametrization see Anderson (1998). We will briefly present next the implications of the Youla-Kučera parametrization in the design of RST controllers. Let assume that a *central controller* has been designed such that the closed loop is asymptotically stable and that we are looking to the redesign of the controller in order to achieve some objectives but without re-computing the *central controller*. Instead we will tune $Q(z^{-1})$ in order to achieve the desired properties of the closed loop. We will examine next some of these situations.

Introduction of Auxiliary Poles (for Robustness)
From Equation E.5 one can see that desired additional auxiliary poles can be introduced by taking

$$Q(z^{-1}) = \frac{1}{\alpha(z^{-1})} = \frac{1}{P_F(z^{-1})} \tag{E.6}$$

Affine Parametrization of the Sensitivity Functions
Using the Q polynomial given in Equation E.2, the sensitivity functions given in Chapter 2, Equations 2.5.14 through 2.5.16 and 2.5.18 will take the form:

$$
\begin{aligned}
S_{yp} &= \frac{AS}{P} = \frac{AS_0}{P} - \frac{q^{-d}B}{P}Q = S_{yp0} - \frac{q^{-d}B}{P}Q \\
S_{up} &= -\frac{AR}{P} = -\frac{AR_0}{P} - \frac{A}{P}Q = S_{up0} - \frac{A}{P}Q \\
S_{yb} &= -\frac{q^{-d}BR}{P} = -\frac{q^{-d}BR_0}{P} - \frac{A}{P}Q = S_{yb0} - \frac{A}{P}Q \\
S_{yv} &= \frac{q^{-d}BS}{P} = \frac{q^{-d}BS_0}{P} - \frac{q^{-d}B}{P}Q = S_{yv0} - \frac{q^{-d}B}{P}Q
\end{aligned}
\tag{E.7}
$$

One distinguishes two cases :

$Q = \beta(z^{-1}); \; \alpha(z^{-1}) = 1$: this allows tuning the sensitivity functions by using the coefficients of $\beta(z^{-1})$ without modifying the closed loop poles

$Q = \beta(z^{-1}) / \alpha(z^{-1})$: this allows tuning the sensitivity functions by using the coefficients of $\beta(z^{-1})$ and by modifying the auxiliary poles using the coefficients of polynomial $\alpha(z^{-1})$

Automatic tuning of the sensitivity functions taking advantage of this affine parametrization can be done by convex optimization. See Boyd and Barratt (1991); Rantzer and Megretsky (1994); Langer and Landau (1999).

Introduction and Tuning of the Internal Model in the Controller Without Changing the Closed Loop Poles
In the context of RST controllers the internal model corresponding to the disturbance model is the polynomial H_S (see Chapter 3), which is a factor of polynomial S. But the internal model can be alternatively introduced using the Youla-Kučera parametrization. In fact one should select β such that

$$
S = S' H_S = S_0 - z^{-d}B\beta
\tag{E.8}
$$

In order to leave unchanged the poles of the closed loop one should take $\alpha(z^{-1}) = 1$ and $\beta(z^{-1})$ should satisfy the following polynomial equations resulting from the condition of Equation E.8 :

$$
S' H_S + -z^{-d}B\beta = S_0
\tag{E.9}
$$

where $z^{-d}B, H_S$ and S_0 are known and β and S' are unknown. The order of the polynomial S_0 is

$$n_{S_0} \le n_{H_S} + n_B + d - 1$$

and therefore

$$n_\beta = n_{H_s} - 1; n_{S'} = n_B + d - 1.$$

This allows determination of the order and coefficients of the polynomial $\beta(z^{-1}) = \beta_0 + \beta_1 z^{-1} + \ldots\ldots$ See Tsypkin (1997); Landau *et al.* (2005).

E.2 Notes and References

For a tutorial on the Youla-Kučera parametrization see:

Anderson B.D.O. (1998) From Youla Kučera to identification, adaptive and nonlinear control, Automatica vol. 34, 1485-1506.

For the use of the Youla-Kučera parametrization in convex optimization see:

Boyd St.P, Barratt C.H. (1991) Linear controller design. Limits of performance, Prentice Hall, Englewood Cliffs.

Langer J., Landau I.D. (1999) Combined pole placement/sensitivity function shaping method using convex optimization criteria, Automatica, vol.35, 1111-1120.

Rantzer A., Megretski A. (1994) A convex parametrization of robustly stabilizing controllers, IEEE Trans. Aut. Control. Vol. 39, pp. 1802-1808.

For the introduction of the internal model in Youla-Kučera parametrized controllers see:

Tsypkin Y.Z. (1997) Stochastic discrete systems with internal models, Journal of Automation and information Sciences, Vol. 29 pp. 156-161.

Landau I.D., Constantinescu A., Rey D. (2005) Adaptive narrow band disturbance rejection applied to an active suspension – an internal model approach, Automatica vol. 41, no.4.

The Adaptation Gain Updating – The U–D Factorization

F.1 The U-D Factorization

The adaptation gain given by Equation 5.2.73 is in principle sensitive to round-off errors, in particular if the number of samples to be processed is very high (typical situation encountered in real time applications). The objective is to preserve the positive definite property of the adaption gain matrix despite round-off errors. This problem is comprehensively discussed in Bierman (1977) where the *U-D* factorization has been developed in order to ensure the numerical robustness of the adaptation gain. To this end the adaptation gain matrix is rewritten as follows:

$$F(t) = U(t)D(t)U(t)^T \tag{F.1}$$

where $U(t)$ is an upper triangular matrix with all diagonal elements equal to 1 and $D(t)$ is a diagonal matrix (with positive elements). This allows the adaptation gain matrix to remain positive definite despite the presence of the rounding errors.

Let

$$G(t) = D(t)V(t)$$
$$V(t) = U(t)^T \Phi(t)$$
$$\beta(t) = 1 + V(t)^T G(t)$$
$$\delta(t) = \frac{\lambda_1(t)}{\lambda_2(t)} + V(t)^T G(t) \tag{F.2}$$

Define:

$$\Gamma(t) = \frac{U(t)G(t)}{\beta(t)} = \frac{F(t)\Phi(t)}{1 + \Phi(t)^T F(t)\Phi(t)} \tag{F.3}$$

The U-D factorization algorithm for adaptation gain updating is given next.

Initialize $U(0)$ and $D(0)$ at time $t=0$. This provides the initial value of the adaptation gain matrix $F(0) = U(0)D(0)U(0)^T$. At time $t+1$, compute $\Gamma(t)$ while updating $D(t+1)$ and $U(t+1)$ by performing the following steps

- Compute $V(t)$, $G(t)$, $\beta_0 = 1$ and $\delta_0 = \lambda_1(t)/\lambda_2(t)$;
- For $j=1$ to n_p (number of parameters) compute

$$\beta_j(t) = \beta_{j-1}(t) + V_j(t)G_j(t)$$

$$\delta_j(t) = \delta_{j-1}(t) + V_j(t)G_j(t)$$

$$D_{jj}(t+1) = \frac{\delta_{j-1}(t)}{\delta_j(t)\lambda_1(t)}D_{jj}(t)$$

$$\Gamma_j(t) = G_j(t)$$

$$M_j(t) = -\frac{V_j(t)}{\delta_{j-1}(t)}$$

       If $j\neq 1$ then
          For $i=1$ to $j-1$ compute

$$U_{ij}(t+1) = U_{ij}(t) + \Gamma_i(t)M_j(t)$$

$$\Gamma_i(t) = \Gamma_i(t) + U_{ij}(t)\Gamma_j(t)$$

          end
        end
      end
- For $i=1$ to n_p do

$$\Gamma_i(t) = \frac{1}{\beta_{n_p}(t)}\Gamma_i(t)$$

      end

A lower bound on the adaptation gain is simply obtained by maintaining the values of the elements of the diagonal matrix $D(t)$ above some specified threshold.

The U-D algorithm is implemented in the function: *udrls.m* (MATLAB®) (see the book website).

F.2 Notes and References

Bierman, G. (1977) Factorization methods for discrete sequential estimation, Academic Press, New York.

Landau I.D., Lozano R., M'Saad M. (1997) Adaptive control, Springer, London, U.K.

G

Laboratory Sessions

In this appendix we propose several laboratory sessions in order to help the reader to become familiar with the techniques presented in this book. The sessions described in this Appendix can be worked out with the help of the MATLAB® and Scilab control design and identification functions and toolboxes available from the book website (*hhtp://landau-bookic.lag.ensieg.inpg.fr*). Additional laboratory sessions can be developed by the reader from the examples and applications presented in Chapters 3, 4, 7, 8 and 9. Other laboratory sessions are described on the book website.

G.1 Sampled-data Systems

Objective
To become familiar with the discrete-time representation of a physical system.

Functions to be used: cont2disc.m (cont2disc.sci) and standard MATLAB®/Scilab functions.

Sequence of operations

1. Study the step response of the first-order discrete-time model corresponding to the ZOH discretization of a first-order continuous-time model $G/(1+\tau s)$:

$$H(z^{-1}) = \frac{b_1 z^{-1}}{1 + a_1 z^{-1}}$$

for $a_1 = -0.2$; -0.5 ; -0.7 ; -0.9 by choosing b_1 such that the static gain $b_1 / (1 + a_1) = 1$ for all models. Superimpose the curves obtained on the same plot for an easy comparison.

2.   Find for all models the equivalent time constant (normalized with respect to the sampling period).

Reminder: the time constant is equal to the time necessary to reach the 63% of the final value.

3.   Question: Does $a_1 = - 0.2$ correspond to a fast sampling of a continuous time first order system, or to a slow one ?
Same question for $a_1 = - 0.9$.

4.   What are the approximated values for a_1 if the sampling periods $T_s = T / 4$ and $T_s = T$ (T = time constant) are chosen?

5.   Study the response in time domain for $a_1 = 0.5$ and 0.7. Give an interpretation of the results.

6.   Study the effect caused by the sampling period and the fractional delay on the discrete time model obtained by discretizing with a ZOH the following system:

$$H(s) = \frac{G.e^{-sL}}{1 + sT}$$

for $G = 1$, $T = 10s$; $L = 0$; $0.5s$; $1s$; $2s$; $3s$ and $T_s = 5s$ and $10s$.

Analyze the properties of poles and zeros of the pulse transfer function for all models.

G.2 Digital PID Controller

Objective
To become familiar with the digital PID and to emphasize the importance of using a three-branched digital controller (RST).

Functions to be used: cont2disc.m (cont2disc.sci), bezoutd.m (bezoutd.sci) and standard MATLAB®/Scilab functions.

Sequence of operations

1.   Find the discrete-time model from the ZOH discretization of

$$H(s) = \frac{G.e^{-sL}}{1 + sT} \quad for\ G = 1, T = 10s,\ L = 3s\ and\ T_s = 5s$$

2. Find the digital PID1[1] controller with a continuous-time equivalent which provides the best performances, corresponding to a second order system with a damping equal to 0.8.
3. Once the PID is computed, perform a simulation of the closed loop in the time domain. Use a rectangular wave as reference.
    Is it possible to obtain a closed loop response faster than the open loop response for this ratio of time delay/time constant equal to 0.3?
4. Find the digital PID1 with no continuous time equivalent that provides a closed loop rise time approximately equal to half of the open loop one.
5. Perform a simulation in the time domain. Explain the possible overshoot from the values of $R(q^{-1})$.
6. In the same conditions as the point 4, compare the digital PID controllers type 1 and 2.

G.3 System Identification

Objectives
Session G.3-1: to become familiar with the recursive parameter identification methods and the model validation procedures.

Session G.3-2: to identify two real systems (a distillation column and a flexible robot arm) directly from I/O data.

   Functions to be used: rls.m, rels.m, oloe.m, afoloe.m, foloe.m, estorderiv.m, vimaux.m, xoloe.m, olvalid.m (or the corresponding Scilab functions) and standard MATLAB®/Scilab functions.

Session G.3-1

Data files
The files to be used are TØ, T1 and XQ, each one containing 256 I/O samples[2]. The files should be centered.
   The input signal is a PRBS generated by a shift register with $N = 8$ cells. The file TØ has been obtained by simulating the following model:

$$A(q^{-1}) \, y(t) = q^{-d} B(q^{-1}) \, u(t) \; ; \quad u \text{ - input, } y \text{ - output}$$

$$A(q^{-1}) = 1 - 1.5 \, q^{-1} + 0.7 \, q^{-2} \; ; \; B(q^{-1}) = 1 \, q^{-1} + 0.5 \, q^{-2} \; ; \; d = 1$$

---

[1] Digital PID 1 and 2 controllers are presented in Chapter 3, Section 3.2.
[2] Files available on the website: *http//:landau-bookic.lag.ensieg.inpg.fr.*

The file T1 has been generated with the same polynomials A and B, but a stochastic disturbance has been added to the model output.

The file XQ is given with no *a priori* information (degree of $A \leq 5$).

Sequence of operations

1. Use structure 1 and the RLS (Recursive Least Squares) method with a decreasing adaptation gain to identify the model corresponding to TØ (with $n_A = 2$, $n_B = 2$, $d = 1$).
2. Try with degrees $B = 4$ and $d = 0$ on the same file TØ using the same algorithm previously used, to see how a pure time delay can be found out (the first coefficient of B is much smaller than the second one).
3. Identify the model corresponding to the file T1 using structure 1 and the recursive least squares method with a decreasing adaptation gain (as for TØ). Note the bias of the identified parameters. Try to validate the model identified.
4. Try other structures and methods to improve the results (S3 then S2).
5. When a satisfactory validation is obtained, compute and plot the step response.
6. Perform a complexity estimation (values of n_A, n_B, d) with the different methods described in Chapters 6 and 7.
7. Identify the model corresponding to XQ.

Session G.3-2

Data files: QXD, ROB2
QXD: this file contains 256 I/O samples that describe the interaction between the heating power on the bottom of a binary distillation column and the concentration of product on the top. The input signal is a PRBS generated by a shift register with eight cells. The sampling period is 10 s. *A priori* knowledge: for the distillation columns the degree of polynomial A is generally not greater than 3. The file has to be centered.

Sequence of operations
Identify the model(s) corresponding to the I/O data set, validate the models identified, and plot the step response(s).

Question 1: what is the value of the plant rise time?

Question 2: has the PRBS been correctly chosen with respect to the rise time?

ROB2: 256 I/O samples corresponding to a very flexible robot arm. Input: motor torque. Output: robot arm end position. The input signal is a PRBS. The sampling frequency is 5 Hz. Any vibration mode over 2.5 Hz has been eliminated by an anti-aliasing filter. *A priori* knowledge: in the frequency region from 0 to 2.5 Hz, there are two vibration modes. The file has to be centered.

Sequence of operations
Identify the model(s) corresponding to the I/O data set, validate the models identified, and plot the frequency responses (≥ 200 points).

Question 1: at what frequencies are the vibration modes? What are the corresponding damping factors?

Question 2: what are the differences between the frequency responses of a validated model and of a model which does not pass the validation test?

G.4 Digital Control (Distillation Column Control Design)

Objective
The purpose of this session is to apply the control algorithms presented in Chapter 3 to the distillation column case study (file QXD), and to evaluate the performances in simulation.

We consider two different models for the distillation column, corresponding to a time delay $d = 1$, and $d = 2$ respectively (the sampling period is 10 s). Digital controllers will be computed for both models, and the controllers robustness will be tested by carrying on crossed simulations (Model 1 with controller 2 and Model 2 with controller 1). The models are the following:

$$d = 1$$
$$\text{M1:} \quad A(q^{-1}) = 1 - 0.5065\,q^{-1} - 0.1595\,q^{-2}$$
$$B(q^{-1}) = 0.0908\,q^{-1} + 0.1612\,q^{-2} + 0.0776\,q^{-3}$$

$$d = 2$$
$$\text{M2:} \quad A(q^{-1}) = 1 - 0.589\,q^{-1} - 0.098\,q^{-2}$$
$$B(q^{-1}) = 0.202\,q^{-1} + 0.0807\,q^{-2}$$

Functions to be used: bezoutd.m, cont2disc.m, omega_dmp.m, fd2pol.m (or the corresponding Scilab functions) and standard MATLAB®/Scilab functions.

Sequence of operations

1. Design a digital controller (with integrator) for the model M1 (then for the model M2) of the distillation column.
   This operation can be carried on by using a deterministic control algorithm well suited to the model used (depending on the zeros position with respect to the unit circle).
   The desired closed loop dynamics (regulation) is specified by a second-order system, with damping factor equal to *0.9*, such that the closed loop rise time be 20% shorter than the open loop rise time. The dynamics specified for the tracking performances is the same as for the regulation.
2. Compute the controller that guarantees the performances specified above and the robustness margins $\Delta M \geq 0.5$, $\Delta\tau \geq 10s = T_s$.

3. Perform cross-simulation tests with both controllers and both models. Give an interpretation of the results.

4. Repeat the sequence of operations to design a controller assuring an acceleration (reduction) of 50% of the closed loop rise time compared to the open loop system, an overshoot less than 5% (both tracking and regulation), and satisfying the robustness margins ($\Delta M \geq 0.5$, $\Delta \tau \geq 10s$). If both performance and robustness cannot be simultaneously satisfied, select the desired closed loop dynamics such that the resulting robustness margins match the specifications. *Question:* is it possible to assure the robustness requirements with this specification for the regulation dynamics (acceleration of 50%)?

5. Perform cross-simulation tests with both controllers and both models. Give an interpretation of the results. Do a comparison with the previous case.

G.5 Closed Loop Identification

Objective
To introduce the closed loop identification algorithms. This is carried on with a simulated closed loop system.

Functions to be used: functions contained in the CLID® toolbox.

Data files: the file simubf.mat contains the external excitation r, the input u and the output y of the plant. The external excitation is a PRBS applied on the reference. The file simubf_rst.mat contains the parameters of the RST controller. The simulated plant model to be identified is characterized by: $n_A = 2, n_B = 2, d = 0$.

Sequence of operations

1. Identify and validate a model with open loop identification techniques (from u to y) ignoring the feedback.

2. Identify plant models with different closed loop identification algorithms and the given digital controller. Validate the model. Compare the various models on the basis of validation results.

3. Compare the models obtained with an open loop identification technique with the models obtained with the closed loop identification algorithms. Check the closed loop poles proximity (between the identified closed loop poles and the computed ones using both open loop and closed loop identified models).

4. Plot the step responses and the frequency responses for the identified models.

G.6 Controller Reduction

Objective

To focus attention on controllers' complexity, and to present an effective solution to the controller implementation in the case of strict constraints imposed by the available resources. The flexible transmission example will be studied and a reduced order controller will be estimated.

Functions to be used: functions contained in the REDUC® toolbox.

Data files: the file flex_prbs.mat contains the external excitation ; the file BF_RST.mat contains the parameters of the RST controller ($n_R = n_S = 5, T = R(1)$); the file BF_mod.mat contains the discrete-time model of the plant.

Sequence of operations

1. Compute the simulated control signal on the basis of the given closed loop system (will be used in the estimation algorithms). Use the appropriate sensitivity function with respect to the reduction algorithm to be used (Closed Input Matching (S_{up}) and Closed Loop Output Matching (S_{yr}).

2. Find a reduced order RST controller with $n_R = n_S = 3, T = R(1)$ starting from the given one. Validate the resulting controller.

3. Impose $n_R = n_S = 2, T = R(1)$ in order to get a digital controller corresponding to a digital PID 2 controller.

4. Compare the results obtained with the two methods (CLIM and CLOM). Give an interpretation.

5. Plot the response for a step reference change and a step disturbance on the output for the closed loop system with the nominal controller and the reduced controllers.

H

List of Functions (MATLAB®, Scilab and C⁺⁺)

All functions listed in the table below are available on the website : *http ://landau-bookic.lag.ensieg.inpg.fr.*

Scilab Function	MATLAB® Function	C⁺⁺ function	Description
bezoutd.sci	bezoutd.m		Bezout equation ($P=AS_0H_S+BR_0H_R$) solution
cont2disc.sci	cont2disc.m		Continuous-time to discrete-time transfer function conversion ($F(s) \rightarrow F(z)$) with Zero Order Hold
fd2pol.sci	fd2pol.m		Second-order discrete-time system computation from second-order continuous-time system natural frequency and damping factor
omega_dmp.sci	omega_dmp.m		Computation of natural frequency and damping factor for a second-order continuous-time system from desired rise time and overshoot
predisol.sci	predisol.m		Predictor equation solution
nyquist_ol.sci	nyquist_ol.m		Nyquist plot
filter22.sci	filter22.m		Resonant zero-pole filter computation
	ppmaster.zip		Pole placement digital control design
	reduc.zip		Reduction of controller complexity
	clid.zip		Closed loop system identification
estorderls.sci	estorderls.m		Complexity estimation with the least squares criterion
estorderiv.sci	estorderiv.m		Complexity estimation with the instrumental variable criterion
nrls.sci	nrls .m		Non-recursive least squares
rls.sci	rls.m		Recursive least squares
	udrls.m		U-D factorized recursive least squares

rels.sci	rels.m		Recursive extended least squares
oloe.sci	oloe.sci		Output error (recursive)
vi_maux.sci	vi_maux.m		Instrumental variable with auxiliary model
foloe.sci	foloe.m		Output error with filtered observations
afoloe.sci	afoloe.m		Output error with adaptive filtered observations
xoloe.sci	xoloe.m		Output error with extended prediction model
	olvalid.m		Open loop model validation
	prbs.m	prbs.c	Pseudo random binary sequence generation
		rst.c	RST controller algorithm

Polynomials ($A, B, R, \ldots$) are considered in MATLAB® environment as vectors of coefficients. In Scilab the object "polynomial" is directly available. However, in order to have uniformity in polynomials processing, polynomials must be defined as vectors of coefficients to be used as inputs of Scilab functions. In the same way, the outputs of the Scilab functions are vectors of coefficients instead of the object "polynomial".

References

Adaptech (1996a) WinPIM⁺TR – System Identification Software, Adaptech, St. Martin d'Hères, France.

Adaptech (1997b) WinPIM-BF – Software for closed loop identification, Adaptech, St. Martin d'Hères, France.

Adaptech (1998) Optreg – Software for automated design of robust digital controllers using convex optimization (for MATLAB®), Adaptech, St. Martin d'Hères, France.

Adaptech (1999a) REDUC – Controller order reduction by closed-loop identification (Toolbox for MATLAB®), Adaptech, St. Martin d'Hères, France

Adaptech (1999b) CLID – Plant model identification in closed loop (Toolbox for MATLAB®), Adaptech, St. Martin d'Hères, France.

Adaptech (2001a) Guidelines for RST controller implementation, Adaptech, St. Martin d'Hères, France.

Adaptech (2001) Guidelines for PRBS integration, Adaptech, St. Martin d'Hères, France.

Adaptech (2004) Wintrac – Software for data acquisition and real time RST digital control, Adaptech, St. Martin d'Hères, France.

Anderson B.D.O., Liu Y. (1989) Controller reduction concepts and approaches, IEEE Trans. on Automatic Control, vol. 34, no. 8, pp. 802-812.

Anderson B.D.O. (1993) Controller reduction: moving from theory to practice, IEEE Control Magazine, vol. 13, pp. 16-25.

Anderson B.D.O. (1998) From Youla Kučera to identification, adaptive and nonlinear control, Automatica vol. 34, 1485-1506.

Anderson B.D.O, Moore J.B. (1971) Linear Optimal Control, Prentice Hall, Englewood Cliffs, N.J.

Aström K.J. (1970) Introduction to Stochastic Control Theory, Academic Press, N.Y.

Aström K.J., Hägglund I. (1995) PID Controllers Theory, Design and Tuning, 2nd edition ISA, Research Triangle Park, N.C., U.S.A.

Aström K.J., Wittenmark B. (1995) Adaptive Control, 2nd edition, Addison Wesley, Reaching, Mass.

Aström K.J., Wittenmark B. (1997) Computer Controlled Systems – Theory and Design, 3rd edition, Prentice-Hall, Englewood Cliffs, N.J.

Bendat J.S., Piersol A.G. (1971) Random Data Analysis and Measurement Procedures, John Wiley.

Béthoux G. (1976) Approche unitaire des méthodes d'identification et de commande adaptative des procédés dynamiques, Ph.D., Institut National Polytechnique de Grenoble, juillet.

Bierman, G. (1977) Factorization methods for discrete sequential estimation, Academic Press, New York.

Bourlès H., Irving E. (1991) La méthode LQG/LTR une interprétation polynomiale, temps continu/temps discret , RAIRO-APII, Vol. 25, pp. 545-568.

Box G.E.P., Jenkins G.M. (1970) Time Series Analysis, Forecasting and Control, Holden Day, S. Francisco.

Boyd St.P, Barratt C.H. (1991) Linear controller design. Limits of performance, Prentice Hall, Englewood Cliffs.

Camacho, E.F., Bordons, C. (2004) Model Predictive Control, 2nd edition, Springer, London, UK.

Candy J.V. (1986) Signal Processing - The Model Based Approach, MacGraw-Hill, N.Y.

Clarke D.W., Gawthrop P.J. (1975) A Self-Tuning Controller , Proc. IEE, vol.122, pp. 929-34.

Clarke D.W., Gawthrop P.J. (1979) Self-tuning Control, Proc. IEEE, vol. 126, pp .633-40.

Clarke D., Mothadi C. (1987) Generalized predictive control, Automatica, vol 23, pp. 137-160.

Clarke D., Mothadi C. (1989) Properties of generalized predictive control, Automatica, vol 25, pp. 859-876.

Doyle J.C., Francis B.A., Tanenbaum A.R.,(1992) Feedback Control Theory, Mac Millan, N.Y.

Dugard L., Landau I.D. (1980) Recursive Output Error Identification Algorithms Theory and Evaluation , Automatica, vol. 16, pp.443-462.

Duong H.N., Landau I.D. (1996) An I.V. based criterion for model order selection, Automatica, vol. 32, no. 6, pp. 909-914.

Eykhoff P. (1974) System Identification Parameter and State Estimation, John Wiley, London.

Fenot C., Rolland F., Vigneron G., Landau I.D. (1993) Open loop adaptive feedback control of depozited zinc in hot dip galvanizing , Control Engineering Practice, vol. 1, no. 5, pp 347-352.

Fenot C., Vigneron G., Rolland F., Landau I.D. (1993) Régulation d'épaisseur de dépôt de zinc à Sollac, Florange , Revue Générale d'Electricité, no. 11, pp. 25-30, Déc.

Forsell U., Ljung L. (1999) Closed loop identification revisited, Automatica, vol. 35, n. 7, pp. 1215-1241.

Franklin G., Powell J.D. (1986) Feedback Control of Dynamic Systems, Addison Wesley, Reading, Mass.

Franklin G.F., Powell J.D., Workman M.L. (1998) Digital Control of Dynamic Systems, 3rd edition, Addison Wesley, Reading, Mass.

Gevers M. (1993) Towards a joint design of identification and control , in Essays in Control (H.L. Trentelman, J.C. Willems, Eds), Birkhäuser, Boston, USA, pp. 111-152.

Goodwin G.C., Payne R.L. (1977) Dynamic System Identification Experiment Design and Data Analysis, Academic Press, N.Y.

Goodwin G.C., Sin K.S. (1984) Adaptive Filtering Prediction and Control, Prentice-Hall, Englewood Cliffs, N.J.

Hogg R., Graig A. (1970) Introduction to Mathematical Statistics, MacMillan, N.Y.

Isermann R. (1980) Practical aspects of process identification, Automatica, vol.16, pp. 575-587.

Isermann R. (Ed) (1981) Special Issue on System Identification, Automatica, vol 17, n° 1.

Kailath T. (1980) Linear systems, Prentice Hall, Englewood Cliffs, N.J.

Karimi A., Landau I.D. (1998) Comparison of the closed loop identification methods in terms of the bias distribution, Systems and Control Letters, vol. 34, pp. 159-167.

Kuo B. (1980) Digital Control Systems, Holt Saunders, Tokyo.

Kuo B.C. (1991) Automatic Control Systems (6th edition), Prentice Hall, N.J.

Kwakernaak H. (1993) Robust control and H_{inf} optimization – a tutorial Automatica, vol.29, pp.255-273.

Landau I.D. (1976) Unbiased recursive identification using model reference adaptive tehniques I.E.E.E., Trans. on Automatic Control, vol AC-20, n°2, pp. 194-202.

Landau I.D. (1979) Adaptive control – the model reference approach, Dekker, N.Y.

Landau I.D. (1981) Model Reference Adaptive Controllers and Stochastic Self-tuning Regulators, A Unified Approach, Trans. A.S.M.E, J. of Dyn. Syst. Meas. and Control, vol. 103, n°4, pp. 404, 416.

Landau I.D. (1982) Near Supermartingales for Convergence Analysis or Recursive Identification and Adaptative Control Schemes, Int J. of Control, vol.35, pp. 197-226.

Landau I.D. (1984) A Feedback System Approach to Adaptive filtering, IEEE Trans. on Information Theory, vol. 30, n°2, pp. 251-262.

Landau I.D. (1986) La Commande Adaptative Un Tour Guidé, in Commande adaptative - Aspects pratiques et théoriques (Landau, Dugard- Editeurs), Masson, Paris.

Landau I.D. (1990) System Identification and Control Design, Prentice Hall, Englewood Cliffs, N.J.

Landau I.D. (1993) Evolution of Adaptive Control, A.S.M.E. Transactions, Journal D.S.M.C., vol. 115, no. 2, pp. 381-391, june.

Landau I.D. (1995) Robust digital control of systems with time delay (the Smith predictor revisited), Int. J. of Control, vol. 62, pp. 325-347.

Landau I.D. (2001a) Identification in closed loop a powerful design tool (better models, simple controllers), Control Engineering Practice, vol. 9, no. 1, pp. 51-65.

Landau I.D. (2001b) Les bases de l'identification des systèmes, in Identification des Systèmes (I.D. Landau, A. Bensançon-Voda, ed), pp. 19-130, Hermes, Paris.

Landau I.D., Constantinescu A., Rey D. (2005) Adaptive Narrow Band Disturbance Rejection Applied to an Active Suspension – An Internal Model Principle Approach, Automatica, Vol.61, n°4.

Landau I.D., Karimi A. (1997a) An output error recursive algorithm for unbiased identification in closed loop, Automatica, vol. 33, no. 8, pp. 933-938.

Landau I.D., Karimi A. (1997b) Recursive algorithms for identification in closed-loop – a unified approach and evaluation, Automatica, vol. 33, no. 8, pp. 1499-1523.

Landau I.D., Karimi A. (1998) Robust digital control using pole placement with sensitivity function shaping method, Int. J. of Robust and Nonlinear Control, vol. 8, pp. 191-210.

Landau I.D., Karimi A. (2001) Identification des modèles de procédé en boucle fermée, in Identification de systèmes (I.D. Landau, A. Besançon-Voda, Eds), pp. 213-244, Hermès, Paris.

Landau I.D., Karimi A. (2002) A unified approach to closed-loop plant identification and direct controller reduction, European J. of Control, vol. 8, no.6.

Landau I.D., Karimi A., Constantinescu A. (2001) Direct controller reduction by identification in closed loop, Automatica, vol. 37, no. 11, pp. 1689-1702.

Landau I.D., Langer J., Rey D., Barnier J. (1996) Robust control of a 360° flexible arm using the combined pole placement / sensitivity function shaping method, IEEE Trans. on Control Systems Tech., vol. 4, no. 4, pp. 369-383.

Landau I.D., Lozano R. (1981) Unification of Discrete-Time Explicit Model Reference Adaptive Control Designs, Automatica, vol. 12, pp. 593-611.

Landau I.D., Lozano R., M'Saad M. (1997) Adaptive Control, Springer, London, UK.

Landau I.D., M'Sirdi N., M'Saad M. (1986) Techniques de modélisation récursives pour l'analyse spectrale paramétrique adaptative, Traitement du Signal, vol.3, pp. 183-204.

Landau I.D., Rolland F. (1993) Identification and digital control of electrical drives, Control Engineering Practice, vol. 1, no. 3.

Langer J., Constantinescu A. (1999) Pole placement design using convex optimization criteria for the flexible transmission benchmark, European Journal of Control, vol. 5, no. 2-4, pp. 193-207.

Langer J., Landau I.D. (1996) Improvement of robust digital control by identification in closed loop. Application to a 360° flexible arm , Control Engineering Practice, vol. 8, no. 4, pp. 1079-1088.

Langer J., Landau I.D. (1999) Combined pole placement / sensitivity function shaping using convex optimization criteria , Automatica, vol. 35, pp. 1111-1120.

Ljung L.,(1999) System Identification - Theory for the User, 2nd edition, Prentice Hall, Englewood Cliffs.

Ljung L., Söderström T. (1983) Theory and Practice of Recursive Identification, MIT Press, Cambridge, Mass.

Lozano R., Landau I.D. (1982) Quasi-direct adaptive control for nonminimum phase systems, Transactions A.S.M.E., Journal of D.S.M.C., vol. 104, n°4, pp. 311-316, décembre.

Macejowski J.M.(2001) Predictive control with constraints, Prentice Hall, N.J.

Mathworks (1998) Identification toolbox for Matlab, The Mathworks Inc., Mass. U.S.A.

Morari M., Zafiriou E. (1989) Robust Process Control, Prentice Hall International, Englewood Cliffs, N.J.

M'Saad M., Landau I.D. (1991) Adaptive Control An overview, Actes du Symposium International IFAC ADCHEM 91, pp.3-11, Toulouse.

Narendra K.S., Taylor J.H. (1973) Frequency Domain Criteria for Absolute Stability, Academic Press, New York.

Ogata K. (1990) Modern Control Engineering (2nd edition), Prentice Hall, N.J.

Ogata K. (1987) Discrete-Time Control Systems, Prentice Hall, N.J.

Phillips, C.L., Nagle, H.T. (1995) Digital Control Systems Analysis and Design, 3rd edition, Prenctice Hall, N.J.

Press W.H., Vetterling W.T., Teukolsky S., Flanery B. (1992) Numerical recipes in C (The art of scientific computing), 2nd edition, Cambridge University Press, Cambridge, Mass.

Prochazka H., Landau I.D. (2003) Pole placement with sensitivity function shaping using 2nd order digital notch filters, Automatica, Vol. 39, 6, pp. 1103-1107.

Rantzer A., Megretski A. (1994) A convex parametrization of robustly stabilizing controllers, IEEE Trans. Aut. Control. Vol. 39, pp. 1802-1808.

Rolland F., Landau I.D. (1991) Pour mieux réguler le PC va vous aider, Mesures, pp. 71-73, december.

Shinskey F.G. (1979) Process Control Systems, McGraw-Hill, N.Y.

Söderström T., Stoica P. (1983) Instrumental variable methods for system identification , Lectures Notes in Control and Information Sciences, Springer Verlag, Berlin.

Söderström T., Stoica P. (1989) System Identification, Prentice Hall International, Hertfordshire.

Solo V. (1979) The Convergence of A.M.L., IEEE Trans. Automatic Control, vol. AC-24, pp. 958-963.

Sung H.K., Hara S. (1988) Properties of sensitivity and complementary sensitivity functions in single-input, single-output digital systems, Int. J. of Cont., vol. 48, n°6, pp. 2429-2439.

Takahashi Y., Rabins M., Auslander D. (1970) Control, Addison Wesley, Reading, Mass.

Tsypkin Y.Z. (1997) Stochastic discrete systems with internal models, Journal of Automation and information Sciences, Vol. 29 pp. 156-161.

Van den Bossche E. (1987) Etude et commande adaptative d'un bras manipulateur flexible, Ph.D., I.N.P.G.

Van den Hof P., Shrama R. (1993) An indirect method for transfer function estimation from closed loop data, Automatica, vol. 29, no. 6, pp. 1523-1528.

Vieillard J.P. (1991) Machine automatique pour la fabrication de cables torsadés téléphoniques, La Lettre d'Adaptech, no. 2, pp. 1-2, Adaptech, St. Martin d'Hères, France.

Vinnicombe G. (1993) Frequency domain uncertainty and the graph topology , IEEE Trans. on Automatic Control, vol. 38, no. 9, pp. 1371-1383.

Voda A., Landau I.D. (1995a) A method for the auto-calibration of P.I.D. controllers, Automatica, no. 2.

Voda A., Landau I.D. (1995b) The auto-calibration of P.I. controllers based on two frequency measurements , Int. J. of Adaptive Control and Signal Processing, vol. 9, no. 5, pp. 395-422.

Wirk G.S. (1991) Digital Computer Systems, 3rd edition, MacMillan, London.

Young P.C. (1969) An Instrumental Variable Method for Real Time Identification of a Noisy Process , Automatica, vol.6, pp. 271-288.

Young P.C., Jakeman A.J. (1979) Refined Instrumental Variable Methods of Recursive Time Series Analysis, Part I, Single-input, Single-output Systems, Int. J. of Control, vol. 29, pp.1-30.

Zames G. (1966) On the input-output stability of time-varying non linear feedback systems, IEEE-TAC, vol. AC-11, April, pp. 228-238, july pp. 445-476.

Zhu K. (1998) Essentials of robust control, Prentice Hall, N.J., U.S.A.

Ziegler J.G., Nichols N.B. (1942) Optimum Settings for Automatic Controllers, Trans. ASME, vol. 64, pp. 759-768.

Index

A

a posteriori prediction error 210
a priori prediction error 208
adaptation error 236
adaptation gain 213
    choice of 221
    constant 225
    constant trace 223
    decreasing 222
    initial 225
adaptive control 331
    direct 331
    indirect 331
air heater
    digital control of a 333
    identification of a 296
algorithm
    gradient 209
    parameter adaptation 206
    recursive estimation 215
aliasing 29
analysis
    frequency domain 2, 38
    time domain 1, 34
anti-aliasing filter 30
anti windup (device) 321
ARMAX 178
analog to digital converter 27
autocorrelation 172
    function 172
    normalized 172

B

backward shift operator 34
bandwidth 8
bias 233

C

closed loop identification 377
closed loop input error 404
    method 405
closed loop output error
    extended 382
    method (identification) 377
    with filtering 381
complexity reduction
    direct 401
    indirect 400
    of controllers 399
computational time delay 319
computer control 25
confidence interval 257
constant trace (see also adaptation
    gain) 223
continuous-time system 1
contraction factor 252
control
    cascade 326
    internal model 129
    position 327

479